T0297441

Technological and Institutional Innovations for Marginalized Smallholders in Agricultural Development

Franz W. Gatzweiler • Joachim von Braun
Editors

Technological and Institutional Innovations for Marginalized Smallholders in Agricultural Development

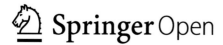

Editors
Franz W. Gatzweiler
Center for Development
  Research (ZEF)
University of Bonn
Bonn, Germany

Joachim von Braun
Center for Development
  Research (ZEF)
University of Bonn
Bonn, Germany

ISBN 978-3-319-25716-7   ISBN 978-3-319-25718-1   (eBook)
DOI 10.1007/978-3-319-25718-1

Library of Congress Control Number: 2015959025

Springer Cham Heidelberg New York Dordrecht London
© The Editor(s) (if applicable) and The Author(s) 2016. The book is published with open access at
SpringerLink.com.
Open Access This book is distributed under the terms of the Creative Commons Attribution-
Noncommercial 2.5 License (http://creativecommons.org/licenses/by-nc/2.5/) which permits any
noncommercial use, distribution, and reproduction in any medium, provided the original author(s) and
source are credited.
The images or other third party material in this chapter are included in the work's Creative Commons
license, unless indicated otherwise in the credit line; if such material is not included in the work's
Creative Commons license and the respective action is not permitted by statutory regulation, users will
need to obtain permission from the license holder to duplicate, adapt or reproduce the material.
This work is subject to copyright. All commercial rights are reserved by the Publisher, whether the whole
or part of the material is concerned, specifically the rights of translation, reprinting, reuse of illustrations,
recitation, broadcasting, reproduction on microfilms or in any other physical way, and transmission or
information storage and retrieval, electronic adaptation, computer software, or by similar or dissimilar
methodology now known or hereafter developed.
The use of general descriptive names, registered names, trademarks, service marks, etc. in this
publication does not imply, even in the absence of a specific statement, that such names are exempt
from the relevant protective laws and regulations and therefore free for general use.
The publisher, the authors and the editors are safe to assume that the advice and information in this book
are believed to be true and accurate at the date of publication. Neither the publisher nor the authors or the
editors give a warranty, express or implied, with respect to the material contained herein or for any errors
or omissions that may have been made.

Printed on acid-free paper

Springer International Publishing AG Switzerland is part of Springer Science+Business Media
(www.springer.com)

Foreword

We face a global food crisis of many dimensions. Food prices for the poor are rising and volatile. About a billion people are chronically hungry. Most shocking of all, 1 in 3 children under the age of five are seriously malnourished and will grow up physically and mentally stunted. At the same time, some two billion people are overweight or obese. Furthermore, we have to feed a growing world population demanding more varied and nutritious diets, including a wide range of livestock products. We will have to produce more food, but on more or less the same amount of land and with the same amount of water.

In recent years, I and a team of experts drawn from Europe and Africa, known as the Montpellier Panel, have been attempting to articulate the concepts, frameworks and practical actions we will need to cope with these challenges. We have argued that a way forward is sustainable intensification, producing more with less, but also using inputs more prudently, adapting to climate change, reducing greenhouse gases, improving natural capital and building resilience. It is a tall order, a challenge far greater than that we faced at the time of the Green Revolution.

An important contribution to the debate is this volume edited by Franz Gatzweiler and Joachim von Braun. Its aim is to improve the understanding of how, when and why innovation can bring about sustainable intensification in agriculture, improving the lives of poor smallholders, a majority of which live in Sub-Saharan Africa and South Asia. It presents contributions from theory, policy and practice to the science of sustainable intensification. The volume explores opportunities for marginalized smallholders to make use of technological and institutional innovations in order to achieve sustainable intensification and improve productivity and wellbeing.

The insightful framework developed by Gatzweiler and von Braun considers the different needs of smallholders in different agro-ecological environments and with different human capabilities. The diversity of strategies in each of the segments improve the targeting of innovations when they need to be people and area specific. The identified strategies also tell us about the type of enabling environment in which innovations can tap unused productivity potential by leveraging human and agro-ecological capital. According to that framework, innovations which are people focused are likely to be more relevant in areas where agro-ecological potentials are low and innovations which create additional value in agricultural productivity are more relevant in areas with unused agro-ecological potentials. That is an important aspect which will make sustainable intensification more focused. For some small-holders, technology will be the dominant innovation for food security, others will need multiple and diverse strategies and build on their social capital, while for others again non-farm income opportunities are the better alternative.

The examples provided in this volume tell us that technological innovation can take diverse forms from high-yielding and stress-tolerant varieties to modified farming practices. Those innovations need to be accompanied by institutional innovations at multiple scales and engage stakeholders from government, local communities and business. Institutional innovation is not only necessary to ensure the access and use of technological innovations but also to create an enabling environment which rewards grassroots innovators for being creative and sharing their knowledge.

The rich collection of contributions from Sub-Saharan Africa and South Asia in this volume are based on original field-based research demonstrating an in-depth understanding of the lives of poor smallholders and the conditions under which they themselves engage in innovation or adopt innovations. It shows by a host of telling examples that a lot more can be done fast and sustainably for and with smallholders by making use of an area-and-people focused targeting concept.

The support of this research by the Bill & Melinda Gates Foundation is a significant indication of the Foundation's commitment to enhance agricultural innovation for poverty reducing actions at scale in marginalized communities and complex diverse agro-ecologies.

The book is a rich source of knowledge for students, scholars and practitioners in the field of science and policy for understanding and identifying agricultural productivity growth potentials for smallholder farmers and development.

Professor, Faculty of Natural Sciences   Gordon Conway
Centre for Environmental Policy
Chair in International Development
Imperial College, UK

Acknowledgements

We would like to express our sincere gratitude to the Bill and Melinda Gates Foundation who supported a research project on "Ex-ante technology assessment and farm household segmentation for inclusive poverty reduction and sustainable productivity growth in agriculture (TIGA)" at the Center for Development Research (ZEF), University of Bonn under Grant No. OPP1038686. We strongly appreciate the close collaboration and intensive exchanges on the topic with Prabhu Pingali and Kate Schneider at the Foundation, which helped us shed light on various aspects of the multifarious topic of this book.

The research carried out and the contributions to this book would not have been possible without the support of our partners and collaborators at the Forum for Agricultural Research in Africa (FARA) in Ghana, the Ethiopian Economics Association (EEA) in Ethiopia, the International Food Policy Research Institute (IFPRI) in India, and BRAC in Bangladesh.

The present volume was accomplished with the cooperation of researchers, funders, practitioners and decision makers from a variety of backgrounds, who collectively took a fresh look at how to overcome the barriers and grasp the opportunities of innovations in agriculture, which would sustainably reduce poverty and marginality. We are grateful to all authors of this volume and those who directly or indirectly contributed to advancing our knowledge on innovations for sustainable agricultural intensification.

Apart from various research colloquia and meetings, findings under this research project had been presented and discussed at the 8th conference of the Asian Society of Agricultural Economics in Dhaka, October 2014.

We express our special gratitude to Arie Kuyvenhoven and Franz Heidhues who provided continuous advice and contributed with insightful comments to the research process.

Moreover, helpful critiques from external anonymous reviewers on an earlier draft of this volume are gratefully acknowledged.

Bonn, Germany   Franz W. Gatzweiler
                                                                       Joachim von Braun

Contents

Contributors

Assefa Admassie Ethiopian Economic Association Yeka, Addis Ababa, Ethiopia

Akhter U. Ahmed International Food Policy Research Institute (IFPRI), Dhaka, Bangladesh

Irene Annor-Frempong Forum for Agricultural Research in Africa (FARA), Accra, Ghana

Samuel Asuming-Brempong Department of Agricultural Economics and Agribusiness, University of Ghana, Legon, Accra, Ghana

Heike Baumüller Center for Development Research (ZEF), University of Bonn, Bonn, Germany

Tina D. Beuchelt Center for Development Research (ZEF), University of Bonn, Bonn, Germany

Ashish Bharadwaj Jindal Global Law School, O.P. Jindal Global University, Delhi, India

Joachim von Braun Center for Development Research (ZEF), University of Bonn, Bonn, Germany

Tilaye Teklewold Deneke Amhara Agricultural Research Institute (ARARI), Bahir Dar, Ethiopia

Manfred Denich Center for Development Research (ZEF), University of Bonn, Bonn, Germany

Stephen Frimpong United Nations University-Institute for Natural Resources in Africa, Accra, Ghana

Franz W. Gatzweiler Center for Development Research (ZEF), University of Bonn, Bonn, Germany

Nicolas Gerber Center for Development Research (ZEF), University of Bonn, Bonn, Germany

Daniel Gulti Agricultural Transformation Agency (ATA), Addis Ababa, Ethiopia

Md. Latiful Haque BRAC Research and Evaluation Division (RED), Dhaka, Bangladesh

Ricardo Hernandez International Food Policy Research Institute (IFPRI), Dhaka, Bangladesh

Mohammad Syful Hoque BRAC Research and Evaluation Division (RED), Dhaka, Bangladesh

Mahabub Hossain BRAC Research and Evaluation Division (RED), Dhaka, Bangladesh

Bekele Hundiea Kotu International Institute of Tropical Agriculture, Tamale, Ghana

Christine Husmann Center for Development Research (ZEF), University of Bonn, Bonn, Germany

Deden Dinar Iskandar Center for Development Research (ZEF), University of Bonn, Bonn, Germany

Abu Hayat Md. Saiful Islam Center for Development Research, University of Bonn, Bonn, Germany

Department of Agricultural Economics, Bangladesh Agricultural University, Bonn, Germany

P.K. Joshi International Food Policy Research Institute (IFPRI), Washington, DC, USA

Simone Kathrin Kriesemer Horticulture Competence Centre, University of Bonn, Bonn, Germany

Arnim Kuhn Institute for Food and Resource Economics, University of Bonn, Bonn, Germany

Manasi Kumar Department of Psychiatry, College of Health Sciences, University of Nairobi, Nairobi, Kenya

Department of Psychology, University of Cape Town, Cape Town, South Africa

Mohammad Abdul Malek BRAC Research and Evaluation Division (RED), Dhaka, Bangladesh

University of Bonn-Center for Development Research (ZEF), Bonn, Germany

Md. Abdul Mazid BRAC International, Dhaka, Bangladesh

Alisher Mirzabaev Center for Development Research (ZEF), University of Bonn, Bonn, Germany

Firdousi Naher University of Dhaka, Dhaka, Bangladesh

Alex Barimah Owusu Department of Geography and Resource Development, University of Ghana, Legon, Accra, Ghana

Evita Pangaribowo Department of Environmental Geography, University of Gadjah Mada, Yogyakarta, Indonesia

Devesh Roy International Food Policy Research Institute (IFPRI), Washington, DC, USA

Vinay Sonkar International Food Policy Research Institute (IFPRI), New Delhi, India

Justice A. Tambo Center for Development Research (ZEF), University of Bonn, Bonn, Germany

Gaurav Tripathi International Food Policy Research Institute (IFPRI), New Delhi, India

Detlef Virchow Center for Development Research (ZEF), University of Bonn, Bonn, Germany

Katinka M. Weinberger Environment and Development Policy Section, United Nations Economic and Social Commission for Asia and the Pacific, Bangkok, Thailand

Tobias Wünscher Center for Development Research (ZEF), University of Bonn, Bonn, Germany

Josefa Yesmin BRAC Research and Evaluation Division (RED), Dhaka, Bangladesh

Chapter 1
Innovation for Marginalized Smallholder Farmers and Development: An Overview and Implications for Policy and Research

Franz W. Gatzweiler and Joachim von Braun

Abstract Smallholders in Asia and Africa are affected by increasingly complex national and global ecological and economic changes. Agricultural innovation and technology shifts are critical among these forces of change and integration with services is increasingly facilitated through innovations in institutions. Here we focus mainly on innovation opportunities for small farmers, with a particular emphasis on marginalized small farm communities. The chapter elaborates on the concept of the 'small farm' and offers a synthesis of the findings of all the chapters in this volume. The contributions have reconfirmed that sustainable intensification among smallholders is not just another optimization problem for ensuring higher productivity with less environmental impact. Rather it is a complex task of creating value through innovations in the institutional, organizational and technological systems of societies.

Keywords Marginality • Poverty • Innovations • Policy • Smallholder farmer

Introduction

The large majority of the world's 570 million small farms are in Asia and Africa, if we define smallness by land size (Lowder et al. 2014), and about 80 % of them actually live in Asia. They are the largest employment category and small business group among the poor. Their businesses use mostly local resources and face local constraints, but at the same time, they are affected by increasingly complex national and global economic changes. These changes are partly inside farming and partly very much outside agriculture, partly domestic and partly international, i.e.:

- returns to **labor** in small scale farming are increasingly determined outside agriculture through more integrated labor markets; opportunity costs of farm

F.W. Gatzweiler (✉) • J. von Braun
Center for Development Research (ZEF), University of Bonn, Bonn, Germany
e-mail: gatzweiler@gmail.com; jvonbraun@uni-bonn.de

© The Author(s) 2016
F.W. Gatzweiler, J. von Braun (eds.), *Technological and Institutional Innovations for Marginalized Smallholders in Agricultural Development*,
DOI 10.1007/978-3-319-25718-1_1

labor are rising, as are aspirations of youth in farming families who do not want to feel relatively deprived;

- agricultural **innovation and technology** shifts are critical among the forces of change; integration with services is increasingly facilitated through innovations in institutions;
- the market value of smallholder **land** is rising because of agricultural price changes and the increasing influence of non-agricultural demand for land use, as well as expected value changes in other capital asset classes;
- **international dynamics** result from changing price levels and volatility, and trade policies defining competitiveness; consumption shifts are among the fundamental drivers;
- domestic **policies**, especially the scale and pattern of investments in public goods, such as infrastructure, innovation systems, and social policy, change the socio-economic framework of small scale farming.

This volume and the overview chapter focuses mainly on innovation opportunities for small farmers. The other above-mentioned important forces of change are touched upon only as a backdrop. Moreover, we focus in particular on marginalized small farm communities.[1] Small farmers have shown strong resilience in the context of economic transformation. They are faced with forces of continuing change in coming decades, including far more integrated and quality-focused agricultural value chains and more complex technological and institutional choices for production, processing and marketing. Policies must be designed to facilitate an integral role for small farm households not only as passive absorbers of change, but as important contributors to development.

Defining Small Farms Comprehensively

Small farms are highly heterogeneous and diverse. Small farmers exhibit specific characteristics and play different (sometimes multifunctional) "roles" in their regions, and these roles differ in significance at different stages of economic development. Most of the literature defines small farms based on the size of their land or livestock holdings (Eastwood et al. 2010), a standard but arbitrary cut-off size being less than 2 ha (World Bank 2003). Land quality and access to resources such as water are also key differentiators of small farms. It is important to capture the institutional and technical characteristics in the definition of *small farm*. Being small is not only about the land or herd size, but also about varied access to markets and natural resources and the degree of commercialization (von Braun and

---

[1] We define marginality as "an involuntary position and condition of an individual or group at the margins of social, political, economic, ecological, and biophysical systems, that prevent them from access to resources, assets, services, restraining freedom of choice, preventing the development of capabilities, and eventually causing extreme poverty" (von Braun and Gatzweiler 2014).

Table 1.1 Defining small farms: concepts and criteria

Concepts	Measurements	Strengths	Deficiencies
1. Land holding (or herd) size	Size in hectares cultivated Number of livestock	Simple accounting of physical characteristics; Important for agrarian societies	Lack of economic valuation of farm enterprise (quality of land, location to markets, etc.); ownership issues neglected
2. Employment	Labor in small farms	Important for economy- wide considerations, and for livelihoods	Returns to labor (especially marginal returns) undefined; economics of multiple job-holdings missing
3. Income	Annual produc-tion and net returns	Integrates with GDP shares; identifies growth and innova-tion performance; a basis for poverty identification in the small farm economy	Highly variable; pricing own consumption of farm products; externalities not captured (eco-systems services)
4. Total eco-nomic value (TEV)	Comprehensive capital stock (assets) account	Identification of wealth; credit worthiness; important for economy- wide consider-ations beyond GDP	Difficulties to value land and human capital (skills); value of inter-farm collective action (as a form of social capital)
5. Societal role	Small-farm communities; villages; local services	Shows collective action (potentials); governance and fiscal settings; public goods investments	Lacks focus on the farm enterprises

Source: von Braun and Mirzabaev (2015)

Mirzabaev 2015). Given the important role of small farms in reducing rural poverty, the definition of small farms ideally should be asset- and income-based (ibid.), not solely area-based.

Actually, a whole dashboard of concepts and related measurable criteria should be applied to identify size, relevance and potential of the small farm economy. Table 1.1 lists such a dashboard of five sets of concepts (land size, employment, TEV, income, socio-economics). The literature is rich in studies on all these five concepts and, to some extent, their inter-linkages. A general international statistical basis, however, exists only for the land-based accounting of "small farms", and even that is quite deficient (Lowder et al. 2014). The definition of farm class sizes for which data are collected is often divergent among countries, making their cross-country comparisons challenging (FAO 2010). Moreover, the discussion of small farms is dominated by crop production, whereas small pastoralists are usually not much taken into consideration, with little attention being paid to small scale horticulturalists and aquaculturalists as well.

Using the area size of a farm alone to identify whether it is small or big may lead to misguided policy actions. For example, 1 ha of irrigated fertile land planted with high value vegetables and fruits and located close to major urban markets could generate much higher total income than, say, 20 ha of rainfed area under subsis-tence crops in remote areas. The same 1 ha of irrigated land may lead to quite divergent incomes depending on whether it is sustainably managed or highly degraded (Nkonya et al. 2011).

Determinants of "Smallness" of Farms

Smallness of farms is largely endogenous. The fundamental insights of Tschajanov (1923) based on empirical analyses of the relationships between labor use and farm size in Russia around the beginning of the twentieth century emphasized that the small farm (including household plots for home production) should not be viewed as just a short-term transitional phenomenon. It is an economic reality and it depends directly on the household utility function and on the underlying economic conditions in product and labor markets, as well as social system risks. The factors that put small farms at an advantage or disadvantage compared to large farms have been debated by economists for years, and there are long-standing debates on the viability and the role of small farms in economic development (Schultz 1964; von Braun and Kennedy 1994; Hazell et al. 2010). The seminal research of Schultz (1964) on the efficiency of small and poor farmers brought many misleading debates, equating small with inefficient, to an end.

Often, small farms are not considered "viable", but concepts of viability need to be carefully assessed in relation to small farms. Economic viability in family farming means the ability and capacity of a farm to 'make a living', say, over the seasons of a year or over the long run. Given the relevance of multiple job holdings on small farms, defining viability purely on the basis of the farm component of the households' total economy is inappropriate, as farm production, labor and capital allocation are optimized in an integrated, inseparable fashion in most instances (Singh et al. 1986). Furthermore, defining small farm viability from an economy-wide perspective would need to be based on considerations of TEV and productivity (innovation) potentials. These "people potentials" in the small farm sector, such as entrepreneurship and expanding human capital, may be much more relevant for growth and development, rather than simply being the economics of land connected to the small farm economy.

The concept of *returns of scale* has been used to probe many of the theories of optimal farm size (Chavas 2001). Empirical studies of this inverse relationship in the 1970s found that, in India, small farms are more technically efficient than large farms (Yotopoulos and Lau 1973; Berry and Cline 1979). Hired labor is the main reason for the lower land productivity of larger farms (Binswanger and Rosenzweig 1986). Family workers are more efficient than hired workers because family members receive a share of the profit and thus pay greater attention to quality of work than hired labor. Family members also require no hiring or search costs, and each family member assumes a share of the risk; however, there are tendencies towards (self-) exploitation of labor in family farming, especially in relation to child labor (ILO 2006) and remuneration of women's work.

In many cases, the small family farm is the optimum size because scale economies that arise from using inseparable inputs (like machinery) are offset by the scale diseconomies that arise from using hired labor (Hayami 1996). Ultimately, the optimal farm size will be the one under which labor productivity of the agricultural sector approaches that of the non-agricultural sector, given the same quality of labor. Transitions to such a state can take a long time due to institutional rigidities, transformation risks, and policies. A simple calculation highlights this: under an assumption of farm closure rates of 5 % per annum (be it through sales or renting out), it would take 45 years to move from an average of 1-ha farms to an average of 10-ha farms. Europe has only managed that process with half such an exit rate. These small farms will be there for many years to come. Radically accelerated and enforced change in farm size usually entails suffering and is economically inefficient.

Small farms require focused developmental attention for several reasons. Firstly, they play key roles in broader economic transformations. Considering that small farms are home to large shares of populations in developing countries, the successes of economic transformation need to take the economy-wide roles of small farms into account. Secondly, the protection and sustainable use of natural resources by small farms is becoming a critical aspect of their productivity. For example, land degradation is found to affect more than three billion people around the world, the majority of whom are small farmers and pastoralists in developing countries (Le et al. 2014), and has serious economic consequences for them (Nkonya et al. 2011). Thirdly, globalization and changes in markets offer new opportunities and competitive threats for small farms. These opportunities and threats need to be evaluated with a view to enabling small farmers to successfully integrate into new value webs or at least partly exit agriculture in favor of nonfarm activities. Fourthly, small farms play a key role in reducing poverty. Most of the poor in the world reside on small farms (von Braun 2011), so what happens on small farms will be decisive in actions against poverty. And among the population, those most affected by food insecurity are the smallholder farmers, because of income and direct production linkages. Therefore, if these farmers were better off, hunger and the sticky problem of child malnutrition would diminish.

Patterns and Change of Small Farms

The world currently has about 570 million farms, if we include small household agricultural production (Lowder et al. 2014). Table 1.2 depicts their estimated size distribution. Approximately 85 % of the world's farms are smaller than 2 ha. About half of small farms are in low or lower middle income countries (Lowder et al. 2014). The majority of farms, including small farms, are located in Asia, particularly China and India. It should be stressed that land quality differs widely among these small holdings.

Average farm size is decreasing in Asia and Africa. However, in those countries which are experiencing farm size decreases, the rate of decrease has decelerated,

Table 1.2 An approximation of world farm size distribution by regions

Region	Land size classes					
	<1 ha	1–10 ha	10–50 ha	50–100 ha	100–500 ha	>500 ha
Asia	78 %	19 %	1 %	–	–	–
Sub-Saharan Africa	62 %	37 %	1 %	–	–	–
Middle East and North Africa	60 %	33 %	7 %	–	–	–
Latin America and Caribbean	17 %	47 %	23 %	6 %	–	–
Europe	–	77 %	15 %	3 %	3 %	–
North America and Australia	–	19 %	32 %	16 %	24 %	9 %

Source: data from Lowder et al. (2014), FAO datasets
Note: Blank cells mean the number of farms under this land size class is less than 0.1 % of the total

whereas some countries, notably China and Vietnam, have begun experiencing a recently increasing trend in their average farm sizes. The share of the active population employed in agriculture is decreasing, albeit at a much slower rate than the share of agriculture in GDP. As a result, even in countries where agriculture plays a minor role in terms of its contribution to GDP, its role in employment is still quite substantial (von Braun and Mirzabaev 2015).

Size and Productivity

Cross-country comparisons show that average farm size is positively associated with agricultural value added per worker. The comparison of agricultural growth rates with changes in farm size, however, does not show a consistent picture for the Asian countries (Fig. 1.1): The overall trend seems to show that increase in farm size is associated with faster agricultural growth, but this seems heavily influenced by just a few countries (Tajikistan, Uzbekistan, South Korea, Vietnam). The results for SSA countries (Fig. 1.2) show little association between farm size changes and the rates of agricultural growth. Most of the Asian and SSA countries have very small average farm sizes, so passing from, say, 0.4–0.8 ha may not necessarily instigate any strong qualitative changes that influence agricultural growth rates.

Several studies found small farms to have higher land productivity than bigger farms due to higher incentives and productivity of family labor (Eastwood et al. 2010), especially in Asia where labor is more abundant than land (Hazell et al. 2010). For example, decreasing returns to scale in agricultural production were found in East Java, Indonesia (Llewelyn and Williams 1996), and Pakistan (Heltberg 1998). However, Fan and Chan-Kang (2005) also indicate that, in certain cases, once the varying degrees of soil fertility and land potential (irrigated vs. rainfed) are taken into account, the diseconomies of scale in land productivity between small and large farms may disappear. Moreover, there is plentiful evidence

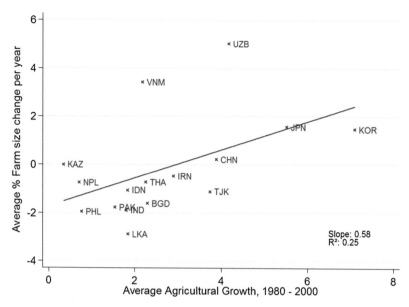

Fig. 1.1 Average annual changes in farm size and average agricultural growth rates in Asia (Sources: von Braunand Mirzabaev (2015), based on word development indicators, World Bank, and Lowder et al. (2014), FAO)

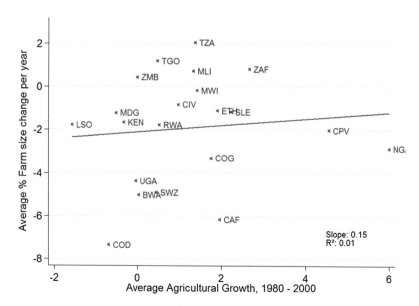

Fig. 1.2 Changes in farm size and average agricultural growth rates in SSA, both in logs (Sources: von Braun and Mirzabaev (2015), based on word development indicators, World Bank, Lowder et al. (2014), FAO)

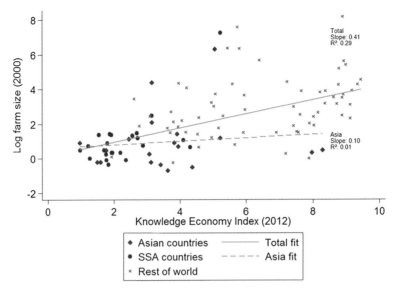

Fig. 1.3 Farm size and knowledge economy (Index, both in logs) (Sources: von Braun and Mirzabaev (2015) based on Lowder et al. (2014), FAO; World Development Bank)

that small farms could be less efficient in terms of labor productivity (ibid.). Wiggins et al. (2010) conclude that the distinct advantages of small farms are present in cases when the main agricultural input is family labor and there is very little use of external inputs, the production being chiefly for home consumption with whatever surpluses exist being sold to small-scale traders. There is no clear-cut answer to the question as to whether small farms perform more productively under what circumstances. Certainly, the performance of small farms is modulated by a variety of accompanying policy, institutional, market and agro-ecological conditions. In fact, the variations in farm size could be explained either by deliberate national policies (Fan and Chan-Kang 2005) or by varied population pressures (Eastwood et al. 2010).

In view of the rising role of innovation (Total factor productivity, TFP) in agricultural growth (Fuglie 2013), we would expect that agriculture grows more and better if accompanied by a strong knowledge society. The comparison of average farm size and the knowledge economy index[2] shows positive association globally, but less so in emerging economies (Fig. 1.3).

---

[2] The Knowledge Economy Index (KEI) measures a country's ability to generate, adopt and diffuse knowledge. It takes into account whether the environment is conducive for knowledge to be used effectively for economic development.

Persisting Rural Poverty and Untapped Potential

Deprivation, hunger and malnutrition remain predominantly rural. Globally, a billion people still live in extreme poverty. Progress in reaching the poorest and most marginalized has been slow, and income inequality continue to increase. The depth of poverty has become less severe, however, most improvements have been achieved in China and India. For other developing countries, the number of people living in extreme poverty today is as bad as it was 30 years ago (Olinto et al. 2013). However, even in China and India, growth has been unevenly distributed, and poverty persists in China's interior and in three of India's states in particular. Moreover, income inequality, the difference between average rural and urban incomes, is increasing, as well as inequality in terms of access to land. The majority of farms in low income countries cover less than 2 ha, and in Sub-Saharan Africa, less than 1 ha (von Braun 2005, von Braun and Mirzabaev 2015).

New efforts need to be made to reach out to those persisting in poverty. For that reason, research into agricultural innovation needs to develop ways for improving the lives of the rural marginalized and poor. Our starting point for re-addressing the topic of agricultural innovation is a new perspective on the lives of the rural poor, by recognizing developmental potential along social and ecological dimensions: capabilities of the rural poor and agro-ecological potential. Identifying those dimensions recognizes that rural poverty is a multi-dimensional phenomena. Agro-ecological potential refers to potential provided by the land and its respective ecosystem services. Since the 1970s, agro-ecological zoning has been used for determining agricultural production potential (Beddow et al. 2010, pp. 8–38)

We take a spatial-, people-, and transaction-specific approach to matching institutional and technological innovations with human capacities and agro-ecological potential. Such an approach recognizes the challenges of adopting technological innovations in complex marginalized social and ecological environments. The conceptual framework we have developed aims to identify suitable strategies and innovations for different segments of marginalized smallholders in agriculture, along a gradient of human capability and agro-ecological potential. Sustainable intensification is thereby achieved through different strategies in specific contexts and can mean increasing crop harvest per area, increasing crop diversity per area and nutrient supply per household member, or increasing income opportunities and household income.

According to Sen (1999), capabilities are realized freedoms for people to do and be what they value. Having capabilities enables people to actually make choices from a set of opportunities, which requires freedom to choose and the availability of options to choose from (which he refers to as functionings). Freedom does not mean "free from any restraint" but rather the possibility to actually choose to be and do. Having fewer choices can sometimes be more enabling (and thereby contribute to wellbeing) than having to make (costly) trade-offs among many choices. The central idea behind the concept of capabilities is to increase freedoms and not only to increase the number of choices. Realized capabilities are "functionings".

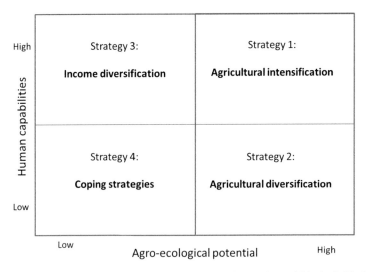

Fig. 1.4 Potential strategies, technological and institutional innovations within the field of tension between human capabilities and agro-ecological potentials

Four broad people-and-land related segments within those two dimensions can be identified (Fig. 1.4): (1) Areas where rural populations have relatively high capabilities and land with relatively high agro-ecological potential (AEP). (2) Areas where the level of human capabilities are relatively high but AEP is low. (3) Areas where human capabilities are relatively low and AEP is high. And (4) where capabilities and AEP are both low.

What our framework essentially shows is that when high capabilities overlap with low agro-ecological potential, the "realized freedoms" cannot be achieved alone from tapping unused agro-ecological potential or closing yield gaps by means of technological innovations in agriculture. Rather, they need to be achieved by alternative income opportunities – either in agriculture related service or business sectors or outside of the agricultural sector.

Within each segment, different types of innovation help the rural poor to improve wellbeing. "Innovation is the process by which inventions are produced – it may involve new ideas, new technologies, or novel applications of existing technologies, new processes or institutions, or more generally, new ways of doing things in a place or by people where they have not been used before" (Juma et al. 2013, p. 2). We refer to institutional and technological innovations as doing things in new ways, on the basis of new sets of rules or organization or by means of technical inventions which are introduced or have been invented by the small-holders themselves.

Innovations for the rural poor include institutional and technological innovations which broaden the set of opportunities for the poor to improve their wellbeing (Conway and Waage 2010). Technological innovations in agriculture can improve wellbeing by increasing efficiency in the production process and reducing labor costs. Institutional innovations can improve the wellbeing of the poor, e.g., through improved access to land, better land use rights, or better income opportunities that do not involve working on the land. The two cannot be entirely separated, however, distinguishing between them shows different opportunity sets for the rural poor to improve their wellbeing.

Innovation Strategies for the Rural Poor

The number of rural poor has persisted, despite successful attempts at various innovation approaches in agriculture, such as the agricultural innovation system (AIS) approach (Leeuwis and Ban 2004; Bergek et al. 2010) or the science-based productivity enhancement approach of the Green Revolution. Many of the remaining rural poor are poor, not only because they are unable to produce more product more efficiently, but because they remain marginalized (von Braun and Gatzweiler 2014). Enhancing agricultural productivity by means of improved seeds, fertilizer and pesticide use, and by reducing post-harvest losses, are definitely among the options for innovation for some rural poor. However, those innovations will not solve the problems of fragmented farm sizes and exploitative relations between landowners and tenant farmers.

Improving the wellbeing of the rural poor will need to be achieved by providing opportunities for increasing capabilities and widening the innovation portfolio by following strategic pathways out of rural poverty. According to our framework (Fig. 1.4), we identified three strategic options and four strategies for each segment of the rural poor. The strategic options are: (1) Intensification, (2) Diversification, and (3) Coping strategies.

The dominant type of productivity to be improved in each segment varies. Innovations which lead to improved land productivity will be favored in segments 1 and 2, where agro-ecological potentials are relatively high. Innovations which lead to improving labor productivity will be favored in segments 1 and 3, where human capabilities are relatively high. In segment 4, intensified efforts for improving both types of productivity need to be made. This segment is typically the domain of development organizations and needs to be embraced by national development and social safety net programs.

Strategy 1 applies to areas with relatively high human capabilities and relatively high agro-ecological potential. This segment is preferable for Green Revolution type of interventions. Land which is agro-ecologically suitable and located in areas with high population densities also shows high opportunity costs of land and labor. People have income opportunities other than working in agriculture, demand for agricultural products increases and land is becoming scarce. Those developments lead to adopting a strategy of intensification, typical for the Green Revolution in Asia and Latin America (Pingali 2012).

Here, a strategy of sustainable agricultural intensification could involve improved access to production means, e.g., high yielding varieties, fertilizer, pesticides, and seeds to enhance productivity through intensification (The Montpellier Panel 2013). In this segment, technological and institutional innovations need to support the aim of increasing yields per area of land.

The optimism for productivity gains in segment 1, however, needs to be accompanied by a word of caution. Decreasing land/labor ratios alone does not automatically suggest similar opportunities for intensification, especially not in marginalized areas where infrastructure development is not a priority, alternative income opportunities are scarce, and property rights do not favor the majority of the poor and are unlikely to change. In those areas, the rural poor are predominantly wage laborers who engage in multiple income generating activities, and even when agricultural productivity increases as a result of intensification, the majority of the rural poor, who might be tenant farmers, do not benefit from that growth (Singh 2012).

This has been shown by Hirway and Shah (2011), who detected a low elasticity of poverty reduction to growth 1993–2005 for the state of Gujarat, India. Despite growth, the state of Gujarat slipped from rank 6 to 12 in rural poverty, from rank 5 to 9 on the Human Development Index (1996–2006), its health and education index decreased by one rank, and it was among the five states in India, together with Bihar, performing worst on the Global Hunger Index (IFPRI 2009). Also, the Gini coefficient of income for Gujarat was 0.47, indicating extremely unequal income distribution (Shukla 2010).

Strategy 2 applies to areas which show low levels of capabilities and high agro-ecological potential, i.e., extreme poverty in areas with high agro-ecological potential. In those areas, a strategy of agricultural diversification (possibly including non-staples and animal production) could enable food security. In this segment, innovations need to support the aim of diversifying and increasing agricultural yields per land area.

Marginalized rural populations in segment 3 are characterized by relatively high capabilities and low agro-ecological potential (less favored or highly marginal). Strategies of income diversification are promising and could include measures which facilitate access to agricultural and non-agricultural markets. Examples for measures taken in this segment include access to micro-credit, social protection, and seeds which are stress tolerant and can cope in harsh environments. In this segment, innovations should support the goal of increasing income opportunities per household.

In segment 4, stallholders have the lowest capabilities and live in areas with low agro-ecological potential, i.e., extreme poverty in harsh environments. Integrated strategies apply which involve access to land, water and public services such as education and health. Examples are integrated rural development approaches or BRAC's program for the ultra-poor in Bangladesh that facilitate readiness for participation in mainstream development initiatives. Innovations in this segment are also people-focused, as in segment 3, and aim to secure livelihoods by diversifying strategies for coping.

Poor smallholders in Sub-Saharan Africa and South Asia are rather vulnerable to technological changes introduced from outside (Holmes and Jones 2009; Farrington et al. 2007). Even the most promising innovations in agricultural technology, which fit the local ecological environment and promise to close yield gaps, are not automatically accompanied by programs which reduce risks, enable smallholders to scale up production levels or secure them the benefits from productivity increases. Tapping agro-ecological potentials by means of agricultural technology innovations will therefore be more sustainable the more the human capabilities of the marginalized smallholders are realized.

Overview: Innovative Intensification and Diversification for and with Marginalized Small Farmers

This volume is structured according to the major innovation strategies we have identified in our framework. Part I (Chaps. 2, 3, 4 and 5) presents theoretical insights into innovation at multiple scales of society. Foundations are laid for understanding sustainable intensification in agriculture as a complex development towards creating value in multiple domains of society, economy and ecology. Contributions in this part of the volume look into how innovation occurs in multi-layered social organization (polycentricity) and how the human psychology of innovation works. A theoretical model for technological adaptions among marginalized smallholders is presented, as well as impacts and trends in innovations for food security.

Gatzweiler (Chap. 2) explains how institutions can be enabling or inhibiting for the rural poor to escape poverty. He shows how rule changes within multi-layered, nested (polycentric) social order can create value horizontally and vertically, either from inside social systems or externally induced, by reducing transaction costs and enabling connectivity, interactions, exchange, and communication. Technological and institutional innovations can change the rule-change calculus, providing incentives to change the sets of rules which keep smallholders marginalized, and thereby better position them in society to escape poverty. Accordingly, sustainably improving the lives of smallholders in Sub-Saharan Africa and South Asia cannot be achieved by improving productivity through technology innovations alone – institutional innovations are required for smallholders to change their marginalized positions in society.

Food and nutrition security among rural smallholders remain in a critical state. Pangaribowo and Gerber (Chap. 3) address the issue by presenting the current situation which, despite overall progress, still gives reason for concern, especially in Sub-Saharan Africa and South Asia. They provide examples of how new platforms and traditional technological innovations (see also Chap. 9) but also institutional innovations enable farmers to collaborate and learn from each other, and thereby can have positive direct or indirect impacts on food and nutrition security.

Little attention has been paid to the states of mind which drive innovation. In Chap. 4, Manasi Kumar and Ashish Bharadwaj look at the psychology behind innovations, identify barriers and processes of innovation diffusion, and explain when internal stimuli for innovation might be more promising as compared to externally inducing innovation as a result of uncertainty perception. They provide answers to the questions of how, when and where creativity and innovation occur among the rural poor and why poverty and deprivations can mar capacities that drive novelty seeking behavior. The insights the authors provide into the psychology of innovation illustrate the circumstances under which an innovation is worthwhile and why it is that, as time passes, the motivational and need structures can change for the worse.

In Chap. 5, Iskandar and Gatzweiler develop an optimization model showing that productivity gains among rural smallholders can be achieved, but are conditional on human and natural capital stocks and transaction costs. Corresponding to the conceptual frame proposed by Gatzweiler and von Braun in Chap. 1, they explain why adjustments in rural infrastructure and institutions to reduce transaction costs is a more preferable investment strategy than adjusting agricultural technologies to marginalized production conditions. After defining the optimization problem for rural households under the poverty and survival line, the authors observe the impact of technology adoption and the transaction cost effects on the income generation capacity in specific segments of the rural poor. Their analysis sheds light on the question of why technology adoption is not the preferred strategy for productivity growth under the presence of high transaction costs which are common amongst poor smallholders.

Part II consists of contributions from the Asian and African Regions which present examples of income and production diversification strategies. The authors present studies on the role of large non-governmental organizations, private businesses and governmental organizations for facilitating income and agricultural diversification strategies, show innovative approaches for encouraging smallholders to make use of local innovations, project the potential impacts from innovations in agriculture on gender, and deliberate on the underused potential of synergy effects in interwoven value chains, so-called value webs, in the bioeconomies of Subsahan Africa and South Asia.

In Chap. 6, Mazid et al. present the BRAC approach to innovation among smallholders in Bangladesh for reducing hunger and improving food security. BRAC applies a mix of multiple approaches, ranging from better adapted and higher yielding crop varieties to improved production processes to agricultural microcredit schemes. The combination of innovations and interventions which address what matters for communities and their involvement in the development process, as well as the provision of extension services, is an encouraging example of diversified agricultural development strategies, which are also being implemented in Africa.

Strong linkages between research and extension organizations in Ethiopia's agricultural innovation system are also the focus of the authors in Chap. 7. With reference to the Ethiopian Agriculture Development Partners Linkage Advisory Council (ADPLAC), their findings show how important it is to institutionalize joint research and extension processes by improving accountability, monitoring and evaluation. The involvement of stakeholders in the innovation process is also crucial.

Opportunities for innovations in the Ethiopian seed system are outlined by Husmann in Chap. 8. The study underlines the importance of transaction costs, as demonstrated by Iskandar and Gatzweiler in Chap. 5. The current government dominated seed system is characterized by transaction costs which make it less attractive for the private sector to invest and meet the demand for improved seeds. Although seed demand assessments are carried out at a local level and passed on to governmental agencies which eventually engage in seed production, the lack of a market price system and of agro-dealers make distribution of seeds inefficient. Governmental control prevents private seed companies from breeding and seed production. Prices are set after a negotiation process between governmental organizations and not based on a farmer's actual willingness to pay. Further, the transaction costs involved in the seed system are carried by the government and not covered by the price of the seed, which is a disincentive to engage in efficient seed provision. Institutional innovations are proposed to improve access to and supply of seeds in Ethiopia.

Mobile communication technology plays an important role in the set of diversification strategies for innovation in agriculture. In Chap. 9, Baumüller investigates the effect of delivering services through mobile phones to smallholders in Kenya by outlining the key factors that have supported the growth of the Kenyan m-services sector. From her case study, the author concludes that, as a result of accessing information on demand and price, some farmers have changed their cropping patterns, whereas it remains inconclusive whether their bargaining power improved as a result of improved access to price information. The author recommends m-services to be embedded in complementary support programs and infrastructure developments to tackle other production and marketing limitations smallholders are facing. The author's findings support an essential argument throughout this volume: in order to improve the quality of life of smallholders in Sub-Saharan Africa and South Asia, technological innovations need to be accompanied by innovations in the broader institutional infrastructure.

In Chap. 10, Wünscher and Tambo present an innovative approach for identifying farmer innovations in upper East Ghana. By means of a farmer innovation contest, innovation behavior of farmers is stimulated. They become more creative, share their knowledge and engage in experimentation. The authors point to the fact that, despite poverty, a farmer's innovative capacity remains part of their capabilities, which can be made use of by changing incentive systems for innovation. Before introducing innovations, this human resource of creativity and innovation can be made better use of, reducing a farmer's dependence on external inputs (Gatzweiler, Chap. 2) and strengthening identity and self-esteem (Kumar, Chap. 4).

The contribution of women in agriculture tends to be undervalued, and innovations affect or bypass women in different ways. In Chap. 11, Beuchelt looks at gender and social equity trade-offs related to the promotion and diffusion of improved technologies for agricultural development. Her analysis underlines the importance of the social context-specificity of innovations in agriculture and calls for ex-ante assessments of potential gender and social effects from the introduction of innovations. Introducing innovations in a more participatory and gender sensitive manner can significantly contribute to meeting the food and nutritional needs of marginalized smallholders.

Productivity in agriculture does not only need to increase, it also needs to be sustainable. Adoption of new technologies depends on whether they are adapted to local conditions. In Chap. 12, Kriesemer, Virchow and Weinberger present an approach for assessing the sustainability and suitability of agricultural innovations for rural smallholders, taking into consideration environmental resilience, economic viability, and social sustainability, as well as technical sustainability. They develop indicators which help smallholders and local extension agents to make decisions which are locally adapted and more sustainable, increasing resilience.

Innovative sustainable land management (SLM) technologies and practices can help in addressing land degradation and improving rural livelihoods, however, they are not generally being adopted at larger scales. Mirzabaev (Chap. 13) looks at factors constraining the adoption of SLM innovations and identifies the major incentives for adoption: access to markets, credit and extension, as well as secure land tenure. SLM technologies alone, however, cannot comprehensively address land degradation. From his review, Mirzabaev concludes that a combination of technological, social and economic changes, achieved through synergies of bottom-up and top-down approaches, has led to successful examples for sustainable land management.

Virchow, Beuchelt, Kuhn and Denich (Chap. 14) look at the potentials of biomass-based value webs for economies in Sub-Saharan Africa by merging value chains into value webs. They apply a multi-dimensional methodology to understand the inter-linkages of value chains as a system of flexible, efficient and sustainable production, processing, trading and consumption, which they have termed "value webs". Their systems approach focuses on alternative uses of raw products, including recycling processes and cascading effects, during the processing phase of biomass utilization. This innovative perspective goes beyond the controversy of food versus non-food cropping systems, and helps identify synergy effects by combining the food and non-food branch of the biomass-based bioeconomy. These can lead to improved food security, employment, urgently needed export earnings, and to the conservation of environmental assets.

Part III presents contributions which show pathways towards sustainable intensification by the adoption of stress tolerant rice varieties, access to and use of improved seeds, and by adjusted crop combinations, such as integrated rice-fish farming practices, growing vegetables, cocoa, ginger and maize in Ghana, or wheat, rice and pulses for the Indian states of Odisha and Bihar. The contributions show where, how and under which circumstances further productivity growth potentials can be made use of by sustainable intensification strategies in agriculture.

In Chap. 15, Ahmed, Hernandez and Naher evaluate the adoption, retention, and diffusion of a set of modern stress-tolerant rice varieties promoted by the Cereal Systems Initiative in Bangladesh. Cultivating stress tolerant rice varieties is especially relevant in Bangladesh, which is one of the countries suffering the worst effects of climate change. The authors found generally higher adoption rates among educated and medium to large farmers but a relatively low adoption of stress-tolerant varieties by marginal farmers, which could be explained by their high risk aversion. The coverage of area under stress-tolerant varieties was found to be very low. High-yielding rice varieties which were introduced three decades ago are now being replaced by varieties which perform better under stressful agro-ecological conditions such as soil salinity, drought or submergence.

In Chap. 16, Malek, Hoque, Yesmin and Haque conducted a household survey in 5855 marginalized households and ask whether only cereal based cropping technologies can improve food and livelihood security of poor smallholders in marginal areas of Bangladesh. The authors carry out a mapping exercise for the identification of marginal rural areas with potential and then carry out an in-depth household analysis. The authors identify unused potential in each district and conclude that adverse agro-ecological conditions and associated perceived risks discourage poor smallholders from taking innovative steps and adopting technology useful for agricultural intensification. Cereal-based technologies are recommended as part of the solution, but should be integrated with other income diversification and agricultural diversification strategies. Intensive crop system, hybrid seeds, water management technologies, non-crop farming, non-farm enterprise and business are the suggested potential technological innovations for the study areas.

In Chap. 17, Islam carries out value chain, partial budgeting and SWOT analysis for assessing competitive performance and identifying the key factors affecting adoption and diffusion of rice-fish technology by indigenous farmers in Bangladesh. The overall quantitative results from gross margin, partial budgeting and gendered employment analysis show positive benefits due to the introduction of rice-fish technology as compared to rice monoculture. Findings indicate that rice-fish systems offer considerable potential for increasing overall agricultural productivity and farm incomes. Those potentials could be realized by government support and better value chain development.

The potentials of technologies for maize, wheat, rice and pulses in marginal areas of Bihar and Odisha, India are assessed in Chap. 18 by Joshi, Roy, Sonkar and Tripathi. Through multi-stakeholder consultations and secondary data analysis, the authors assess the potential of those technologies in terms of their agro-ecological suitability, as well as required complementary inputs. Maize and pulses are identified as crops that farmers aspire to cultivate. Their analysis also reveals that, in both states, there is a general, significant lack of awareness about agricultural technology. There is also a dissonance between expert technology evaluations and the valuations of likely adopters. Hybrid rice and varietal improvements in wheat and organic/semi organic farming are among the technologies with potential in the study area. Adoption of technologies which require more complementary inputs has been more difficult. The authors call for a holistic approach, taking the entire process from information to adoption support into account.

Chapter 19 presents an assessment of technological innovations for marginalized smallholders in Ghana. The authors explore community-based technologies that have the greatest potential for reducing poverty and vulnerability. Their findings show that the dominant technologies with the potential to reduce smallholder farmers' poverty in specific areas are inorganic fertilizers for Afigya-Kwabre; zero tillage for Amansie-West; storage facilities for Atebubu-Amantin; marketing facilities for Kintampo South; improved varieties for Gonja East; and pesticides for the Tolon Districts.

This study also underlines the importance of complementary government interventions and strengthening of extension services for marginalized farmers.

Hundie and Admassie (Chap. 20) assess the potential impact of yield increasing crop technologies on productivity and poverty based on data collected from two districts of Ethiopia: Basoliben in the Amhara region and Halaba in the Southern Nations, Nationalities and Peoples' region. The use of improved seeds, chemical fertilizer use and rowplanting techniques are considered to be promising technologies. The authors used the partial budget approach to analyze the potential impacts of those technologies. Results show that the mean per capita net benefit per day from technology adoption would be enough to lift the "better-off-poor" households out of poverty; not so, however, for the ultra-poor who require other livelihood strategies for improving their wellbeing.

Conclusions and Implications for Policy and Research

Sustainable intensification in agriculture is a response to the locally and globally increasing demand for food and non-food agricultural products, the need to maintain the functions of ecosystems and the stagnating availability of productive land. The contributions to this volume have reconfirmed that sustainable intensification is not just another optimization problem for ensuring higher productivity with less environmental impact. Rather it is a complex task of creating value through innovations in the institutional, organizational and technological systems of societies. Sustainable intensification is therefore not only a challenge to be met by and within the agricultural sector alone, but by society at large.

Opportunities for creating value by means of institutional, organizational and technological innovations can be grasped in two broad dimensions: in the dimension of human capabilities and that of agro-ecological potential. Whereas technological innovations help to close yield gaps, improve efficiency of production, and make full use of agro-ecological potentials, institutional and organizational innovations create the societal environment which enables people to make full use of their capabilities.

Depending on the specific context in which innovations are sought, strategies towards sustainable intensification will need to be more people and/or area-focused. These strategies include:

1. Intensifying crop production and minimizing environmental impact by making use of improved varieties and technologies adjusted to changing environmental and climatic conditions.
2. Diversifying agricultural crop production and production techniques to reduce external inputs and risks of failure and maintain agro-biodiversity.
3. Diversifying income opportunities and facilitating exit strategies, as well as enabling private business opportunities.

4. Providing basic educational, food, and health services for the most deprived, including them in social safety nets, and connecting them to communication and transport infrastructure.

Common lessons from the contributions to this volume confirm those conclusions and underline the importance of strategies which involve technological and institutional innovations. They can be summarized as follows:

1. Innovations for sustainable intensification in agriculture can be created in multiple value domains of society. Improving the wellbeing of smallholders means facilitating a change to positions in social and economic systems in which they are less marginalized. Being inclusive and sensitive to gender and social inequalities is not just fairer but also contributes to improving productivity.
2. Creating linkages and incentives which facilitate learning and encourage exchange of knowledge about innovations in agriculture. Such linkages unleash productivity potentials in the creation of value chains and value webs, stimulate private business engagement and improve collaboration between research and extension organizations. Apart from the pivotal role of smallholders themselves, entrepreneurs and governments need to collaborate to achieve sustainable intensification.
3. Technological innovations in various domains, such as mobile phones, stress-tolerant, high-yielding varieties and quality seeds, sustainable land management technologies and integrated farming, need to come together, rather than being pursued in isolation. They have the potential to boost productivity of smallholders. Actual productivity increases, however, depend on the extent of adoption. Adoption among poor smallholders tends to improve with increasing levels of education, increasing farm size and decreasing risk aversion.
4. Maintaining the sustainability of agricultural technologies for productivity growth requires a two-tiered dynamic approach: making technologies people-ready and making people technology-ready. Adjusting introduced technologies to the local context and local capabilities is as important as improving education and skills.

By investing more in farms, and by increasing efficiency of farming, a large portion of poverty and malnutrition could be reduced. Small farms play multifunctional roles in development (HLPE 2013). Importantly, public policies need to regard small farms as a part of a broader development solution. Policy support should be aimed at promoting the dynamism within the family farm sector itself, but also enhancing the dynamic interactions and integration of the family farm sector into the rest of the economy. All three options for small farm transformation need public policy attention, not just a smallholder growth strategy.

Small farmers can play key roles in fostering rural growth. It can no longer be assumed that the millions of small farmers will remain a peacefully suffering community in the future. Information and access to political influence through elections, farmer organizations, and more decentralized political systems are changing the context. Governments need to recognize and uphold smallholder rights, including the right to food, the right for self-organization, and the right to land and gaining equitable access to common pool natural resources (HLPE 2013). Crucially, these rights should be equally enjoyed by both men and women. There will be multiple futures for small farms. Appropriate policies facilitating the improvements of marginalized small farm communities need to be adjusted to the specific local and country contexts. Innovations play a key role in this.

Open Access This chapter is distributed under the terms of the Creative Commons Attribution-Noncommercial 2.5 License (http://creativecommons.org/licenses/by-nc/2.5/) which permits any noncommercial use, distribution, and reproduction in any medium, provided the original author(s) and source are credited.
The images or other third party material in this chapter are included in the work's Creative Commons license, unless indicated otherwise in the credit line; if such material is not included in the work's Creative Commons license and the respective action is not permitted by statutory regulation, users will need to obtain permission from the license holder to duplicate, adapt or reproduce the material.

References

Beddow JA, Parday PG, Koo J, Wood S (2010) The changing landscape of global agriculture. In: Alston J, Babcock B, Pardey P (eds) The shifting patterns of agricultural production and productivity worldwide. Iowa State University, The Midwest Agribusiness Trade Research and Information Center (MATRIC), Ames

Bergek A, Jacobsson S, Hekkert M, Smith K (2010) Functionality of innovation systems as a rationale for and guide to innovation policy. In: Smits RE, Kuhlmann S, Shapira P (eds) The theory and practice of innovation policy: an international research handbook. Edward Elgar, Cheltenham/Northampton, pp 115–144

Berry RA, Cline WR (1979) Agrarian structure and productivity in developing countries. John Hopkins University Press, Baltimore/London

Binswanger HP, Rosenzweig MR (1986) Behavioural and material determinants of production relations in agriculture. J Dev Stud 22(3):503–539

Chavas JP (2001) Structural change in agricultural production: economics, technology and policy. In: Gardner BL, Rausser GC (eds) Handbook of agricultural economics, vol 1, Part 1. Elsevier, Amsterdam/London, pp 263–285

Conway G, Waage J (2010) Science and innovation for development. UK Collaborative on Development Sciences (UKCDS), London

Eastwood R, Lipton M, Newell A (2010) Farm size. In: Pingali PL, Evenson RE (eds) Handbook of agricultural economics, vol 4, Elsevier. Amsterdam, London, pp 3323–3397

Fan S, Chan-Kang C (2005) Is small beautiful? Farm size, productivity, and poverty in Asian agriculture. Agric Econ 32(s1):135–146

FAO (2010) Characterisation of small farmers in Asia and the Pacific. Asia and Pacific Commission on agricultural statistics, twenty-third session, Siem Reap, 26–30 Apr 2010

Farrington J, Holmes R, Slater R (2007) Linking social protection and the productive sectors, ODI briefing paper 28. Overseas Development Institute, London

Fuglie K, Nin-Pratt A (2013) Agricultural productivity: a changing global harvest. 2012 Global food policy report. International Food Policy Research Institute, Washington DC, pp 15–28

Hayami Y (1996) The peasant in economic modernization. Am J Agric Econ 78(5):1157–1167

Hazell P, Poulton C, Wiggins S, Dorward A (2010) The future of small farms: trajectories and policy priorities. World Dev 38(10):1349–1361

Heltberg R (1998) Rural market imperfections and the farm size—productivity relationship: evidence from Pakistan. World Dev 26(10):1807–1826

Hirway I, Shah N (2011) Labour and employment under globalisation: the case of Gujarat. Econ Polit Weekly 46(22):57–65

HLPE (2013) Investing in smallholder agriculture for food security. A report by the high level panel of experts on food security and nutrition of the committee on world food security, Rome

Holmes R, Jones N (2009) Gender inequality, risk and vulnerability in the rural economy: refocusing the public works agenda to take account of economic and social risks, background report for the state of food and agriculture 2010. Overseas Development Institute, London

IFPRI (2009) India state hunger index, in report on global hunger index. International Food Policy Research Institute, Washington, DC

ILO (2006) Tackling hazardous child labour in agriculture: guidance on policy and practice. International programme on the elimination of child labour. International Labour Organisation, Geneva

Juma C, Tabo R, Wilson K, Conway G (2013) Innovation for sustainable intensification in Africa, the Montpellier panel. Agriculture for Impact, London

Le QB, Nkonya E, Mirzabaev A (2014) Biomass productivity-based mapping of global land degradation hotspots, ZEF discussion papers 193. Center for Development Research, Bonn

Leeuwis C, Ban A (2004) Communication for rural innovation: rethinking agricultural extension. Oxford Blackwell Science, Oxford

Llewelyn RV, Williams JR (1996) Nonparametric analysis of technical, pure technical, and scale efficiencies for food crop production in East Java, Indonesia. Agric Econ 15(2):113–126

Lowder SK, Skoet J, Singh S (2014) What do we really know about the number and distribution of farms and family farms in the world? Background paper for the state of food and agriculture 2014. Food and Agriculture Organization of the United Nations, Rome

Nkonya E, Gerber N, Baumgartner P, von Braun J, De Pinto A, Graw V, Kato E, Kloos J, Walter T (2011) The economics of desertification, land degradation, and drought: toward an integrated global assessment, ZEF discussion papers on development policy No 150. Center for Development Research, Bonn

Olinto P, Beegle K, Sobrado C, Uematsu H (2013) The state of the poor: where are the poor, where is extreme poverty harder to end, and what is the current profile of the world's poor? World Bank, Washington, DC

Pingali P (2012) Green revolution: impacts, limits and the path ahead. Proc Natl Acad Sci U S A 109:12302–12308

Schultz TW (1964) Transforming traditional agriculture. Yale University Press, New Haven

Sen A (1999) Development as freedom. Harvard University Press, Cambridge

Shukla R (2010) The official poor in India summed up. IJHD 4(2):301–328

Singh S (2012) The woes of rural wage labour. The limitations of inclusiveness. Institute of Economic Growth, New Delhi

Singh I, Squire L, Strauss J (1986) Agricultural household models: extensions, applications, and policy. Johns Hopkins University Press, Baltimore

The Montpellier Panel (2013) Sustainable intensification: a new paradigm for African agriculture. Agriculture for Impact, London

Tschajanov AV (1923) Die Lehre von der bäuerlichen Wirtschaft: Versuch einer Theorie der Familienwirtschaft im Landbau. P. Parey, Madison

von Braun J (2005) Agricultural economics and distributional effects. Agric Econ 32(s1):1–20. doi:10.1111/j.0169-5150.2004.00011.x

von Braun J (2011) Food security and the futures of farms. In: Lindquist AK, Verba T (eds) Report from the Bertebos conference on "Food security and the futures of farms: 2020 and towards 2050", Falkenberg, 29–31 Aug 2010

von Braun J, Mirzabaev A (2015) Small farms: changing structures and roles in economic development. ZEF discussion paper, No. 204. Center for Development Research (ZEF), Bonn

von Braun J, Gatzweiler FW (2014) Marginality. Addressing the nexus of poverty, exclusion and ecology. Springer, Dordrecht/Heidelberg/New York/London von Braun J, Kennedy E (eds) (1994) Agricultural commercialization, economic development, and nutrition. The Johns Hopkins University Press, Baltimore/London

Wiggins S, Kirsten J, Llambí L (2010) The future of small farms. Worl Dev 38(10):1341–1348

World Bank (2003) Reaching the rural poor: a renewed strategy for rural development. World Bank, Washington, DC

Yotopoulos PA, Lau LJ (1973) A test for relative economic efficiency: some further results. Am Econ Rev 63(1):214–223. doi:10.2307/1803137

Part I
Innovation for the Rural Poor:
Theory, Trends and Impacts

Chapter 2
Institutional and Technological Innovations in Polycentric Systems: Pathways for Escaping Marginality

Franz W. Gatzweiler

Abstract There is increasing consensus that institutional innovations are just as important for development as technological innovations. Polycentric systems are social systems of many autonomous decision centers operating under an overarching set of rules. The rural poor hold positions in polycentric systems, which are marginalized as a result of poverty, exclusion and degraded environments. Horizontal and vertical position changes by means of technological and institutional innovations within polycentric systems create escape routes from marginality. Productivity growth in agriculture through technological innovations is one way to enhance the wellbeing of the rural poor. Sustainable productivity growth, however, also requires institutional innovations. This contribution shows pathways for escaping marginality by means of technological and institutional innovations in polycentric systems.

Keywords Institutional innovations • Technological innovations • Polycentric systems • Marginality • Poverty

Innovation in Polycentric Systems

The rural poor are in positions in which they have limited options for selecting, with more or less control, from a set of alternative actions in light of available information about the general structure and outcomes that may be affected by benefits and costs assigned to actions and outcomes. This general situation is what Elinor Ostrom has referred to as an 'action situation' (Ostrom 2005, p. 189), and it serves as a starting point for describing how the rural poor can escape marginality in polycentric systems.

Polycentricity is "a social system of many decision centers having limited and autonomous prerogatives and operating under an overarching set of rules." (Aligica 2014, p. 37). Because polycentric systems "...take each other into account in

F.W. Gatzweiler (✉)
Center for Development Research (ZEF), University of Bonn, Bonn, Germany
e-mail: gatzweiler@gmail.com

© The Author(s) 2016
F.W. Gatzweiler, J. von Braun (eds.), *Technological and Institutional Innovations for Marginalized Smallholders in Agricultural Development*,
DOI 10.1007/978-3-319-25718-1_2

competitive relationships, enter into various contractual and cooperative undertakings or have recourse to central mechanisms to resolve conflicts. . ." they are not chaotic, but "function in a coherent manner", and therefore, Vincent Ostrom defines them as "systems" (Ostrom et al. 1961, p. 831). Ostrom (2005, p. 11) also refers to polycentricity as institutions being organized in multiple hierarchies of nested part-whole structures (see also Kiser and Ostrom 1982).

The concept of polycentricity has been used to refer to a particular type of multi-level organization (Aligica 2014; Aligica and Tarko 2012; Ostrom 1972, 1999a, b). Within polycentric order, rules govern action situations at multiple, nested hierarchical levels. Within the nested hierarchy of governance, at the operational level of decision-making, rules directly shape the outcomes of actions, for example, rules defining the amount and timing of fertilizer application on a field. Operational level rules are contained in and affected by the rules at a higher collective-choice level of decision-making.

Rules at the collective-choice level govern action situations, the outcomes of which are rules which define the operational-level rules. The collective-choice rules are affected by and contained in yet a higher level of decision-making, the constitutional choice level. Action situations in polycentric systems are therefore coupled. The ways in which systems of rules at each level match or complement rule systems at other levels has been addressed by the literature on institutional complementarity. Institutional complementarity (Aoki 2001; Gagliardi 2013; Amable 2000; Hall and Gingerich 2004) has also been referred to as fit, match and interplay of institutions at multiple scales complementing and reinforcing each other in order to improve the robustness or other performance criteria of social or ecological systems (Folke et al. 2007; Vatn 2012; Young 2002; Young et al. 2008). Such complementarities are specific to different types of technological innovations, types of actors and the strategies they follow, and the agro-ecological environments in which they operate (Gatzweiler 2014).

"Polycentric systems tend to enhance innovation, learning, adaptation, (. . .) and the achievement of more effective, equitable, and sustainable outcomes at multiple scales, . . ."(Ostrom 2010, p. 552). Technical and institutional innovation occurs by movement and respective changes of interaction patterns within polycentric systems. Interaction patterns are defined by sets of rules, here referred to as institutions. Among others, institutions define what actors do, how they perform actions and in which positions. Institutions thereby define the actors' action possibilities in action situations – the actors' functioning. Functioning is what people are actually able to be and do – the realized capabilities, people's wellbeing or quality of life (Nussbaum and Sen 1993).

Marginality in Polycentric Systems

Marginality is an involuntary position of an individual or group at the margins of social, political, economic, ecological, and biophysical systems, that prevents them from access to resources, assets, and services, restraining freedom of choice, preventing the development of capabilities, and causing extreme poverty (von Braun and Gatzweiler 2014). The marginality framework has been developed to explore the social, economic, and ecological spaces in which poor smallholders can advance and tap unused capabilities and agro-ecological potentials.

Being marginalized means having a marginal position in polycentric systems. From an institutional perspective, being marginalized means being excluded from institutional frameworks which enable innovation of technological or institutional infrastructures. With reference to the picture of a bicycle chain, being marginalized means being constrained in "changing gears" and creating value at higher levels of polycentric order. Institutions of marginality cannot always be directly identified by specific attributes or forms of rule configurations in which they appear; rather, they can be detected indirectly ex-post by (1) the ways in which decisions are made, (2) the types of behavior and actions they enable or inhibit, and (3) the effects they have on society and the environment.

Those institutions which define the everyday lives of the marginalized do not enable them to improve their wellbeing vertically by creating value in systems at a higher polycentric order.[1] Institutions of marginality consist of strong horizontal value-creating institutions, e.g., strong bonding social capital institutions, like traditions, habits or life-long contracts. Survival networks and social networks protect the marginalized from life-endangering extremes, but also prevent them from having access to resources, rights and services outside their immediate social and physical environment.

Being marginalized means being excluded, especially from vertical value creation, and having to carry the costs which are externalized to secure private benefits of other, less marginalized actors. Further, geographical remoteness and being socially and culturally confined prevents structural coupling with higher level systems of governance in the polycentric order, such as collective-choice or constitutional level decision-making, for example, by claiming citizen rights, or by an attempt to act outside the boundaries of tradition. Being trapped in marginality is determined by situations in which actors are less well connected to higher orders of decision-making, have relatively low potential and low levels of resilience (Carpenter and Brock 2008). Although change is needed for improving livelihoods, it is risky.[2] Therefore innovation often occurs among marginalized groups when potential and connectedness may be low but resilience is sufficiently high to absorb risks of failure.

---

[1] Operational, collective-choice and constitutional levels of decision making are levels of increasing polycentric order.

[2] To explain how institutions of marginality work, scholars have attempted to set up dichotomies of different types of institutions. North et al. (2007, 2009) explains how limited access orders work in contrast to open access orders. Acemoglu and Robinson (2012, 2008) identify extractive versus inclusive institutions in explaining why some nations prosper while others fail. Hagedorn (2008) refers to segregative in contrast to integrative institutions for explaining sustainable resource use in agriculture. Despite the different use of terms, all those theories contribute to explaining how institutions of marginality work: they limit access and property rights, create poverty and inequality, limit control of a community's rights to change it's own institutions, free the economic and political elites from accountability, and limit access to resources, rights, freedoms and opportunities. And, as Acemoglu and Robinson (2008, 2012) argue, most of the time, they are intentionally designed to extract resources and limit rights of a majority in favor of those of a political and economic elite.

Pathways Out of Marginality: Institutional and Technological Innovations

Escaping marginality requires innovation in institutional and technological infrastructures. Technological innovations in agriculture refer to products or machines which are new in the production process, whereas institutional innovations are changes in sets of rules which define peoples' positions and actions in action situations.

Institutions are the sets of rules which emerge from the attempt to structure social interactions. Social interactions, in turn, are shaped by institutions. Within institutional structures, behaviors in action situations[3] are motivated and enabled by reducing the costs involved in agreeing on why, when, and which actions are carried out, and by whom. The costs of agreeing on sets of rules can be substantial, and because of the sunk cost effect, they can actually be a reason for avoiding behavioral changes (Ostrom 2005, p. 58; Janssen et al. 2003; Janssen and Scheffer 2004). High investments for establishing political, social or economic arrangements which are cemented in institutional arrangements result in a resistance against reform, especially in periods of uncertainty and ecological vulnerability (Pakandam 2009).

Institutions enable and constrain actions within action situations. The boundaries of the action situation itself are defined by institutions, as well as membership, authority, and a variety of other rules which specify the scope of outcomes, the information available, or how costs and benefits are allocated (Ostrom 2005). Desired behavior is motivated by institutions, while undesired behavior is sanctioned. While some institutions have emerged spontaneously, without purposeful design, and have eventually become habits or traditions, others are the result of purposeful design. Either they are determined and imposed or they are the result of continuous deliberations and "struggles" to find and improve sets of rules which serve the purpose of the actors within or in charge of an action arena. Variations and combinations of both top-down and bottom up processes of institutional evolution are common (van den Bergh and Stagl 2003; Arnold 1980; Richerson and Boyd 2001; Farrel and Shalizi undated).

As technology always comes with rules on how to use the technology, who is to use it, and who has the rights to the outcomes from its use, institutional fit or complementarity is relevant for the adoption of technological innovations. Technological innovations can change action situations by changing roles and rights of actors related to the use of technology. Local technological innovations can

1. evolve from specific local institutional contexts (Fig. 2.1a), or
2. be developed elsewhere and introduced into specific local institutional contexts (Fig. 2.1b).

---

[3] An action situation involves "participants in positions who must decide among diverse actions in light of the information about how actions are linked to potential outcomes and the costs and benefits assigned to actions and outcomes" (Ostrom et al. 1994, p. 29).

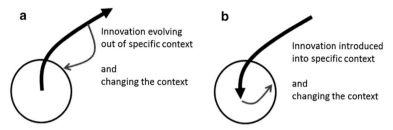

Fig. 2.1 Institutional or technological innovations evolving from (**a**) or introduced into (**b**) local contexts

When local technological innovations evolve from specific constraining contexts, they primarily intend to reduce workload or improve the working and living conditions of the marginalized. Grassroot innovators (Gupta et al. 2003) can improve their immediate working and living conditions despite being economically poor. This is not only because they possess intellectual capabilities and can be ingenious, but also because of their desire and motivation to lift the constraints of their work and living conditions.

The creativity of innovators is triggered by the adverse and constrained living and working conditions in which they are working. Although creativity usually declines under pressure (Amabile et al. 2002; Gunkel 2010), marginalized innovators attempt to improve their work and living conditions through innovations which make work processes easier, less dangerous and open opportunities. This seemingly adverse context for creativity results from tensions, conflicts and dilemmas which need to be resolved (Lewis et al. 2002; King et al. 1991) by means of creating a technological or institutional solution for an existing interaction pattern (status quo) which is perceived as being too costly or burdensome. The attempt to improve contextual fit or match (Bledow 2009; Young 2002) brings forth innovation. The enabling and inhibiting characteristics of institutions are also a potential source of conflict when actors are no longer willing to carry the costs of being constrained (see following chapter).

Alternatively, innovations which have been developed outside local contexts, once adopted, can change the specific local context with regards to who uses the technology, how it is made use of, and how costs and benefits from the use of it are allocated. In this case, the motivation to develop an innovation which mainly serves the interests of local communities is not always straightforward, as the innovator and the user are two different actors with different attributes and interests. Although innovators outside the local context can be motivated by noble goals (e.g., to help the poor increase yields by improved seeds), the incentive to implement a technology which has been costly to develop is high, and the hidden costs, like ecological externalities or additional costs required to accompany the innovation (e.g. fertilizers, pesticides, in the case of high yielding varieties), are less visible. From the perspective of the prospective users of the innovation, externally developed technologies are attractive because the costs for development are carried by an external innovator who can hide them in the per unit costs of the purchased technology over a long time period. Technologies externally developed for marginalized smallholders are not necessarily less sustainable than technologies developed from grassroot actors themselves, especially when the potential users of the technology have been part of the innovation process in innovation systems (Nyikahadzoi et al. 2012).

Enabling and Inhibiting Functions of Institutions

When do actors undertake efforts to change sets of rules which govern action situations? In other words, when do they undertake the effort to change institutions? Whereas institutions are the preconditions for continuity of interactions in changing and uncertain environments, institutions themselves frequently change and adapt to changing environments. Institutional innovation is also a result of the tension between the constraining and enabling functions of institutions.

When costs related to the constraining functions of institutions are perceived as too high, people undertake efforts to change them so that the benefits which can be captured by the enabling functions of institutions prevail. An incentive to change institutions can occur when the expected benefits from an old set of rules are less than the anticipated benefits from a new set of rules plus all the costs involved in the process of changing the rules – that is what Elinor Ostrom has referred to as the "rule change calculus" (2005, p. 245). When the old set of rules is perceived as being too costly, or when a set of institutions provide for too many constraints, actors may perceive the need for freedoms and institutional innovation. This need for change can only be perceived if an improvement of status quo is imaginable and when processes of structural coupling and cognition take place, as described above.

The variety of institutional arrangements in different action situations thereby "...provides the raw material for adaptation" (Axelrod and Cohen 1999, p. 32). Institutional arrangements define the decisions and actions people can take within action situations, as well as the relationships between actors interacting in those situations. They also define the linkages between action situations at various levels of the polycentric order. Accordingly, institutional innovations, here understood as rule changes or rule creations, can occur within or between coupled action situations, or they lead to the creation of rules (e.g., monitoring) which link action situations. Changing the relative frequency of types of rule configurations leads to institutional innovation.

Variety in rule configurations can be a source for innovation and wealth creation, however, "...beyond some level, variety ... can also be a source of debilitating inconsistency and conflict" (Axelrod and Cohen 1999, p. 122). This is the case when rule configurations in interaction domains provide space for too much individual wealth generation without contributing to collective (societal) wealth generation and without constraints from higher level institutions. The "tragedy of the commons" is a popular example (Hardin 1968).

Per definition, institutions have *enabling* and *inhibiting* functions. They enable interactions and functions within the subsystems they regulate, and thereby are an incentive for favorable behavior. They inhibit and create disincentives for unfavorable behavior. The rules created within a system or an organization can be perceived as enabling or constraining – depending on the cognition and aspirations of individuals within the organization. A constant adjustment process between internal and external organizational structures takes place, comparing status quo with how a system should or could be organized. This adjustment process has been referred to as *structural coupling*, or *cognitivism* (Rizzello 1999, p. 79) – a process that is structure-determined and structure-determining and that leads to structural congruence between two or more systems (Gatzweiler and Hagedorn 2013). This process of matching internal representations with external (changing) environments (Wexler 2008, p. 15), or structural coupling, is important for adaptation and survival in social and ecological systems.

Hagedorn (2008, p. 12) explains that institutions '...are always (more or less) integrative and segregative at the same time. On the one hand, they are integrating the interests of those who want to be protected against adverse effects into the decision-making of other actors by exposing these to corresponding constraints. On the other hand, they secure reasonable room to manoeuvre for a decision-maker, even if this may require that not all adverse effects are kept away from other actors.' Institutions can thereby function as incentives or disincentives; they can reward or punish (Hodgson 2006, p. 2). They are never universal and always specific, and they are inclusive for some and exclude others.

How much freedom people have in the actions they can take and how constrained people are in the decisions they can make is defined by institutions. The adjustment process between enabling and constraining functions of institutions reoccurs in the discourse of freedom and organization. Vincent Ostrom (1983, p. 1) has defined freedom as "...the capacity to act on the basis of one's own considered judgment" and as "... the availability of adaptive potential." In that sense, development gains can be made when "...constraints (are built) into people's relationships..." and when "human beings (...) take advantage of each other's capabilities and pursue opportunities for joint advantage through teamwork."

In order to be able to tap those capabilities, it is necessary to "...establish a basis for creating stable expectations about their relationships with one another" (ibid., p. 3). Wealth creation under freedom can only be realized in the presence of organization. On the other side, organization requires rules, rulers and those being ruled. Grasping the benefits of ordered relationships is only possible if sanctions can credibly be imposed on defectors. If those who rule are excluded from the rules imposed on the ruled, organization can become a threat to freedom. "The prerogatives of rulership become a threat to freedom when organization provides unequal opportunities for some to exploit and oppress others" (ibid., p. 5).

Horizontal and Vertical Value Creating Institutions in Polycentric Order

Institutional innovations which are enabling create value. Value creating institutions are part of polycentric order and therefore value can be created horizontally and vertically within polycentric systems. Enabling institutions can therefore be regarded as sets of rules which increase the set of opportunities for escaping marginality. They define the type of actions actors can take in action situations. Having access and being connected to those institutions enables actors to create value or, as in the case of public service provision, benefiting from value created by the community. Being fully included in the systems in which value is created means being able to make use of low institutional entropy (Auster 1974, 1983), e.g., a CEO running an organization, a government collecting taxes from organizations, or an individual taking a position in a company after studying at university. Low institutional entropy means ordered interactions which produce outcomes that are beneficial. Institutional entropy is a component of what has been referred to as social entropy (Bailey 1990, 2006) and social capital (Putnam 2000). In the polycentric organization of social systems, social capital can be built by bonding and bridging, and similarly, institutional innovations can be the result of vertical or horizontal value creation.

Innovations from horizontal value creation result from a combination of actions (or materials) within a given biophysical or social environment. Horizontal value creation in social systems (bonding) happens by improving social and exchange relationships, or enforcing trust and solidarity. Assembling parts to build tools and machines or recycling waste for new uses are examples of horizontal value creation in physical systems. The marginalized benefit from horizontal value creation by being integrated into and benefiting from community activities and local networks and by making use of simple technological solutions. Rotating savings and credit associations in Indonesia (Varadharajan undated) are an example. Creating value through strong ties and identities within the community is an important social safety net, but it can also be a disincentive to break out of the bonds and create value vertically.

Vertical value creation in polycentric systems happens by connecting to actors and networks outside one's own networks or by making use of the rules and policies at a higher level of decision-making. Linking to knowledge and social systems beyond the local level opens opportunities for value creation by having access to external assets and information and by making rules instead of only obeying them. Gupta et al. (2003) provide the example of the Gujarat Grassroots Innovation Augmentation Network (GIAN). GIAN adds value to local innovations by facilitating the shift from innovators to entrepreneurs, access to risk capital, and technological knowhow or design input which transforms the innovation into a commercial product. GIAN also files patents on behalf of the innovators or provides support in building business networks.

Another example of successful innovation which specifically takes vertical value creation into consideration is the spread of the shallow tubewells for (boro) rice irrigation during the dry season in Bangladesh in the 1980s. The expansion of boro rice production from 10 % in the 1960s to 60 % of the total rice production by 2008 was triggered by a change in policy which liberalized the procurement and marketing of irrigation equipment in the 1980s. The availability of lower cost machines from China and the elimination of import duties and standardization requirements led to the rapid spread of this technology and an increase in productivity (Hossain 2009).

Vertical and horizontal value creation reduces institutional entropy and serves the purpose of reducing transaction costs (Iskandar and Gatzweiler 2014). Being able to produce, exchange, transact and communicate more efficiently can be achieved by changing the set of rules which structure various types of exchange and interactions. Very often, this has been referred to as "enabling institutional environment" (Joshi and Moore 2000). North (1990, p. 125) lists three types of institutional innovations which have led to increasing economic performance and "the rise of the Western World". They relate to lowering the costs of mobility, capital and information, and spreading the costs of risk in economic exchange relationships.

Horizontal and vertical value creation are not always equally important. Depending on the situation of the marginalized, it may be more important to create strong networks and innovate horizontally through bonding, especially if transcending accustomed behaviors is a threat to survival and social integrity. In those cases, bonding is good for adapting, getting along and building survival networks. In other situations, vertical value creation is an opportunity for getting ahead, connecting to actors at different locations of the value chain, making decisions on rights which secure the benefits from innovation and building mobility networks (Briggs 1989; Briggs et al. 2010; Dominguez and Watkins 2003).

Social and physical technology (Beinhocker 2006, p. 241)[4] mutually influence each other and co-evolve (van de Ven and Garud 1994; Lewin et al. 1999). The use and diffusion of physical technology very often depends on an enabling institutional environment. Institutions not only provide freedoms and incentives to come up with new ideas and innovate, but also enable diffusion and make use of technological innovations once they are available. Using a new technology will not be considered as long as the costs, risks and modes of use and maintenance are not institutionalized. In order to create benefits from the use of new technology, it might be necessary to change policies or create new organizations, like machine rings, and define property rights related to the use of the technology and the outcomes from said use. Physical technologies may also be a trigger and driver

---

[4] Social technology is referred to by Beinhocker (2006, pp. 242, 261) as, "The rule of law, the existence of property rights, well-organized banking system, economic transparency, lack of corruption and other social and institutional factors." Physical technology are the tools, techniques and machines, and the transmission of the knowledge as to how to produce and make use of them.

for institutional change, as opposed to situations in which a general favorable institutional environment for enabling the use and diffusion of a technology must be achieved first. The use of mobile phones, e.g., has created new forms of social relationships and rules of how to communicate, interact and behave in those social networks (Srivastava 2005; Huberman et al. 2008).

In multi-level polycentric orders, value can be created at horizontal and vertical levels. Value creation at horizontal levels refers to the governance of action situations at one level of decision-making. For example, farmers can cooperate and help each other during the harvest season, or achieve improved productivity resulting from the use of technologies, like improved seeds or fertilizers. Such innovations are the result of doing things differently within existing institutional arrangements. They are adapted to and fitted into status quo social structures without fundamentally changing them. Vertical value creation happens when associations, organizations, networks or cooperatives are established which are specifically designed to change previous cost/benefit streams and interaction patterns. Additional value is created by linking to higher levels of organization, e.g., by improving access to rights and resources and inclusion at higher decision-making levels.

Institutional innovation is a continuous process of incremental changes. "Incremental change comes from the perceptions of the entrepreneurs in political and economic organizations that they could do better by altering the existing institutional framework at some margin." (North 1990, p. 6), or it is a response to failure, crisis and constraints, presuming information flows and feedback. North further elaborates (1990, p. 108) that, "The success stories of economic history describe the institutional innovations that have lowered the costs of transacting and permitted capturing more of the gains from trade ...". The existence of transaction and interaction costs and a situation of incomplete information, together with the desire to increase gains from interacting, is therefore a necessary precondition for innovation.

The condition of perfect fit, i.e., a situation in which transaction costs for existing interaction patterns are at a minimum, where actors have the same information and mental models of how the world around them works, and a situation which is perceived as being too expensive to change behavior (e.g., because of contracts), is also referred to as the lock-in effect (North 1990, p. 7). Such lock-in occurs in a state of equilibrium. In such a state, the economy performs in an endless round of activities – something Schumpeter (Greiner and Hanusch 1994, p. 262) termed the "circular flow" – a situation in which demand and supply correspond, and interest rates and savings are equal to zero. Profits and losses are zero too, assuming that there is also no time preference. Time does not play a role because production and consumption are synchronized. For given prices, all economic agents have their economic transactions fulfilled. This Walrasian system[5] is in a state of unending stasis, in which individual tastes, techniques and resources are fixed and no innovation occurs.

Doner (2010) makes the vulnerability argument which explains institutional innovation as a response to failures, pressures or crises. The vulnerability argument says that institutional innovations occur in vulnerable systems, systems in which political leaders are confronted with popular discontent and external pressures. They are under pressure by the need for maintaining living standards, avoiding public unrest, maintaining national security, the need for foreign exchange and acting within scarce budget constraints (Doner et al. 2005). At the level of the individual smallholder, the continued absence of support from outside and the remote and marginalized situation can lead individuals to become creative and find technological solutions for their locality (Gupta et al. 2003).

Schumpeter's (1934) notion of creative destruction can be regarded as creating value through new institutional arrangements and leaving behind those which have hindered innovation and suppressed entrepreneurial growth and development. The entrepreneur breaks out of the circular flow model in equilibrium and diverts capital to novel uses in order to be creative and make use of market niches. In order to finance innovation, he/she diverts labor (from the circular flow) to novel uses. However, because resources are already optimally allocated in the equilibrium state, this diversion leads to instability of the equilibrium state in industrial capitalism, which explains outbursts of violence and catastrophes in the history of capitalism (Schumpeter 1939, 1942).

From a polycentricity perspective, the institutions of marginalized communities are not well linked to higher level orders of decision-making and value creation in specific action situations. The pathways do not exist through which information can flow and the monitoring of the performance of actions could occur. Segmented sections of rural society interact more among themselves than with political and economic elites, and their rights to participate and self-organize are constrained. The linkages between rural populations and decision-making elites are weak and interactions do not take place within viable organizational units which could build up sufficient pressures for triggering institutional reform. The coherence and functioning of nation-states in Africa (Gatzweiler 2013, p. 4) are insufficient and the positions of political leaders are too secure for building up the pressures needed to trigger substantial institutional reforms, as explained in the vulnerability argument.

Without explicitly mentioning polycentricity, Gupta et al. (2003, p. 981) refer to similar dynamics hindering institutional innovation by transcending constraints at

---

[5] Leon Walras (1834–1910) describes a closed circular system of exchange between producers and consumers, firms and households. Humans, in that model, act with regards to themselves and according to a very particular conception of rationality. Their preferences are exogenous, i.e., independent of social context and the preferences of others. This self-regarding feature of economic agents who are not influenced by others is a central building block of the Walrasian model. Without it, the mathematical proof of the efficiency of competitive equilibrium collapses.

one institutional level of the nested hierarchy: "A lot of people have learnt to adapt and adjust to a constraint rather than transcend it. In the case of women based technological problems, this constraint has been a consequence of cultural institutions, which prevented them from acquiring blacksmithy or carpentry tools. Women are very creative in coping with the constraints. ..."

Conclusions

Improving the wellbeing of marginalized smallholders means changing their position in polycentric systems to where they have more influence on deciding what they are and what they do. This can be achieved through technological and institutional innovations which emerge from social systems or are introduced to them from outside. Pathways for improving the wellbeing of marginalized smallholders can be found by making use of the enabling functions of institutions and creating value horizontally and vertically.

When connectedness to higher level decision-making is low and the marginalized have little control over the institutions which inhibit them from unfolding their potential, and when social system resilience is sufficiently high to absorb risks from experimentation, innovation will occur, leading to more favorable positions at multiple scales in polycentric systems.

Although appropriate technology innovations need to fit into local social and ecological context in order to be adopted, the process of institutional and technological innovation is triggered by situations of misfit creating costs carried by the economically and socially marginalized. If those costs become too high, decisions will be made to change the status quo and innovation occurs. The trigger for changing institutions will occur when the calculus for institutional innovation is positive, which means when expected benefits from an old set of rules are less than the anticipated benefits from a new set of rules plus all the costs involved in the process of changing the rules.

Technological innovations need to be accompanied by institutional changes if progress towards productivity growth is to be sustainable. Technological innovations often improve efficiency (horizontal value creation), e.g., by reducing labor costs, but they do not always enable marginalized smallholders to create value vertically, e.g., by improving access to rights, services or higher decision-making levels. A sole focus on technology innovations in agriculture can therefore improve agricultural productivity growth but make little progress towards sustainable intensification. Donors, investors and development organizations supporting technological innovations in agriculture therefore need to better understand how technological and institutional innovations change the marginal position of poor smallholders in polycentric systems.

In order to facilitate escape from marginality, a broader concept of value creation in polycentric systems is required that reflects the multiple dimensions in which rural smallholders are currently marginalized but also shows diverse pathways down which they could progress.

Open Access This chapter is distributed under the terms of the Creative Commons Attribution-Noncommercial 2.5 License (http://creativecommons.org/licenses/by-nc/2.5/) which permits any noncommercial use, distribution, and reproduction in any medium, provided the original author(s) and source are credited.

The images or other third party material in this chapter are included in the work's Creative Commons license, unless indicated otherwise in the credit line; if such material is not included in the work's Creative Commons license and the respective action is not permitted by statutory regulation, users will need to obtain permission from the license holder to duplicate, adapt or reproduce the material.

References

Acemoglu D, Robinson J (2008) The role of Institutions in growth and development. Working paper No 10. Commission on Growth and Development, The World Bank, Washington, DC

Acemoglu D, Robinson J (2012) Why nations fail, the origins of power, prosperity, and poverty. Crown Business, New York

Aligica P (2014) Institutional diversity and political economy. The ostroms and beyond. Oxford University Press, New York

Aligica P, Tarko V (2012) Polycentricity: from Polanyi to Ostrom, and beyond. Governance 25(2):237–262

Amabile T, Hadley CN, Kramer SJ (2002) Creativity under the gun. Special issue on the innovative enterprise: turning ideas into profits. Harv Bus Rev 80(8):52–61

Amable B (2000) International specialisation and growth. Struct Chang Econ Dyn 11(4):413–431

Aoki M (2001) Toward a comparative institutional analysis. MIT Press, Cambridge, MA

Arnold RA (1980) Hayek and institutional evolution. J Libert Stud 4(4):341–352

Auster RD (1974) The GPITPC and institutional entropy. Public Choice 19(1):77–83

Auster RD (1983) Institutional entropy, again. Public Choice 40(2):211–216

Axelrod R, Cohen MD (1999) Harnessing complexity. Free Press, New York

Bailey KD (1990) Social entropy theory. State University of New York Press, New York

Bailey KD (2006) Living systems theory and social entropy theory. Syst Res Behav Sci 23(3):291–300

Beinhocker E (2006) The origin of wealth. Evolution, complexity and the radical remaking of economics. Harvard Business School Press, Boston

Bledow R (2009) Managing innovation successfully: the value of contextual fit. Dissertation, Justus-Liebig University, Gießen

Briggs XS (1989) Brown kids in white suburbs: housing mobility and the multiple faces of social capital. House Policy Debate 9(1):177–221

Briggs XS, Popkin SJ, Goering J (2010) Moving to opportunity. The story of an american experiment to fight ghetto poverty. Oxford University Press, New York

Carpenter SR, Brock WA (2008) Adaptive capacity and traps. E&S 13(2):40

Dominguez S, Watkins C (2003) Creating networks for survival and mobility: social capital among African-American and Latin-American low-income mothers. Soc Probl 50(1):111–135

Doner RF (2010) Explaining institutional innovation. Case studies from Latin America and East Asia. Social Science Research Council, New York

Doner RF, Ritchie BK, Slater D (2005) Systemic vulnerability and the origins of developmental states: Northeast and Southeast Asia in comparative perspective. IO 59(2):327–361

Farrel H, Shalizi C (undated) Evolutionary theory and the dynamics of institutional change. http://
    iis-db.stanford.edu/evnts/6595/Farrell-Shalizi_4.3.pdf. Accessed 21 Feb 2014
Folke C, Pritchard L, Berkes F, Colding J, Svedin U (2007) The problem of fit between ecosystems
    and institutions: ten years later. Ecol Soc 12(1):30. http://www.ecologyandsociety.org/vol12/
    iss1/art30/. Accessed 10 June 2015
Gagliardi F (2013) A bibliometric analysis of the literature on institutional complementarities.
    Paper presented at the 2nd GROE Meeting, Hitchin, 20–22 Sept 2013. [online] http://www.
    uhbs-groe.org/user/image/gagliardi.pdf
Gatzweiler FW (2013) Institutional and livelihood changes in East African forest landscapes:
    decentralization and institutional change for sustainable forest management in Uganda, Kenya,
    Tanzania and Ethiopia. Peter Lang GmbH, Internationaler Verlag der Wissenschaften,
    Frankfurt
Gatzweiler FW (2014) Value, institutional complementarity and variety in coupled socio-
    ecological systems. Ecosyst Serv 10:137–143
Gatzweiler FW, Hagedorn K (2013) Biodiversity and cultural ecosystem services. In: Levin SA
    (ed) Encyclopedia of biodiversity, vol 1, 2nd edn. Academic Press, Waltham, pp 332–340
Greiner A, Hanusch H (1994) Schumpeter's circular flow, learning by doing and cyclical growth. J
    Evol Econ 4(3):261–271. doi:10.1007/BF01236372
Gunkel J (2010) Forms of work satisfaction and creativity (Formen der Arbeitszufriedenheit und
    Kreativität). Dissertation, Fakultät für Wirtschaftswissenschaften, Technische Universität
    München, Munich
Gupta AK, Sinha R, Koradia D, Patel R, Parmar M, Rohit P, Patel H, Patel K, Chand VS, James TJ,
    Chandan A, Patel M, Prakash TN, Vivekanandan P (2003) Mobilizing grassroots' technolog-
    ical innovations and traditional knowledge, values and institutions: articulating social and
    ethical capital. Futures 35(9):975–987. doi:10.1016/S0016-3287(03)00053-3
Hagedorn K (2008) Segregating and integrating institutions – a dichotomy for nature related
    institutional analysis. In: Schäfer C, Rupschus C, Nagel U-J (eds) Enhancing the capacities of
    agricultural systems and producers, proceedings of the second green week scientific confer-
    ence. Margraf-Publishers, Weikersheim
Hall PA, Gingerich DW (2004) Varieties of capitalism and institutional complementarities in the
    macroeconomy: an empirical analysis. Max-Planck-Institut für Gesellschaftsforschung, Köln
Hardin G (1968) The tragedy of the commons. Science 162(3859):1243–1248
Hodgson G (2006) What are institutions? J Econ Iss 40(1):1–25
Hossain M (2009) The impact of shallow tubewells and boro rice on food security in Bangladesh.
    International Food Policy Research Institute, Washington, DC
Huberman BA, Romero DM, Wu F (2008) Social networks that matter: Twitter under the
    microscope. http://arxiv.org/pdf/0812.1045v1.pdf. Accessed 24 Apr 2014
Iskandar DD, Gatzweiler FW (2014) An optimization model for technology adoption of margin-
    alized smallholders: theoretical support for matching technological and institutional innova-
    tions. ZEF Working paper 136. Center for Development Research, University of Bonn, Bonn
Janssen M, Scheffer M (2004) Overexploitation of renewable resources by ancient societies and
    the role of Sunk-cost effects. Ecol Soc 9(1):2
Janssen MA, Kohler TA, Scheffer M (2003) Sunk-cost effects and vulnerability to collapse in
    ancient societies. Curr Anthropol 44(5):722–728. doi:10.1086/379261
Joshi A, Moore M (2000) Enabling environments: do anti-poverty programmes mobilise the poor?
    J Dev Stud 37(1):25–56. doi:10.1080/713600057
King N, Anderson N, West MA (1991) Organizational innovation in the UK: a case study of
    perceptions and processes. Work Stress 5(4):331–339. doi:10.1080/02678379108257031
Kiser LL, Ostrom E (1982) The three worlds of action: a metatheoretical synthesis of institutional
    approaches. In: Ostrom E (ed) Strategies of political inquiry. Sage, Beverly Hills
Lewin AY, Long CP, Carroll TN (1999) The coevolution of new organizational forms. Organ Sci
    10(5):535–550. doi:10.1287/orsc.10.5.535

Lewis MW, Welsh MA, Dehler GE, Green SG (2002) Product development tensions: exploring contrasting styles of project management. Acad Manage J 45(3):546–564. doi:10.2307/3069380

North DC (1990) Institutions, institutional change and economic performance. Cambridge University Press, Cambridge

North DC, Wallis JJ, Webb SB, Weingast BR (2007) Limited access orders in the developing world: a new approach to the problems of development, World Bank policy research working paper series. World Bank, Washington, DC

North DC, Wallis JJ, Weingast BR (2009) Violence and social orders. A conceptual framework for interpreting recorded human history. Cambridge University Press, New York

Nussbaum M, Sen A (eds) (1993) The quality of life. Clarendon, Oxford

Nyikahadzoi K, Pali P, Fatunbi LO et al (2012) Stakeholder participation in innovation platform and implications for Integrated Agricultural Research for Development (IAR4D). IJAF 2(3):92–100. doi:10.5923/j.ijaf.20120203.03

Ostrom V (1972) Polycentricity. Originally prepared for delivery at the 1972 annual meeting of the American political science association, Washington, DC, Hilton Hotel, 5–9 Sept 1972. Revision published in: McGinnis (eds) Polycentricity and local public economies: readings from the workshop in political theory and policy analysis (1999), pp 119

Ostrom V (1983) Freedon and organization. Workshop in political theory and policy analysis working paper W83-16. Bloomington

Ostrom V (1999a) Polycentricity (1). In: McGinnis M (ed) Polycentricity and local public economies. Readings from the workshop in political theory and policy analysis. University of Michigan Press, Ann Arbor, pp 52–74

Ostrom V (1999b) Polycentricity (2). In: McGinnis M (ed) Polycentricity and local public economies. Readings from the workshop in political theory and policy analysis. University of Michigan Press, Ann Arbor, pp 119–138

Ostrom E (2005) Understanding institutional diversity. Princeton University Press, New Jersey

Ostrom E (2010) Polycentric systems for coping with collective action and global environmental change. Glob Environ Chang 20(4):550–557. doi:10.1016/j.gloenvcha.2010.07.004

Ostrom V, Tiebout CM, Warren R (1961) The organization of government in metropolitan areas: a theoretical inquiry. Am Polit Sci Rev 55(4):831. doi:10.2307/1952530

Ostrom E, Gardner R, Walker J (1994) Rules, games, and common-pool resources. The University of Michigan Press, Ann Arbor

Pakandam B (2009) Why Easter Island Collapsed: an answer for an enduring question. Working papers No 117/09. London School of Economics, London

Putnam RD (2000) Bowling alone, the collapse and revival of American community. Simon and Schuster, New York

Richerson PJ, Boyd R (2001) Institutional evolution in the holocene: the rise of complex societies. In: Runciman G (ed) The origin of human social institutions. Proceedings of the British Academy 110, Oxford University Press, Oxford, pp 197–204

Rizzello S (1999) The economics of the mind. Elgar, Cheltenham

Schumpeter JA (1934) The theory of economic development. Harvard University Press, Cambridge

Schumpeter J (1939) Business cycles: a theoretical, historical, and statistical analysis of the capitalist process. McGraw-Hill, London

Schumpeter JA (1942) Capitalism, socialism and democracy, Harper, 1975 [orig. pub. 1942], New York

Srivastava L (2005) Mobile phones and the evolution of social behaviour. Behav Inform Technol 24(2):111–129. doi:10.1080/01449290512331321910

Van de Ven A, Garud R (1994) The coevolution of technical and institutional events in the development of an innovation. In: Baum JAC (ed) Evolutionary dynamics of organizations. Oxford University Press, Oxford, pp 425–443

Van den Bergh J, Stagl S (2003) Coevolution of economic behaviour and institutions: towards a theory of institutional change. J Evol Econ 13(3):289–317. doi:10.1007/s00191-003-0158-8

Varadharajan S (undated) Explaining participation in Rotating Savings and Credit Associations (RoSCAs): Evidence from Indonesia, http://www.microfinancegateway.org/gm/document-1.9.26314/23451_file_Arisan_May.pdf. Accessed 28 Apr 2014

Vatn J (2012) Can we understand complex systems in terms of risk analysis? Proc Inst Mech Eng Part O J Risk Reliab 226(3):346–358

von Braun J, Gatzweiler FW (2014) Marginality, addressing the nexus of poverty, exclusion and ecology. Springer, Dordrecht/Heidelberg/New York/London

Wexler BE (2008) Brain and culture: neurobiology, ideology, and social change. MIT Press, Cambridge, MA/London

Young OR (2002) The institutional dimensions of environmental change: fit, interplay, and scale/ Oran R. Young. Global environmental accord. MIT Press, Cambridge

Young OR, King LA, Schroeder H (2008) Institutions and environmental change: principal findings, applications, and research frontiers. MIT Press, Cambridge

Chapter 3
Innovations for Food and Nutrition Security: Impacts and Trends

Evita Pangaribowo and Nicolas Gerber

Abstract Achieving food and nutrition security (FNS) is a priority in developing countries. One of the key routes to achieve a resilient global food system and improved FNS requires a reorientation of relevant policies. Among them, policies associated with the creation, adoption and adaptation of technologies, knowledge and innovations and with their related institutional adjustments are key factors to counter the complex and evolving challenges of the global food system. In line with this notion, the objectives of this chapter are severalfold. First, we discuss the main features of innovations for FNS. Second, we describe the impact of innovations on FNS using the examples of new platform and traditional technology. Third, this chapter elaborates on the views of a variety of stakeholders concerning the impacts of technological and institutional innovations, as well as the future priorities of FNS innovation.

Keywords Food and nutrition security • Food system • Resilience • Innovations • Policies

Introduction

According the UN Secretary-General *Ban Ki-moon, food* and nutrition security (FNS) are the foundations of a decent *life*.[1] The UN Universal Declaration of Human Rights stated that "everyone has the rights to a standard of living adequate for the health and well-being of himself and his family, including food" and mandated food as a human right. One of the key routes to achieve a resilient global

---

[1] http://www.un.org/waterforlifedecade/food_security.shtml.

E. Pangaribowo (✉)
Department of Environmental Geography, University of Gadjah Mada, Yogyakarta, Indonesia
e-mail: evitahp@ugm.ac.id

N. Gerber
Center for Development Research (ZEF), University of Bonn, Walter Flex Strasse 3, 53113 Bonn, Germany
e-mail: ngerber@uni-bonn.de

© The Author(s) 2016
F.W. Gatzweiler, J. von Braun (eds.), *Technological and Institutional Innovations for Marginalized Smallholders in Agricultural Development*,
DOI 10.1007/978-3-319-25718-1_3

41

food system and improved FNS requires a reorientation of relevant policies. Among them, policies associated with the creation, adoption and adaptation (a process called diffusion) of technologies, knowledge and innovations and with their related institutional adjustments (Juma and Yee-Cheong 2005) are key factors for countering the complex and evolving challenges of the global food system.

FNS worldwide is currently in an alarming state, despite global progress towards the achievement of Millennium Development Goal number 1. The steep rise in food prices in 2007–2008 and the volatility of food prices in the following period have negatively impacted the poor in particular, and some studies have shown an important reduction in calorie intake and an increase in poverty rates in general (Webb 2010). It is recognized that the overall impact of the high food prices on welfare depends on the status of the target groups or the time horizon of the analysis (e.g., net food buyers versus sellers, short term versus long term impacts) (Swinnen 2011). Notwithstanding, the episode of high and volatile food prices of 2007–2008 has definitely slowed down progress in terms of decreased malnutrition (von Braun and Tadesse 2012) and hampered achievements in the fight against food insecurity. Further, many countries (mostly of low middle income) are currently experiencing a triple burden of under- or malnutrition: undernourishment, overnourishment and hidden hunger. Undernourishment, or hunger, is effectively the insufficient intake of energy and proteins, which has been directly linked to diseases and premature death, as well as poor physical development. The UNICEF framework of undernutrition (Black et al. 2008) laid out how the lack of household access to and use of nutritious foods, health care, water and sanitation services are among the major drivers of undernutrition. Overnourishment is the excessive intake of dietary energy, resulting in overweight, obesity and chronic diseases, as well as with increasing risks of non communicable diseases (NCD). Overnourishment is driven by many factors, including the globalization of trade, finance, change of information and cultures, change of lifestyles and physical activity patterns, and demographic shifts – in particular, urbanization (Hawkes et al. 2005; Popkin et al. 2012). The third burden, hidden hunger, is a situation when people suffer from micronutrient deficiency. The major driver of micronutrient deficiencies is lack of access to and consumption of nutrient dense foods such as fruit and vegetables. In the low and middle income countries, people are mostly suffering from iron, zinc, vitamin A, iodine and folate deficiency (Muthayya et al. 2013). Iron deficiency is one of the leading causes of maternal mortality. Particularly for children, the triple burden of malnutrition has devastating effects on later life, including physical and cognitive development. Under- and overnourishment cannot coexist in the same individual, but can be observed in the same household. Micronutrient deficiency can coexist with under- or overnourishment in an individual and in a household.

The chapter primarily aims to discuss the main features of innovations for FNS, as well as present their impact pathways. A consultation with several stakeholders of the food (innovation) system about the impacts of innovations on FNS, now and in the future, illustrates the plurality of views about the necessity to invest in different types of innovations for FNS, thus helping to identify priorities for action in the field of FNS and innovation. This consultation suggests that, although technological innovation is important for increasing agricultural production, insti-tutional factors such as farmers' collective action should be well supported in directing future science policy for agriculture and FNS. Understanding the impacts of innovations on FNS and the priorities for innovation in the future requires better knowledge on the current state of FNS, which is discussed in the next section.

Current FNS Situation

FAO reported last year that, in the period of 2011–2013, around one in eight people in the world were estimated to be suffering from chronic hunger, a situation where people do not have enough food to perform an active life (FAO et al. 2013). Even though this figure was slightly improved compared to the previous period, substantial efforts are needed to meet the Sustainable Development Goal No. 2 of ending hunger by 2013. The efforts should account for regional differences, although globally, Sub-Saharan Africa and South Asia still rank highest as the homes of malnourishment (Fig. 3.1[2]). As we can see from Fig. 3.1, Sub-Saharan Africa and South Asia have the highest prevalence of stunting among children under-five, both at 38 %. The consequences of stunting for later life are enormous. Victora et al. (2008) pointed out that stunted children were associated with low human capital and higher risk of adult diseases. Apart from stunting, underweight is also more prevalent in South Asia and Sub-Saharan Africa than in other parts of the world, at a rate of 33 % and 21 %, respectively. The consequences of underweight are also severe. Empirical studies show that being underweight in childhood was positively associated with low adult body-mass index, intellectual performance and work capacity (Martorell 1999; Victora et al. 2008). For wasting, it is also evident that the situation in South Asia is alarming. In that region, around one in six children is suffering from wasting. Wasting indicates current or acute malnutrition and children suffering from wasting have a higher mortality risk.

Recent studies revealed that many developing countries have experienced a multiple burden of malnutrition where undernutrition (mainly stunting) and overnutrition (overweight and obesity) coexist in the same population or household (Hawkes et al. 2005; FAO 2006). UNICEF reported that 7 % of children under-five were overweight in 2012, and this number represents a 43 % increase from 1990. Overnutrition is becoming an alarming signal in developing countries, as obesity and diet-related chronic diseases are increasing in developing countries

---

[2] Stunting refers to the proportion of children aged under-five falling below minus 2 standard deviations (moderate and severe) from the median height-for-age of the WHO growth standard. Underweight refers to the proportion of children aged under-five falling below minus 2 standard deviations (moderate and severe) from the median weight-for-age of the WHO growth standard. Wasting refers to the proportion of children aged under-five falling below minus 2 standard deviations (moderate and severe) from the median weight-for-height of the WHO growth standard. Overweight refers to the proportion of children aged under-five falling above 2 standard deviations from the median weight-for-height of the WHO growth standard.

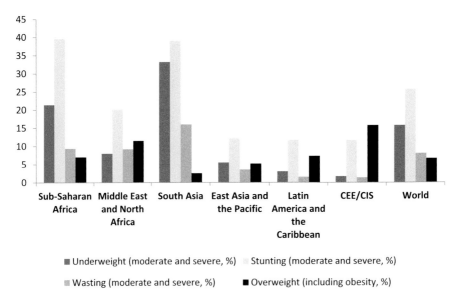

Fig. 3.1 Global Nutritional Status, 2012 (Source: Childinfo.org, UNICEF)

(Shetty 2012). Optimizing the window of opportunities for preventing undernutrition and overnutrition from pre-pregnancy to the first 1000 days of life is strongly needed. Gómez et al. (2013) add another burden, the so-called micronutrient malnutrition or 'hidden hunger', which owes its name to the fact that the symptoms of the problems are not always visible. Hidden hunger is a condition in which people suffer from a chronic deficiency of micronutrients or essential vitamins and minerals. Currently, it is estimated that two billion people suffer from chronic deficiency of micronutrients. India, Afghanistan and many countries of Sub-Saharan Africa have an alarming situation of micronutrient deficiency where iron, vitamin A, and deficiency are highly prevalent in school children (Muthayya et al. 2013). Table 3.1 presents the countries most affected by multiple micronutrient deficiencies, several of them being high on the list of countries with high prevalence of under- and overnutrition. Micronutrient deficiency has huge consequences for later life. A study by Lozoff et al. (2013) shows that a chronic iron deficiency is associated with lower level of educational attainment (not completing secondary school and not pursuing further education/training). Chronic iron deficiency is also associated with poorer emotional health and more negative emotions in later life. Muthayya et al. (2013) estimated that micronutrient deficiencies contribute to 1.5–12 % of the total Disability Adjusted Life Years (DALY).

Despite those above challenges the world is facing nowadays, we should be hopeful about the future. Innovations and FNS-related policies are among the potential ways to address those problems. The rest of the chapter discusses the trend and impact of innovations in reducing malnutrition and enhancing FNS.

Table 3.1 Top 20 countries affected by hidden hunger

Rank	Country	Hidden hunger index score	Deficiency prevalence (%)		
			Zinc	Iron	Vitamin A
1	Niger	52.0	47.0	41.8	67.0
2	Kenya	51.7	35.8	34.5	84.4
3	Benin	51.3	44.7	39.1	70.7
4	Central African Republic	51.0	43.0	42.1	68.2
5	Mozambique	51.0	47.0	37.4	68.8
6	Sierra Leone	50.0	37.4	37.9	74.8
7	Malawi	49.7	53.2	36.6	59.2
8	India	48.3	47.9	34.7	62.0
9	Burkina Faso	48.3	44.5	45.8	54.3
10	Ghana	47.7	28.6	39.0	75.8
11	São Tomé and Príncipe	47.7	29.3	18.4	95.6
12	Afghanistan	47.7	59.3	19.0	64.5
13	Democratic Republic of the Congo	47.7	45.8	35.7	95.6
14	Mali	46.0	38.5	40.7	58.6
15	Liberia	45.3	39.4	43.4	52.9
16	Côte d'Ivoire	44.0	40.1	34.5	57.3
17	Gambia	43.7	27.6	39.7	64.0
18	Chad	43.3	44.8	35.6	50.1
19	Madagascar	43.0	52.8	34.2	42.1
20	Zambia	42.0	45.8	26.5	54.1

Source: http://reliefweb.int/sites/reliefweb.int/files/resources/Hidden_Hunger_Index_Executive_Summary.pdf. The Hidden Hunger Index is the average, for preschool children, of three deficiency prevalence estimates: stunting (as a proxy for zinc deficiency, as recommended by the International Zinc Nutrition Consultative Group), iron-deficiency anemia and vitamin A deficiency. The three components were equally weighted (Hidden Hunger score = [stunting (%) + anemia (%) + low serum retinol (%)]/3)

The Main Features of Technological and Institutional Innovations for FNS

It is argued that both 'hardware' and 'software' are needed for societies to develop and ultimately to prosper (Woodhill 2010). According to Woodhill (2010), 'hardware' refers to technological innovations while 'software' represents institutional innovation and arrangements. Along with this notion, this chapter classifies innovations for FNS into two main types: technological and institutional innovations. The features of technological innovations are closely related to the sources of technology. Following Conway and Waage (2010), the sources of technology are categorized as conventional, traditional, intermediate and new platforms for technology.

"Conventional technologies" are produced by industrialized countries through the application of modern knowhow in physics, chemistry, and biology. They are available in regional or global markets as a packaged form. The conventional technologies were normally developed in the form of agricultural inputs, such as fertilizer, high yielding varieties and irrigation tools, globally known as the tools of the Green Revolution. The original aim of conventional technological innovations is to diffuse knowledge to farmers to increase agricultural production through the transfer of knowledge embedded in the products (Dockes et al. 2011).

Traditional technologies are defined as technologies which have been developed by the local communities to meet their local needs. This type of innovation is derived from the traditional practices, generally shaped over a period of time by communities in developing countries and proven to be effective as complements to conventional technologies. Several traditional technologies, particularly agricultural systems, have been promoted and recognized globally. As a traditional technology is invented and adopted by local people, this technology is also referred to as indigenous technical knowledge (Conway and Waage 2010). In the farming system, a traditional technology is characterized by a low use of inputs, reflecting the (lack of) opportunities available to smallholder farmers (Meyer 2010). A (controversial) example of farming practice rooted in age-old agricultural practices is the system of rice intensification (SRI). SRI has been widely adopted globally in the last decade beyond its country of origin, Madagascar (Uphoff and Kassam 2008).

Intermediate technologies are a mix between conventional and traditional technologies (Conway and Waage 2010). The application of such technologies is supported by an institutional change so that they can provide a full range of benefits to small farmers. As examples, Polak et al. (2003) listed three types of affordable small-plot irrigation systems which developed from the mix of conventional and traditional technologies, including the treadle pump, low cost drip irrigation, and the low cost sprinkler system. These low-cost irrigation technologies enable poor farmers to have access to water and, at the same time, to reduce production costs. The treadle pump is one of the successful intermediate technologies, developed in Bangladesh during the 1980s (Namara et al. 2010). The objective of the treadle pump development was threefold: a high and sustainable agricultural output, low cost technology, and simplicity of production, installation and use. In support, a variety of mass marketing actions were implemented in the 1980s by an international non-profit organization, International Development Enterprises (IDE) (Hierli and Polak 2000; Namara et al. 2010). Currently, the treadle pump has been adopted across Africa and Asia (Kay and Brabben 2000).

The new platform technologies applied in fostering FNS include information and communication technologies (ICT) for the agricultural sector, biotechnology, and nanotechnology. ICT have been widely applied for enhancing better market access, as well as empowering local farmer organizations. Many risks and uncertainties normally faced by smallholder farmers before, during and after production can be overcome via mobile phone information, accordingly boosting their production. The mobile phone services are on several fronts, ranging from providing market and price information to knowledge sharing, insuring crop production to monitoring children's nutrition status. In the application of biotechnology, biofortification is among the most cost-effective ways of improving nutritional outcomes.

Institutional innovations involve social and political processes in which the actors of innovation contribute to a larger action by combining inherited practices, technologies and institutions to address their interest (Hargrave and van De Ven 2006). Institutions are defined as the rules of society or organizations that support the people or members by helping them form and deal with their expectations about each other so that they achieve common objectives (Ruttan and Hayami 1984; World Bank 2002). As mentioned earlier, innovation is a process involving various institutional arrangements and inter-agent coordination. In the FNS-related areas, more specifically in the agricultural sector, institutional innovations have emerged in the form of the coordination of actions and interests of farmers, markets, and policymakers. As mentioned above, the downsides of the Green Revolution are mainly due to the related social policies, not to the technologies themselves. Therefore, institutional innovation plays a substantial role in accompanying technological innovation and making it beneficial for the people.

One of the innovative institutions related to FNS are the Farmer Field Schools (FFS) (Braun et al. 2006). Originated in Indonesia, FFS have long been recognized as an initiative to address the challenge of pest management and the heterogeneous ecological aspects of farming activities. Nevertheless, FFS have also been implemented in other fields, such as resource management (Nepal), adoption of agricultural technologies (Kenya), and diffusion of knowledge (Mexico). Despite the small budgets needed to sustain the FFS, a great number of international and national NGOs have been involved thoroughly in FFS since the early 1990s. A good practice in FFS is the involvement of FFS alumni in Indonesia and the Philippines as full-time FFS facilitators. Apart from pest management and farming practices, the FFS alumni were also trained with new skills, such as computer and entrepreneurial development (Braun et al. 2006; Braun and Duveskog 2008).

IFAD (2007) outlines the importance of institutional innovations in facilitating access to natural resources and local governance, access to productive assets and markets, access to information and knowledge, and increasing political capital. The World Development Report 2008 on Agriculture for Development (World Bank 2008) documents several focus areas of institutional innovations, including new mechanisms to increase land tenure security for smallholder farmers, financial and services access, risk mitigation and management, as well as efficient input markets.

The Impacts of Innovations

This chapter features two types of technology, including new platform and traditional technology, as well as institutional innovations and their contribution to the enhancement of FNS. The new platform technology is now the focus of policies, as this type of technology has profound long-term implications, particularly in the context of FNS. The spikes in food and energy prices in 2007–2008 have triggered the increase in input costs which negatively affected the supply responses from the producer side. Thus, the introduction of the new platform technologies in the agricultural sector plays a role for both producers and consumers. In the more globalized market, the new platform technologies benefit producers and consumers in their involvement in the supply chain through better access to market information. With the challenge of climate variability, new platform technologies offer small-holder farmers tools for decision-making, including on what and when to grow. In addition, traditional technology often contributes to improving agricultural technology. Low income farmers have limited access to modern technology, thus traditional technologies benefit them most, as they are most accessible and affordable. This chapter also provides an overview as to how the institutional innovations through community-based innovation impact FNS.

Analyzing the impacts of innovations on FNS cannot be separated from the FNS dimensions: availability, accessibility, utilization and stability. Following Masset et al. (2012) and Webb (2013), the impact of innovation and agricultural interventions are channeled through multiple pathways, both direct and indirect. The indirect pathway is chiefly linked to the accessibility dimension, while the direct pathways are associated with the availability dimension of FNS. While the indirect pathway goes through income, the direct pathways are channeled through food production and improved food quality, more diverse diet composition, food prices, non-food spending, and intrahousehold resource allocation. The latter can be impacted through three channels: women's control over resources; women's time and caring practices; and improved women's nutrition and health. It is recognized that the pathways through intrahousehold resource allocation are still poorly explored, particularly innovations that target the three channels altogether (Webb 2013). Our chapter focuses on several types of innovation, including the new platform technology through ICT and biofortification, traditional technology exemplified here by home gardens, and institution innovation through community-based actions. The first new platform technologies through ICT and biofortification are chosen, as these two technologies are projected to be among the priority of public investment in agricultural knowledge systems (IAASTD 2009). On the other hand, traditional technology is sometimes overlooked in term of its contribution to FNS. This chapter highlights the long contribution of traditional technology through home gardens that have been providing households with rich and diversified diets. In terms of institutional innovation through community-based institutions, this chapter outlines the impact of institutional arrangement in facilitating small-holder farmers to increase their voice and have better access to markets and services. We focus on FFS as one of the most established institutional innovations.

The Impact of New Platform Technology

The new platform technologies applied in fostering FNS include information and communication technologies (ICT) for the agricultural sector, biotechnology, and nanotechnology. This chapter focuses on two new platform technologies: ICT and biofortification. ICT have been widely applied for enhancing better market access, as well as empowering local farmer organizations. In the application of biotechnology, food fortification is among the most cost-effective ways to improve nutritional outcomes.

ICT

ICT cover technologies used to handle information and communication, including internet, radio, television, video, digital cameras, and other hardware and software. In the former era, ICT in developing countries mainly served as an entertaining gizmo and a means of communication. The modern application of ICT has provided more services through most areas of development, including agriculture, education and health. The mushrooming of ICT applications in many developing countries provides an opportunity to transfer knowledge through the private and public information systems (Aker 2011). One of the ICT applications is the widespread and varied use of mobile phones. Over the past decade, mobile phone subscriptions have increased considerably in developing countries (ITU 2011). The greatest benefits of mobile phones are the significantly reduced communication and information costs, geographic coverage and the convenient use of the technology (Aker and Mbiti 2010). As more and more people, particularly the poor, have enjoyed the benefits of mobile phones, a number of innovators in developing countries have taken the opportunity to use it in various aspects of local life (Conway and Waage 2010). Since early 2007, there have been a number of applications through mobile phones for farming, health, banking, and advocacy.

The impact pathways of ICT on FNS are mainly through improved agricultural production and access to market-related information, which accordingly increases farmers' income. ICT support farmers by improving agricultural productivities through information on the precise input use and environmentally-friendly agricultural production. One notable example of the role of ICT in agricultural production is through software for plant nutrient application rate. Pampolino et al. (2012) found that the use of Nutrient Expert for Hybrid Maize (NEHM) software increased the yield and economic benefits of farmers in Indonesia and the Philippines through the provision of information on nutrient application rate. For farming activities, the expanding use of mobile phones supports farmers' access to information on agricultural extension services, markets, financial services and livelihood support (Donner 2009), translating to better access to extension services, better market links and distribution networks, and better access to finance (World Bank 2011). Ultimately, the mobile phone applications for farmers will improve farmers' income, lower transaction and distribution costs for input suppliers, improve traceability and quality standards for buyers, and create new opportunities for financial institutions. In more detail, Aker (2011) highlights the significance of mobile phones for agricultural services adoption and extension in developing countries through improved access to private information, farmer's management of input and output supply chains, facilitation of the delivery of other services, increased accountability of extension services, and increased communication linkages with research systems. The perspective of the private sector (Vodafone Group and Accenture 2011) also emphasizes the potential solution offered by mobile applications in improving data visibility for supply chain efficiency. Based on the review of 92 mobile applications for agriculture and rural develpoment, Qiang et al. (2011) found that the major service provided by the application is information provision. It is also found that only a few of the applications are already sustainable, while 33 % of them are at the concept proof stage and 55 % are at the scaling-up phase.

Along the agricultural chain, the introduction of the new platform technologies in the agricultural sector through ICT plays a significant role for both producers and consumers. In developing countries, most smallholder farmers act as producers and consumers at the same time, and ICT offer a unique opportunity for rural farmers to access market information, weather, and extension services. Several empirical studies reveal that ICT have a significant impact both on producer and consumer welfare (Jensen 2007). Arguably, there are several potential channels for ICT to affect accessibility: by increasing farmer's profitability, and thus income, ICT can enable a farmer to improve consumption, whilst at the same time enabling them to save and accumulate resources. Labonne and Chase (2009) assessed the positive impact of ICT on per capita consumption. In India, Reuters Market Light (RML) provides services in terms of agricultural information dissemination over mobile phones. However, Fafchamps and Minten (2012) found that RML had a small effect on crop grading and no significant impact on prices received by farmers. There is also no significant difference in crop losses resulting from rainstorms. Fafchamps and Minten (2012) argued that a small number of subscribers and slow take-up rate might play a part as the underlying factor. Another study in India found that internet kiosks and warehouses supplied through the e-Choupal program reduced the price dispersion, thus benefiting both producers and consumers (Goyal 2010).

Biofortification

In the area of FNS, biotechnology plays a supportive role through tissue culture in the quest for more effective and beneficial traits and genetic engineering technology. Genetic engineering has been used widely but has mostly concentrated on increasing resistance to environmental stresses, pests, and diseases. However, recent developments in biotechnology have moved in another direction: high yield crops and more nutritious crops and animal products. In order to bring some of these benefits to the poor, who typically lack access to nutritious foods, such as fruits, vegetables, and animal source foods (fish, meat, eggs, and dairy products) and rely heavily on staple foods, there is a need for staple-related biotechnology.

One of the new platform technologies in this area is biofortification, a process of introducing nutrients into staple foods. Biofortification can be conducted through conventional plant breeding, agronomic practices such as the application of fertilizers to increase zinc and selenium content, or transgenetic techniques (Bouis et al. 2011). The smallholder farmers cultivate a large variety of food crops developed by national agricultural research centers with the support of the Consultative Group on International Agricultural Research (CGIAR). One of the global initiatives for biofortification is known as HarvestPlus.[3] Biofortification provides a large outreach, as it is accessible to the malnourished rural population which is less exposed to the fortified food in markets and supplementation programs. By design, biofortification initially targets the more remote population in the country and is expanded later to urban populations. To be successful, i.e., to improve people's absorption and assimilation of micronutrients, biofortification should meet several challenges, some of which require additional accompanying interventions: successful breeding in terms of high yields and profitability, making sure nutrients of biofortified staple foods are preserved during processing and cooking, the degree of adoption and acceptance by farmers and consumers, and the coverage rate (the proportion of biofortified staples in production and consumption) (Nestel et al. 2006; Meenakshi et al. 2010; Bouis et al. 2011). The development of biofortification is outlined in Table 3.2. In the case of food processing, Meenakshi et al. (2010) estimated that the greatest processing losses are in the case of cassava in Africa, where the loss of vitamin A during the cooking process is between 70 % and 90 %. For other staple crops such as sweet potato and rice, the processing loss can be anticipated, as both staple foods are consumed in boiled form.

Biofortification has been implemented in several countries of Asia and Africa (Table 3.3). A number of crops are biofortified, including rice, wheat, maize, cassava, pearl millet, beans, and sweet potato, depending on the national context. Biofortification is found to be cost-effective in terms of the moderate breeding costs, which amount to approximately 0.2 % of the global vitamin A supplementation (Beyer 2010), while the benefit is far higher than the cost.[4] Compared with other types of interventions, such as supplementation and food fortification, biofortification seems more cost-effective.[5] Nevertheless, biofortification is not without its limitations, as it might not be viable for application in all plants. For instance, from a breeding perspective, the breeding system of some plants is very complex (Beyer 2010). In Uganda, banana is the primrary staple food, accounting for a per capita per year consumption of nearly 200 kg. However, the vitamin and

---

[3] HarvestPlus is a part of the CGIAR research program on Agriculture for Nutrition and Health under the coordination of the International Center for Tropical Agriculture (CIAT) and the International Food Policy Research Institute (IFPRI).

[4] See the detail example in Bouis et al. (2011).

[5] HarvestPlus provides an example as to how much $75 million (US) is worth for supplementation, fortification and biofortification, respectively. That amount of money could buy vitamin A supplementation for 1 year for 37.5 million pre-school children in the South Asian countries of Bangladesh, India and Pakistan; likewise, it could be used for iron fortification for 1 year for 365 million persons, accounting for 30 % of the population in Bangladesh, India, and Pakistan. Contrastingly, the same amount could finance the cost of developing and disseminating iron and zinc fortified rice and wheat for all of South Asia, a crop which would continue to thrive year after year.

Table 3.2 HarvestPlus pathway to impact

Stage	Activity
Discovery	Identifying target populations and staple food consumption profiles
	Setting nutrient target levels
	Screening and applying biotechnology
Development	Crop improvement
	Gene by environment (GxE) interactions on nutrient density
	Nutrient retention and bioavailability
	Nutritional efficacy studies in human subjects
Delivery	Releasing biofortified crops
	Facilitating dissemination, marketing, and consumer acceptance
	Improved nutritional status of target populations

Source: Bouis et al. (2011)

Table 3.3 Target crops, nutrients, countries, and release dates

Crop	Nutrient	Country	Year of release
Bean	Iron	DR Congo, Rwanda	2012
Cassava	Vitamin A	DR Congo, Nigeria	2011
Maize	Vitamin A	Nigeria, Zambia	2012
Pearl millet	Iron	India	2012
Rice	Zinc	Bangladesh, India	2013
Sweet potato	Vitamin A	Mozambique, Uganda	2007
Wheat	Zinc	India, Pakistan	2013

Source: http://www.harvestplus.org/content/crops

mineral content of the banana is very low. Producing bananas fortified with micronutrients is challenging, as the conventional breeding of a banana is less viable and takes more processing. Another limitation of biofortification is the fact that the potential benefits of biofortified staple foods are uneven across staple food groups, as the need of micronutrients varies along the lifecycle of the crop (Bouis et al. 2011).

The Impact of Traditional Technology Through the Home Garden

Traditional technologies are defined as technologies which have been developed by the local communities to meet their local needs. This type of innovation is derived from traditional practices, generally shaped over a period of time by communities in developing countries and proven to be effective as complements to conventional technologies. Several traditional technologies, particularly agricultural systems, have been promoted and recognized globally. As a traditional technology is invented and adopted by local people, this technology is also referred to as indigenous technical knowledge (Conway and Waage 2010). In the farming system, a traditional technology is characterized by low use of inputs, reflecting the (lack of) opportunities available to smallholder farmers (Meyer 2010). A (controversial) example of a farming practice rooted in age-old agricultural practices is the system of rice intensification (SRI). SRI has been widely adopted globally in the last decade beyond its country of origin, Madagascar (Uphoff and Kassam 2008).

Another example of traditional technology globally applied are home gardens. Home gardens represent a traditional agricultural practice that has been applied mostly in rural areas, acting as food buffer stock for smallholders. Apart from that, home gardens provide more benefits, including wealth generation, bargaining power in labor markets, post-harvest storage, non-agricultural income generating activities, and access to credit (Hanstad et al. 2002). It should be noted that the traditional technology of the home garden is also considered to be a viable and effective way to improve micro-nutrient consumption. Vegetables and fruits are both important sources of vitamins and minerals. Some vegetables, including well-known types such as tomato (*Solanum lycopersicum*), cabbage (*Brassica oleracea*), and onions (*Allium cepa*), as well as traditional local vegetables, such as moringa (*Moringa oleifera*), kangkong (*Ipomoea aquatic*), perilla (*Perilla frutescens*), anemone (*Nymphoides hydrophylla*), bitter gourd (*Momordica charantia*) and jute mallow (*Corchorus olitorius*), are available in most Southeast Asian countries and are normally grown in home gardens (Table 3.4). Those traditional vegetables are rich in micronutrients. For example, tomato contains more β-carotene, vitamin E and iron but has lower antioxidant activity compared to cabbage (Yang and Keding 2009). However, compared to commercially-available tomatoes, even moringa can have 38 times the amount of β-carotene, 24 times the amount of vitamin C, and 17 times the amounts of vitamin E, folates and iron (Hughes and Keatinge 2011).

Recently, home gardening has been used as a sustainable strategy that can address multiple micronutrient deficiencies through dietary diversification (Cabalda et al. 2011). At the same time, home gardens also serve as an integrated agro-ecosystem (Soemarwoto et al. 1985; Kehlenbeck et al. 2007; Galluzzi et al. 2010). In Java,[6] home gardens (*pekarangan*) are well-developed and charac-terized with great diversity relative to their size[7] (Soemarwoto et al. 1975, 1985). The structure of home gardens in Java varies from place to place, ranging from 80 to 179 plant species (Soemarwoto et al. 1985). More importantly, Javanese home gardens contribute primarily to vitamin A and C provision, 12.4 % and 23.6 %, respectively, of the recommended dietary allowance (Arifin et al. 2012), and to 20 % of household income (Stoler 1978). In Cambodia and Nepal, 31–65 %, respectively, of household income is derived from revenue from sale of poultry raised in the home garden (Mitchell and Hanstad 2004). In the Philippines, a study found that having a home garden is positively associated with diversity in the children's diet and the frequency of vegetable consumption (Cabalda et al. 2011).

---

[6] Java is one of the principal islands and the most densely populated in Indonesia.

[7] The size of *pekarangan* normally takes at least 120 m^2 (Arifin et al. 2012) or covers 10–15 % of the cultivatable area (Mitchell and Hanstad 2004).

Table 3.4 Nutritional contents per 100 g of selected staples, traditional fruits and vegetables in Southeast Asian Home Garden Households

Crop	Protein (g)	Vitamin A (mg)	Vitamin C (mg)	Calcium (mg)	Iron (mg)
Wheat	11.6	0	0	68	2.8
Rice (white, polished, cooked)	2.2	0	0	7	0.4
Rice (white, polished, raw)	6.8	0	0	19	1.2
Pearl millet, combined varieties, raw	5.7	0	1	18	13.1
Custard apple	1.17–2.47	0.007–0.0018	15–44.4	17.6–27	0.42–1.14
Mangosteen	0.5–0.6	n/a	1–2	0.01–8	0.2–0.8
Persimmon	0.7	n/a	11	6	0.3
Wax apple	0.5–0.7	0.003–0.008	6.5–17	5.6–5.9	0.2–0.82
Jackfruit (pulp)	1.3–1.9	n/a	8–10	22	0.5
Rambutan	0.46		30	10.6	
Durian	2.5–2.8	0.018	23.9–25	7.9–9	0.73–1
Moringa (leaves)	8.6	19.7	274	584	10.7
Okra (fruit)	1.8	0.4	37	44	0.9
Kangkong (leaves)	2.4	0.4	40	220	2.5
Common cabbage	1.7	0.4	49	52	0.7
Mungbean (grain)	23.8	0.02	15	55	2.8
Tomato	0.9	0.2	30	9	0.6
Sweet pepper	4.4	2.5	93	188	2.1
Bird's nest fern (*Asplenium australasicum*)	2.8	n/a	Very high	Low	Low
Anemone (*Nymphoides hydrophylla*)	0.7	Medium	Low	Low	Low
Sesbania (*Sesbania grandiflora*) leaves	8	Very high	Very high	High	Very high
Chinese cedar (*Toona sinensis*)	6.3–9.8	Medium	Very high	High	High

Source: Hughes and Keatinge (2011), compiled from Australian Custard Apple Growers Association (ACAGA) (2011), Lim (1996), Morton (1987), Yang and Keding (2009), Lin et al. (2009), Institute of Nutrition, Mahidol University (2014), Stadlmayr et al. (2010)

Children from households with a home garden are more likely to consume more vitamin A (vegetables) and have a more diverse diet.

Future Trends and Priorities of FNS Innovation: A Stakeholder Survey

The stakeholder survey aims to collect a range of opinions, stakeholder attitudes and understandings of the impacts of innovations on FNS, as well as of the trade-offs of innovations in terms of FNS, socio-economic or environmental impacts. The results provide general directions that can be used in building scenarios for FNS innovations and their impacts in the future, based on the inferred likelihood of innovation creation and development, as well as adoption. The questionnaire is designed as a simple, non-technical survey in order to appeal to respondents with various educational and professional backgrounds. The number of respondents is 42, and the survey was constructed to approach a limited number of stakeholders with a key interest in FNS, agriculture and natural resources. The professional background of the respondents is fairly diverse: almost 40 % of the respondents work with NGOs, 25 % are from the public sector and academia, 17.1 % are from international agencies (i.e., FAO), 7.3 % are from the private sector, 7.3 % are farmers and the rest are from the general public. The survey was conducted online in February 2013.

General FNS Awareness

The first part of the survey assesses the general awareness of the respondents to FNS issues. The respondents were asked whether they had previously heard the term 'food and nutrition security' (FNS), what FNS means, and to list five priorities (multiple choice) for improving FNS. The majority of the respondents (almost 95 %) were aware of the expression "FNS". This high percentage is not surprising, as almost a quarter of the respondents report FNS as their field of expertise. However, it is interesting to see how the respondents defined FNS. The survey provided a closed question with six definitions, namely:

- everyone has enough food,
- stable food supply in the future,
- all food is safe to eat,
- well-functioning food distribution,
- consumption of high quality of food, and
- ensuring consumption of healthy food through hygienic cooking preparation.

Ninety percent of the respondents chose 'everyone has enough food' and 'stable food supply in the future'. Around 78 % of the respondents indicated that FNS should encompass the consumption of quality food (i.e., micronutrients, calorific content). The stakeholders' perception of FNS is paralleled by The United Nations High Level Task Force on Global Food Security (HLTF) through their Comprehensive Framework for Action (CFA). The framework defines food and nutrition security as a condition in which all people, at all times, have physical, social and economic access to sufficient, safe, and nutritious food which meets their dietary needs and food preferences for an active and healthy life.

To understand the future priorities of FNS innovations, respondents were prompted with a list of types of innovation and asked to choose five of them. The most common answers are as follows (% of respondents):

- promoting a sustainable and diversified agricultural sector (71.8 %),
- improving farmer's skill (69.2 %),
- empowering farmers through collective action (53.8 %),
- income generating programs (51.3 %), and
- increasing agricultural crop production (46.2 %).

This result suggests that, although technological innovation is important for increasing agricultural production, institutional factors through farmer's collective action should be given more emphasis in directing future science policy for agriculture and FNS.

We also asked respondents to rank the relevance of the FNS dimensions, availability, accessibility, utilization and stability, in the context of developing countries and how the relevance of these dimensions may change with time. Around 80 % of the respondents agreed that accessibility in the present and in the future is highly relevant for developing countries. Almost 70 % of the respondents reported that utilization, both in the present and in the future, is highly relevant for developing countries. It is interesting that the availability dimension was seen as less relevant. In comparison, about 58 % of the respondents stated that availability in the present and in the future is highly relevant for developing countries. Thus, stakeholders consider that the future FNS innovations should go beyond the availability dimension, as FNS problems in developing countries are more complex. Many developing countries are entrenched with a dual, sometime triple burden of malnutrition, where undernutrition and overnutrition (overweight and obesity) coexist in the same population or household (Hawkes et al. 2005; FAO 2006), often compounded by a deficiency in micronutrients. Overnutrition in particular is mainly a result of a change in information and culture, and a change in lifestyles and physical activity patterns, as well as of the globalization of trade and finance (Hawkes et al. 2005; Popkin et al. 2012). Tackling these drivers of obesity may indeed require more innovations of the institutional type than presently exist.

Agricultural Innovations and FNS

We prompted respondents with a list of agricultural innovations (generic or specific).[8] First, the respondents were asked about their familiarity with the type of innovation provided in the survey. Among the innovations, FFS was the most familiar to the respondents (75 %), followed by local farmer organization empowerment (67 %), and farmer extension services (52 %). The respondents assessed GM

---

[8] The list of innovation included ICT, farmer extension services, FFS, empowering local farmer organization, rural micro-finance schemes, supply chain management, animal breeding programs, new/modern seed varieties, adapted inputs for small scale farming, food fortification programs, new/integrated water management, and GM crops.

crops as the least relevant technological innovation for FNS in developing countries. The results are affected by and, indeed, are consistent with the stated future priorities for innovations (Fig. A1, Appendix). Our survey also asked respondents to rank the innovations according to their environmental friendliness. FFS was seen as the most likely to be environmentally friendly (80 %), followed by new/integrated water management (77 %), empowering local farmer organizations and farmer extension services (both at 46 %). Around 40 % of the respondents reported that adapted inputs for small scale farming and GM crops are likely to have a negative impact on the environment (Fig. A2, Appendix).

Perceptions of the economic sustainability of innovations was also queried. While around 71 % of the respondents saw that FFS is economically sustainable, they were more likely to state that empowering local farmer organization is the most economically sustainable innovation for FNS. Similarly, this type of innovation is seen by the respondents (almost 70 %) as the most widely applicable beyond the original/experimental setting (Fig. A3, Appendix). The issue of trade-offs between the FNS, environmental, social, and/or economic impacts of innovations was also examined. The respondents rated institutional innovations such as FFS and local farmer organization (55 % and 50 %, respectively) as the most likely to have trade-offs. On the other hand, ICT, supply chain management, and food fortification (30 %, 30 %, and 20 %, respectively) were seen as the least likely to have trade-offs between environmental, social, and/or economic aspects (Fig. A4, Appendix). Finally, respondents were asked (closed question) about the two main barriers to the adoption of innovation. For all types of innovations, respondents stated that limited farmer's access and lack of education and training are the two main barriers to adoption (Fig. A5, Appendix).

Conclusions

FNS continues to be an important challenge in developing countries. Volatile food prices have had mixed effects, but overall have slowed down the progress of achieving FNS. Even though many low middle income countries are now reducing hunger, they are currently experiencing a triple burden of malnutrition, experiencing undernutrition, overnutrition and 'hidden hunger' at the same time. Furthermore, it is estimated that two billion of the world's population suffer from hidden hunger, a chronic deficiency of essential micronutrients. To address these problems, a strong performance in FNS-related sectors, including agriculture and health, is urgently required. In addition to that, policies associated with the diffusion of technologies, knowledge and innovations, as well as institutional arrangements, are key factors in countering the complex and evolving challenges of FNS-related problems.

This study aims to review the role of technological and institutional innovation in FNS-related areas, discuss the main features of innovations for FNS, and describe the impact of innovation on FNS using the examples of new platform and traditional technology. Innovations have contributed to countering the challenges to FNS from the drivers of hunger and poverty, such as rising population, environmental pressures and price fluctuations. In many developing countries, where the small-holder farmers are the main target group, many factors hindering the achievement of FNS are related to the increasing demand for and lack of access to food.

Drawing from two types of technologies, innovations impact FNS through multiple pathways, directly and indirectly. The direct pathways perform through improved food production, which might lead to improved food quality through more diverse diet composition. However, FNS innovations should not only emphasize the supply side or the accessibility of food, but also focus on alleviating "hidden hunger". Our analysis shows that new-platform technologies can be directed at improvements in nutrition outcomes for the whole population and for the poorest. Biofortification provides opportunities to smallholder farmers to access and grow more nutritious food crops with rich micronutrient content. In addition, traditional technology through home gardens has proven to be an effective way to enhance the quality of nutritionally deficient diets through the locally grown vegetables and fruits of smallholder farmers.

Finally, the stakeholder survey pointed out that innovations for FNS should address various challenges, including climate change and environmental issues, energy and water availability, globalization of trade, finance, change in lifestyles and physical activity patterns, and demographic shifts. Based on the results of the stakeholder survey, appropriate 'software' through innovative institutions is recognized by several stakeholder groups as one of the most viable and effective FNS innovations. The survey also raised concerns about the role of institutional innovation in enabling developing countries to achieve FNS with lower environmental impacts. In a situation in which the agricultural sector routinely encounters new challenges and uncertainties, it is critical to refine the farming systems to increase resource use efficiency. Therefore, the new technologies for agricultural production should focus on precision farming, new crop varieties that have better nutritional quality, and diversified traditional crop systems for high-value horticulture.

Open Access This chapter is distributed under the terms of the Creative Commons Attribution-Noncommercial 2.5 License (http://creativecommons.org/licenses/by-nc/2.5/) which permits any noncommercial use, distribution, and reproduction in any medium, provided the original author(s) and source are credited.

The images or other third party material in this chapter are included in the work's Creative Commons license, unless indicated otherwise in the credit line; if such material is not included in the work's Creative Commons license and the respective action is not permitted by statutory regulation, users will need to obtain permission from the license holder to duplicate, adapt or reproduce the material.

Appendix

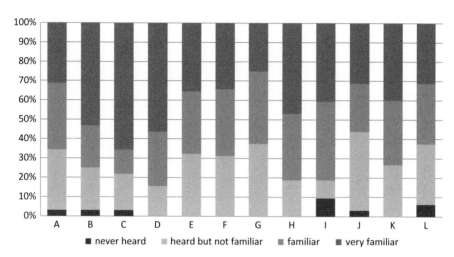

Fig. A1 How familiar are you with the following agricultural types of innovation? (Source: Authors' compilation based on survey)

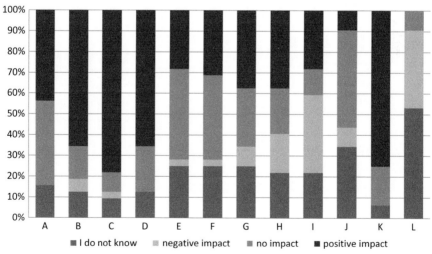

A. ICT for Agriculture	G. Animal breeding programs
B. Farmer extension	H. New/modern seed varieties
C. FFS	I. Adapted inputs for small scale farming
D. Empowering local farmer organization	J. Food fortification
E. Rural micro-finance scheme	K. New/integrated water management
F. Supply chain management	L. GM Crops

Fig. A2 In your opinion, are the following types of innovation likely to be environmentally friendly? (Source: Authors' compilation based on survey)

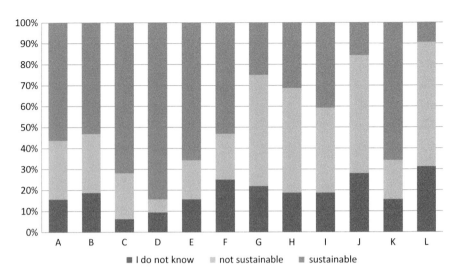

Fig. A3 In your opinion, are the following types of innovation economically sustainable (i.e., under market conditions, without the help of donor/public money) over the longer term, beyond the first implementation program?

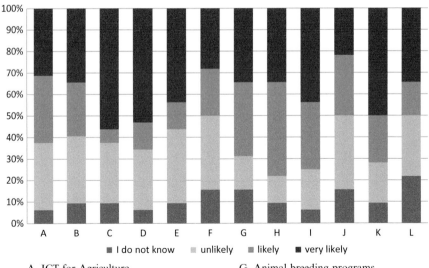

A. ICT for Agriculture
B. Farmer extension
C. FFS
D. Empowering local farmer organization
E. Rural micro-finance scheme
F. Supply chain management

G. Animal breeding programs
H. New/modern seed varieties
I. Adapted inputs for small scale farming
J. Food fortification
K. New/integrated water management
L. GM Crops

Fig. A4 Do you foresee trade-offs between environmental, social, and/or economic impacts in the following types of innovation? (Source: Authors' compilation based on survey)

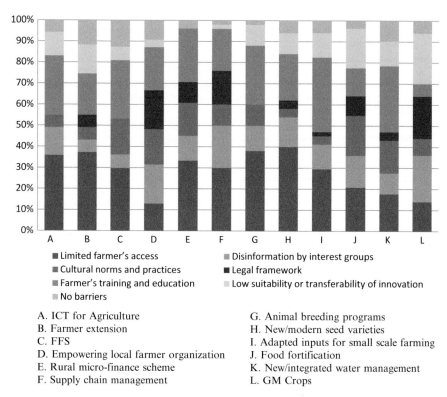

A. ICT for Agriculture
B. Farmer extension
C. FFS
D. Empowering local farmer organization
E. Rural micro-finance scheme
F. Supply chain management

G. Animal breeding programs
H. New/modern seed varieties
I. Adapted inputs for small scale farming
J. Food fortification
K. New/integrated water management
L. GM Crops

Fig. A5 Please state, in your opinion, up to two of the main barriers to fostering innovation in each of the following areas (Source: Authors' compilation based on survey)

References

Aker JC (2011) Dial "A" for agriculture: a review of information and communication technologies for agricultural extension in developing countries. Agric Econ 42(6):631–647. doi:10.1111/j. 1574-0862.2011.00545.x

Aker JC, Mbiti IM (2010) Mobile phones and economic development in Africa. Working paper No 211. Center for Global Development, Washington, DC

Arifin HS, Munandar A, SchultinK G, Kaswanto RL (2012) The role and impacts of small-scale, homestead agro-forestry systems ("pekarangan") on household prosperity: an analysis of agro-ecological zones of Java, Indonesia. Int J AgriSci 2(10):896–914

Australian Custard Apple Growers Association (ACAGA) (2011) http://www.custardapple.com. au/index.php?option=com_content&view=article&id=48&Itemid=55. Accessed 10 May 2014

Beyer P (2010) Golden rice and "golden" crops for human nutrition. N Biotechnol 27(5):478–481. doi:10.1016/j.nbt.2010.05.010

Black RE, Allen LH, Bhutta ZA, Caulfield LE, de Onis M, Ezzati M, Mathers C, Rivera J (2008) Maternal and child undernutrition: global and regional exposures and health consequences. Lancet 371(9608):243–260. doi:10.1016/S0140-6736(07)61690-0

Bouis HE, Hotz C, McClafferty B, Meenakshi JV, Pfeiffer WH (2011) Biofortification: a new tool to reduce micronutrient malnutrition. Food Nutr Bull 32(1):31–40

Braun A, Duveskog D (2008) The farmer field school approach – history, global assessment and success stories, background paper for the IFAD Rural Poverty Report 2011. International Fund for Agricultural Development, Rome

Braun A, Jiggins J, Röling N, van den Berg H, Snijders P (2006) A global survey and review of farmer field school experiences, Report prepared for the International Livestock Research Institute. Endelea, Wageningen

Cabalda AB, Rayco-Solon P, Solon JAA, Solon FS (2011) Home gardening is associated with Filipino preschool children's dietary diversity. J Am Diet Assoc 111(5):711–715. doi:10.1016/j.jada.2011.02.005

Conway G, Waage J (2010) Science and innovation for development. Collaborative on Development Science, London

Dockes A, Tisenkopfs T, Bock B (2011) Reflection paper on AKIS, Collaborative Working Group Agricultural Knowledge and Innovation Systems, WP 1. Available at: http://ec.europa.eu/research/agriculture/scar/pdf/akis-wp1-final.pdf. Accessed 13 Jan 2013

Donner J (2009) Mobile-based livelihood services in Africa: pilots and early deployments. In: Fernandez-Ardevol M, Hijar AR (eds) Communication technologies in Latin America and Africa: a multidisciplinary perspective. Internet Interdisciplinary Institute (IN3), Barcelona

Fafchamps M, Minten B (2012) Impact of SMS-based agricultural information on Indian farmers. World Bank Econ Rev 26(3):383–414. doi:10.1093/wber/lhr056

FAO (2006) The double burden of malnutrition: case studies from six developing countries. Food and Agriculture Organization of the United Nations, Rome

FAO, IFAD, WFP (2013) The state of food insecurity in the world 2013. The multiple dimensions of food security. Food and Agriculture Organization of the United Nations, Rome

Galluzzi G, Eyzaguirre P, Negri V (2010) Home gardens: neglected hotspots of agro-biodiversity and cultural diversity. Biodivers Conserv 19(13):3635–3654. doi:10.1007/s10531-010-9919-5

Gómez MI, Barrett CB, Raney T, Pinstrup-Andersen P, Meerman J, Croppenstedt A, Carisma B, Thompson B (2013) Post-green revolution food systems and the triple burden of malnutrition. Food Policy 42:129–138. doi:10.1016/j.foodpol.2013.06.009

Goyal A (2010) Information, direct access to farmers, and rural market performance in central India. Am Econ J 2(3):22–45. doi:10.1257/app.2.3.22

Hanstad T, Brown J, Prosterman R (2002) Larger homestead plots as land reform? International experience and analysis from Karnataka. Econ Polit Wkly 37(29):3053–3062

Hargrave TJ, van De AH (2006) A collective action model of institutional innovation. Acad Manage Rev 31(4):864–888. doi:10.5465/AMR.2006.22527458

Hawkes C, Eckhardt C, Ruel M, Minot N (2005) Diet quality poverty, and food policy: a new research agenda for obesity prevention in developing countries. SCN News 29:20–22

Hierli U, Polak P (2000) Poverty alleviation as a business. Swiss Agency for Development and Cooperation, Bern

Hughes J d'A, Keatinge JDH (2011) The Nourished Millenium: how vegetables put global goals for healthy, balanced diets within reach. In: Holmer R, Linwattana G, Nath P, Keatinge JDH (eds) High value vegetables in southeast Asia: production, supply, and demand, Proceeding SEAVEG 2012. Available at: http://203.64.245.61/fulltext_pdf/EB/2011-2015/eb0197.pdf. Accessed 20 Jan 2012

IAASTD (2009) Agriculture at a crossroads: synthesis report – international assessment of agricultural knowledge, science and technology for development. Island Press, Washington, DC

IFAD (2007) Innovation strategy. International Fund for Agricultural Development, Rome

Institute of Nutrition, Mahidol University (2014) ASEAN food composition database, electronic version 1, Thaliand. Available at http://www.inmu.mahidol.ac.th/aseanfoods/composition_data.html. Accessed 10 June 2014

ITU (2011) The world in 2011: ICT facts and figures. International Telecommunication Union, Geneva

Jensen R (2007) The digital provide: information (technology), market performance, and welfare in the South Indian fisheries sector. Q J Econ 122(3):879–924. doi:10.1162/qjec.122.3.879

Juma C, Yee-Cheong L (2005) Innovation: applying knowledge in development, UN Millennium Project, Task Force on Science, Technology, and Innovation. Earthscan, London

Kay M, Brabben T (2000) Treadle Pumps for irrigation in Africa, Knowledge synthesis report 1, International Program for Technology and Research in Irrigation and Drainage, Rome

Kehlenbeck K, Arifin H, Maass B (2007) Plant diversity in homegardens in a socio-economic and agro-ecological context. In: Tscharntke T, Leuschner C, Zeller M, Guhardja E, Bidin A (eds) Stability of tropical rainforest margins. Springer, Berlin/Heidelberg, pp 295–317

Labonne J, Chase RS (2009) The power of information: the impact of mobile phones on farmer's welfare in the Philippines. Policy research working paper No 4996. World Bank, Washington, DC

Lim TK (1996) Durian 1. Characteristics and cultivars. Agnote 639 No. D29, Available at http://www.nt.gov.au/d/Primary_Industry/Content/File/horticulture/639.pdf. Accessed 10 May 2014

Lin LI, Hsiao YY, Kuo CG (2009) Discovering indigenous treasures: promising indigenous vegetables from around the world. AVRDC – The World Vegetable Center Publication No 09–720. AVRDC – The World Vegetable Center, Shanhua, Taiwan

Lozoff B, Smith JB, Kaciroti N, Clark KM, Guevara S, Jimenez E (2013) Functional significance of early-life iron deficiency: outcomes at 25 years. J Pediatr 163(5):1260–1266. doi:10.1016/j.jpeds.2013.05.015

Martorell R (1999) The nature of child malnutrition and its long-term implications. Food Nutr Bull 20(3):288–292

Masset E, Haddad L, Cornelius A, Isaza-Castro J (2012) Effectiveness of agricultural interventions that aim to improve nutritional status of children: systematic review. BMJ 344:d8222. doi:10.1136/bmj.d8222

Meenakshi JV, Johnson NL, Manyong VM, DeGroote H, Javelosa J, Yanggen DR, Naher F, Gonzalez C, García J, Meng E (2010) How cost-effective is biofortification in combating micronutrient malnutrition? An ex ante assessment. World Dev 38(1):64–75. doi:10.1016/j.worlddev.2009.03.014

Meyer R (2010) Low-input intensification in agriculture: chances for small-scale farmers in developing countries. GAIA – Ecol Perspect Sci Soc 19(4):263–268

Mitchell R, Hanstad T (2004) Small homegarden plots and sustainable livelihoods for the poor. LSP working paper 11. Food and Agriculture Organization of the United Nations, Rome

Morton J (1987) Fruits of warm climates. Morton JF, Miami

Muthayya S, Rah JH, Sugimoto JD, Roos FF, Kraemer K, Black RE (2013) The global hidden hunger indices and maps: an advocacy tool for action. PLoS One 8(6):e67860. doi:10.1371/journal.pone.0067860

Namara RE, Hanjra MA, Castillo GE, Ravnborg HM, Smith L, van Koppen B (2010) Agricultural water management and poverty linkages. Agric Water Manag 97(4):520–527. doi:10.1016/j.agwat.2009.05.007

Nestel P, Bouis HE, Meenakshi JV, Pfeiffer W (2006) Biofortification of staple food crops. J Nutr 136(4):1064–1067

Pampolino MF, Witt C, Pasuquin JM, Johnston A, Fisher MJ (2012) Development approach and evaluation of the nutrient expert software for nutrient management in cereal crops. Comput Electron Agric 88:103–110. doi:10.1016/j.compag.2012.07.007

Polak P, Adhikari D, Nanes B, Salter D (2003) Transforming rural water access into profitable business opportunities. In: Butterworth J, Moriarty P, van Koppen B (eds) Water, poverty and the productive uses of water at the household level, proceeding of an international symposium, Johannesburg

Popkin BM, Adair LS, Ng SW (2012) Global nutrition transition and the pandemic of obesity in developing countries. Nutr Rev 70(1):3–21. doi:10.1111/j.1753-4887.2011.00456.x

Qiang CZ-W, Kuek SC, Dymond A, Esselaar S (2011) Mobile applications for agriculture and rural development. World Bank, Washington, DC

Ruttan VW, Hayami Y (1984) Toward a theory of induced institutional innovation. J Dev Stud 20 (4):203–223. doi:10.1080/00220388408421914

Shetty P (2012) Public health: India's diabetes time bomb. Nature 485(7398):S14–S16

Soemarwoto O, Soemarwoto I, Karyono, Soekartadireja EM, Ramlan A (1975) The Javanese home garden as an integrated agro-ecosystem. In: Proceeding of an international congress on the human environment, science council of Japan, Kyoto, Available at http://archive.unu.edu/unupress/food/8F073e/8F073E08.htm. Accessed 7 Jan 2013

Soemarwoto O, Soemarwoto I, Karyono, Soekartadireja EM, Ramlan A (1985) The Javanese home garden as an integrated agro-ecosystem. Food Nutr Bull 7(3):44.47

Stadlmayr B, Charrondiere UR, Addy P, Samb B, Enujiugha VN, Bayili RG, Fagbohoun EG, Smith IF, Thiam I, Burlingame B (2010) Composition of selected foods from West Africa. Food and Agriculture Organization of the United Nations, Rome

Stoler A (1978) Garden use and household economy in rural Java. Bull Indones Econ Stud 14(2):85–101. doi:10.1080/00074917812331333331

Swinnen J (2011) The right price of food. Dev Policy Rev 29(6):667–688. doi:10.1111/j.1467-7679.2011.00552.x

Uphoff N, Kassam A (2008) The System of Rice Intensification (SRI), STOA project agricultural technologies for developing countries paper. Food and Agriculture Organization of the United Nations, Rome

Victora CG, Adair L, Fall C, Hallal PC, Martorell R, Richter L, Sachdev HS (2008) Maternal and child undernutrition: consequences for adult health and human capital. Lancet 371 (9609):340–357. doi:10.1016/S0140-6736(07)61692-4

Vodafone Group and Accenture (2011) Connected agriculture: the role of mobile in the food and agriculture value chain, Available at: http://www.vodafone.com/content/dam/vodafone/about/sustainability/2011/pdf/connected_agriculture.pdf. Accessed 10 Feb 2013

von Braun J, Tadesse G (2012) Global food price volatility and spikes: an overview of costs, causes, and solutions. ZEF discussion papers on development policy No 161. Center for Development Research, Bonn

Webb P (2010) Medium – to long-run implications of high food prices for global nutrition. J Nutr 140(1):143S–147S. doi:10.3945/jn.109.110536

Webb P (2013) Impact pathways from agricultural research to improved nutrition and health: literature analysis and research priorities. http://www.fao.org/fileadmin/user_upload/agn/pdf/Webb_FAO_paper__Webb_June_26_2013_.pdf. Accessed 10 Oct 2013

Woodhill J (2010) Capacities for institutional innovation: a complexity perspective. IDS Bull 41 (3):47–59. doi:10.1111/j.1759-5436.2010.00136.x

World Bank (2002) World development report 2002: building institutions for markets. World Bank, Washington, DC

World Bank (2008) World development report 2008: agriculture for development. World Bank, Washington, DC

World Bank (2011) ICT in agriculture: connecting smallholders to knowledge, networks, and institutions. World Bank, Washington, DC

Yang RY, Keding GB (2009) Nutritional contributions of important African Indigenous vegetables. In: Shackleton CM, Pasquini M, Drescher AW (eds) African indigenous vegetables in Urban agriculture. Earthscan, Virginia

Chapter 4
Psychology of Innovation: Innovating Human Psychology?

Manasi Kumar and Ashish Bharadwaj

Abstract Innovation, creativity and novelty-seeking are being driven by particular states of mind and unique, differentiated socio-cultural needs. This chapter identifies the conditions that drive innovation and when the capacities that enable innovation might get marred in individuals. The focus here is on understanding the behavioral characteristics of the inventor and the psychological mechanisms that guide innovation. Creativity could be a starting point for innovation; the question as to whether this is a necessary condition, and further whether it is a sufficient or insufficient one, is looked into from a managerial, legal and, most importantly, psychological standpoint. A number of perspectives from within psychology that have attempted to address the dynamics that guide creativity and innovation are discussed. Finally, the chapter poses questions that are a primer for addressing psychosocial quandaries around innovations as a mechanism for change for the rural poor.

Keywords Creativity • Socio-economic needs • Behavior • Psychological mechanisms • Innovation diffusion

Creative Process, Marginality and the Need to Innovate/Renovate

All innovations start from a creative moment, and before we unpack the psychology of innovation, it is important to understand creativity a little better. Creativity and innovation have an intertwined fate, as they refer to both a (creative) *product* and the *processes* involved in this creatively-derived product (Legrenzi 2005).

M. Kumar (✉)
Department of Psychiatry, College of Health Sciences, University of Nairobi,
Nairobi 00100 (47074), Kenya

Department of Psychology, University of Cape Town, Cape Town, South Africa
e-mail: manni_3in@hotmail.com

A. Bharadwaj
Jindal Global Law School, O.P. Jindal Global University, Delhi 131001, India
e-mail: abharadwaj@jgu.edu.in

© The Author(s) 2016
F.W. Gatzweiler, J. von Braun (eds.), *Technological and Institutional Innovations for Marginalized Smallholders in Agricultural Development*,
DOI 10.1007/978-3-319-25718-1_4

Creativity, some say, is the novel development of ideas, and the kind of transformation implied in a creative process amounts to *making the familiar strange and the strange familiar*. Psychology offers divergent and convergent modes of thinking. Convergent thinking in character is socially guided, more conventional and puts the usual ways of problem-solving into practice; divergent thinking, by contrast, refers to modes of thinking in which problems and solutions are both thought of differently. In a divergent mode, creativity is akin to wanting to invent, innovate, and discover; the urge to change or find unusual solutions to different or even, at times, to the same problems. Other than divergent thinking, for a creative product to come to fore, we should not forget that there is a 'creator' too (imaginary, real or metaphorical, like the post-structuralist ideas of Jacques Lacan!) with an interesting mind and an aptitude for knowledge and innovation. Legrenzi (2005, p. 6) talks about two primary conditions of hierarchies and gradations under which human creation takes form:

(a) every scientific or technological solution, discovery or innovation being creative in respect to another that is less creative and (b) a work of art that produces pleasure or joy (or another emotion) and recreates that emotion each time one comes into contact with it. There's a lasting feeling in reliving an emotion and sensing that the product, process or person is being more creative than something or someone else did before!

Within the fields of creativity, there are different views about the factors and conditions under which creativity thrives in society or in an individual. Some argue that creativity in childhood leads to innovation in adulthood, so there's a human developmental perspective provided here (Bergland 2013). Others argue that creativity emanates out of freedom and choice (Legrenzi 2005), while still others allude to 'optimal marginality' as a thriving condition for intellectual creativity. It is an old debate that a certain kind of marginality gives insight that leads to innovative practices (McLaughlin 2001). An example from within the field of psychoanalysis might link to similar instances in other disciplines. Eric Fromm, a psychoanalyst who was interested in the human condition and social change, challenged mainstream Freudian ideas, and looking at how his 'marginal' position changed the Freudian discourse, we can identify his resourcefulness (in terms of influences from Marxist critical theory, social work and social sciences in general), his ability to engage with and bring in alternative sources of cultural capital, and his unique emotional energy that stimulated the alternative discourse generated around identity and selfhood (McLaughlin 2001) that led to a shift. Similarly, Darwin, Freud and Marx's sojourns and splendid isolations became an active space for creativity that led to change in worldview and praxis.

Marginality is not only an intellectual concept; it is a multidimensional one, involving people at multiple levels of being and functioning (von Braun and Gatzweiler 2014). Gatzweiler and Baumüller (2013), in their work on marginality, linked poverty, ecology and developmental discourse to propose that it was the involuntary position and condition of an individual or group at the margins of social, political, economic, ecological, and biophysical systems that prevented them from accessing resources, assets, services, restrained freedom of choice, prevented the development of capabilities, and eventually caused extreme poverty (p. 30). In this sense, any discourse on marginality begs the questions: How can the poor be creative or innovate as a way of recovering their rights and voices? Can creativity be infused, socially generated or so beyond their reach that more privileged others need to secure it for this group? Can the under-privileged be incentivized to innovate in the first place?

Innovation as the 'Lava' from the Fount of Creativity: Few Behavioral Characteristics

Creativity, one can then argue, is a quality of persons, processes, or products – all three are intertwined in a creative moment (Amabile 1996). Persons have a quality to generate new ideas, and processes of thought and behavior can then lead to products that bring in something unusual and out of the box. Kirton (1994) connected adaptation[1] with creativity to distinguish different cognitive style preferences (called the Kirton Adaptation-Innovation Inventory or KAI). His research revealed that adapters and innovators – both creative – tend to have different cognitive styles which, depending on the context, may act as an advantage and or a disadvantage. Adapters tend to give more weight to structure while innovators tend to assign it less importance.[2] Some differentiation is offered between general creativity and entrepreneurial creativity. In many ways, general creativity could be static and offered in silos that may not do justice to entrepreneurial creativity. Notions of creativity emanating from an eccentric personality, someone who may be essentially highly intelligent or altruistic, or even with a deep flair for the creative arts, might be a misnomer. Entrepreneurial creativity, in this way, is akin to innovation. Amabile (1996) defines it as an activity in which numerous new combinations are tried out, a sort of 'creative destruction' (a'la Schumpeter 1934) within a particular industry which routinely brings the entire system into an unstable equilibrium.

Creativity and innovation have a few components integral to themselves, such as

(a) expertise (includes memory for factual knowledge, technical proficiency, and special talents in the target work domain) (West 2002; Amabile 1996)
(b) creative thinking (alluding to this extra bit of novelty, out of box way of solving problems and finding solutions)

---

[1] Adaptation, in this context, has been defined as the act of adjusting to fit into a specific set of environmental conditions through conformity, agreement, and compliance to acclimatize to an environment to personal advantage (Cohen 2011, p. 9).

[2] See Cohen (2011) for a discussion on how creative adaptation and adaptiveness are related to cognitive style, development of expertise and chance factors.

(c) intrinsic task motivation (this decides what the person will actually do, as opposed to what he or she is capable of doing; curiosity, deep interest, commitment and a sense of challenge drive motivation)
(d) Group task characteristics (difficulty of the task, elements of conflict vs. cooperation, presence of solution multiplicity, presence of awareness of a common task, unity of product and organization, formulation of goals, etc.) (West 2002)
(e) Diversity and knowledge in team members (diversity of knowledge and skills promotes team innovation, creative/informational decision-making, could also pose as a hindrance)
(f) External demands (threat of uncertainty, inhibited creativity at the very early stages of innovation, severity or challenge in demands, time constraints, competition, etc.) (West 2002).

Groups and organizations are settings where these factors of production come to life. Creativity could be a starting point for innovation; this is a necessary but not a sufficient condition. From a managerial standpoint, innovation is the introduction of technologically new products or processes or the improvement of existing products or processes (Ventura et al. 2011). Some others would define innovation as the successful implementation of new ideas in an organizational setting (Amabile 1996; Adams et al. 2006). Entrepreneurship is inextricably linked to innovation. Innovation, as West (2002) defines it, is the intentional introduction and application within a job, work team, or organization of ideas, processes, products or procedures which are new to that job, work team or organization, and which are designed to benefit the job, work team or organization.

Cognitive styles linked with certain attributes exhibited by an individual during the idea implementation stage may influence potentially disruptive innovations led by a group of individuals whose combined efforts exceed those of individual contributors. In their empirical study testing this claim, Spektor et al. (2011) came to the conclusion that "team performance mediated the effect of the cognitive styles on innovation" (p. 740). According to them, inclusion of creative and conformist team players improved the team's radical innovation and inclusion of team players who pay more attention to details hindered it. OECD/Eurostat (2005) define innovation in a more comprehensive way as "(...) implementation of a new or significantly improved product (good or service), or process, a new marketing method, or a new organizational method in business practice, workplace organization or external relations." Innovation, being a multidimensional concept, is inextricably linked with the degree of novelty and creativity, type of process or product innovation, nature of incremental, radical or disruptive innovation, and the technological or non-technological source of innovation. From a legal standpoint, although different types of intellectual property (IP) rights sanction protection for myriad intellectual, creative and artistic creations, the much larger base of ideas and technologies under the open innovation paradigm is gaining momentum. Innovators of technological inventions tend to rely both on IP and non-IP measures to protect their creations.[3] The fact that regimes protecting intellectual property embodied in an innovation predate the psychoanalysis and modern economic analysis of Sigmund Freud and Adam Smith, respectively, lends credibility to this legal instrument used in various forms of innovation.[4]

A number of perspectives from within psychology have tried to address the dynamics that guide creativity and innovation. The German School of *Gestalt* (organized form and a school of thought offering formal conditions to understand psychology of perception) offered us an understanding of how the whole is much more important than its constituent parts- the saying 'beauty lies in the eye of the beholder' could be apt here, as the coming together of a whole 'object' (as opposed to its constituent parts) in a simpler/congruent way is entirely driven by perceptions. Therefore, creativity consists of producing numerous variants with the aim of gradually arriving at the essential. Important works of art, scientific innovations and architectural products are good examples of Gestaltian ideas. Psychoanalysis, as developed by Freud, demonstrated how the domain of desire is dominated by our unconscious mental life, which, in turn, guides conscious behavior. The layered nature of the conscious as discovered by Freud meant that desire shaped creativity and all acts of creation.

Psychoanalytic ideas gave impetus to understanding 'the creator' and the transformation that the work of creation, as well as its creator, went through. As a refined theory of motivation, psychoanalysis helps us grasp the symbolism, instincts and desires that shape the work of creativity, however, Legrenzi (2005) argues that the instinct-based explanations could be circular and may miss something vital. Behaviorism in psychology emerged as a way of tackling the 'subjectivity' and providing findings from observable behaviors in controlled situations. Rigorous experimentation helped us understand that creativity also has another dynamic embedded within it, that is, the urge to 'reproduce creativity intended as the ability of solving problems', perfecting a model of learning through trial and error. Meanwhile, learning theorists like Pavlov and Skinner offered creativity as the ability to reproduce ordered sequences through trial and error, positive and negative reinforcements, a description very different from that offered by Gestaltian psychologists such as Wertheimer or Kohler or from the discourse of the unconscious as extended by Freud (or the psychological functionalism of economists who convert desires into preferences!).

Creativity is not merely trial and error; it is *reproductive as well as productive*, in both phenomenological and perceptual senses. Creativity also involves certain visualizations, solutions at a glimpse, restructuring and reinterpreting the situation. That the creative act is also restructuring (the mind and the end result/solution) was

---

[3] Non-IP protection measures of innovation, on the other hand, are non-statutory alternative mechanisms which include tacit knowledge (uncodified, internalized knowledge and know-how), learning effect advantages, lead-time (first mover advantage) and secrecy.

[4] However, for the same reason, one could be skeptical in treating intellectual property as an infallible system for understanding innovation, as well as a robust metric for measuring the outcome of an inventive activity.

the main lesson drawn from the limitations of behavioral discourse in psychology. In the process of defining creativity, in a nutshell, the knowledge psychology offered was that not all forms of learning are through trial and error; that humans and animals alike tend to invent new strategies to reach a goal, particularly when prior learning doesn't come in handy; and thirdly, that there's a thin line dividing creative and non-creative acts and solutions. With the development of organizational/social psychology as a field of its own, psychologists broadly offered two methods for addressing creativity and innovation in 'rethinking' products or creative enterprises: free association (Freudian technique of tapping into the unconscious) is a method of generating new ideas, and brainstorming is meant to enable a unique exchange of ideas aimed at influencing 'single' solution-oriented thoughts. *Free association* is more intra-psychic, being a process that takes place within an individual, and *brainstorming* is an inter-individual process. Both aim to tap into intersubjective elements to come up with unusual imaginative solutions.

To explain the scientific understanding of the inner essence of individual innovation, Shavinina and Seeratan (2003) attempted to answer the question – 'Why do innovative ideas emerge in human minds?' Developmental and cognitive mechanisms, according to them, are the most important for understanding the conception of individual innovation, which the authors dissect into developmental foundation of innovation, its cognitive basis, its intellectual manifestations, its metacognitive manifestations and its extra-cognitive manifestations (p. 31). From a neurophysiological standpoint, according to Vandervert (2003), "innovation is a recursive neurophysiological process that constantly reduces thought to patterns, thus constantly opening new and more efficient design spaces." (p. 27).

Psychology Behind Innovations

It is important to understand the psychological mechanisms that guide innovation. It would be apt to say that 'while not all change leads to innovation, all innovations are about change' (West and Farr 1990, p. 11), and the change then concerns the individuals who inspired a transformation of ideas towards implementation of these ideas in an organization or work context. We know the difference between creativity and innovation by now. Creativity is about generation of new ideas and innovation refers to the practice of these ideas in shaping a product, process or both. Rank et al. (2004) say that one is about idea generation and the other refers to idea implementation. In that sense, creativity is highly novel, though innovation needs to be maneuvered in a way that it is suitable and acceptable in a social context, and therefore, it is an inter-individual social process, while creativity is, thus, more of an intra-individual cognitive process (Andersen and King 1990 cited in Rank et al. 2004).

Psychologists generally allude to a definition of creativity which describes it as a process that generates an idea (or product) and essentially embodies the twin features of newness or novelty and appropriateness or social value

(Csikszentmihalyi 1996; Sawyer 2006; Gruber and Wallace 1999). It is safe to assume that, in an innovative product design, there is a correlation between the typicality of that design and positive reaction to it (Faerber and Carbon (2013). Further, there is a (positive) causal relationship between familiarity, on one hand, and typicality and positive reaction on the other hand. Novelty (one of the essential requirements for getting design patent protection) has long been considered by product design experts to be at one end of the spectrum, with typicality being at the other end.[5] Building on previous studies,[6] the experiment conducted by Faerber and Carbon (2013) revealed that "humans lacking a visual familiarity towards innovative designs also dislike them because they need time and, most importantly, elaboration to appreciate them" (p. 318). There are other differences in both concepts that are important to understand. Just as openness to experience as a personality trait enables creativity (Schweizer 2006), introversion enables intuition and judgment and the thinking through of ideas to take place, while a reflective and moderate state of orientedness (as opposed to action-oriented) helps in the generation of ideas in creative thinking.

With regards to innovations, extraversion is more beneficial to people who need to sell these ideas to other stakeholders, along with an action-state orientedness that helps to plunge into action and make change that is very goal-directed. High–arousal negative affect could impinge creativity, but seems to be a productive condition for innovating in response to frustrating deficiencies. However, positive arousal also enables energizing other innovations. West (2002) found that individuals and teams are more likely to innovate if the environment is uncertain and threatening. Higher demands, such as competitiveness between organizations, high project urgency (though conducive to creative thinking), and other demands such as high time pressure and competition within groups, is detrimental (Amabile 1996). Charismatic leadership is conducive to project implementation (Rank et al. 2004). Innovations also vary: there are technological versus administrative innovations (Legrenzi gives the example of development of 'zipper versus that of 'Post-its'), evolutionary versus revolutionary innovations, creativity types with internal versus external drivers for engagement, or even specified versus self-discovered problem types (Rank et al. 2004).

Personal initiative is a key factor here; it is about persisting in the face of repeated challenges, and also about proactive behaviors. Initiative as a driver moderates the relationship between innovation and outcomes, thus becoming an important variable to be considered. Research also puts 'voice behavior'- expression and articulation of innovation to others – as a mediator variable between creativity and innovation. Another area that creativity and innovation researchers point to is understanding the cultural differences in innovation and harmonizing

---

[5] Whereas Hekkert et al. (2003) suggest a linear relationship between novelty and typicality which determines positive reaction to the innovative product design, Blijlevens et al. (2012) assumes a non-linear relationship between typicality, per se, and positive reaction.

[6] Faerber et al. (2010), Leder and Carbon (2005) and Carbon and Schoormans (2012).

cross-cultural challenges that teams and organizations face in a modern world. Uncertainty, power, collectivism, intellectual or emotional autonomy, etc., have different meanings for different cultures. It has been found that intellectual auton- omy may be beneficial for creativity cultures which take pride in this value for 'the desirability of individuals independently pursuing their own ideas and intellectual directions', and there are also instances when high amounts of intellectual auton- omy might impede innovation, especially when disagreement concerning ideas and the impulse to be territorial takes over the act of adopting and building on a selected idea (Schwartz 1999 cited in Rank et al. 2004, p. 524). Impact of leadership and low/high uncertainty avoidance cultures are two other thematics that need to be addressed in the context of cross-cultural differences in innovations.

Choi et al. (2011) have conceptualized relationships between cognition and emotions involving innovation using appraisal theory of emotion and affective events theory. In line with Roseman et al. (1990) and Weiner (1986), they suggest that cognitive appraisal of innovation by an employee leads to emotional reactions, which, in turn, explains employees' implementation behavior (p. 108). Their work highlights the crucial role of emotional and cognitive processes in operationalizing innovation and in implementation of innovative outputs. Kaufmann (2003) adds fuel to the affect-creativity relationship by indicating that "tasks of creative think- ing may be particularly mood sensitive and that the main stream argument that positive mood unconditionally and reliably facilitates creativity is characterized as a case of premature closure" (Kaufmann 2003, p. 131).

Even though creativity and innovation conceptually overlap, the differentiating factor, according to Patterson et al. (2005), is novelty. She explains creativity as being concerned with generating new and original ideas, whereas she defines innovation as something which *also* includes use of these novel and original ideas that results in something new and socially useful. Howells (1995) applied a socio- cognitive approach to the process of technological innovation and presented "tech- nological knowledge as socially distributed cognitive knowledge" (Howells 1995, p. 888). He clarifies each step of the long and complex process ("cognitive ensemble") of "linked cognitions", starting from ideation to the final creation of the innovative product. He writes that, "it is the ensemble that makes a project and which can be judged as a 'good idea' worthy of a degree of development." (Howells 1995, p. 891).

Modelling Creativity in Innovation Management

Schweizer (2006) extends a model where creativity is the first step in the novelty generation process. In her model, individual neurocognitive and personality traits guide individual behaviors that, in turn, guide individual motivation, which then informs the behavior of others. More recently, personality theorists have given great thought to creativity and how it shapes behavior and personality (Ventura and Cruz Ventura et al. 2011). Costa and McCrae's five factor theory of personality

(1992) talks about a high score on 'openness to experience' as being a predictor for a creative and healthy personality (Schweizer 2006). The other personality traits included in the big five theory are: conscientiousness, extraversion, agreeableness and neuroticism. 'Innovation is conceived as a means of changing an organization, either as a response to changes in the external environment or as a pre-emptive action to influence the environment. Hence, innovation is here broadly defined to encompass a range of types, including new product or services, new process technology, new organization structure or administrative systems, or new plans or programmes to organization members' (Damanpour 1996, p. 1326 cited in Baregheh et al. 2009). It also refers to successful exploitation of new ideas (UK Department of Trade and Industry 1998 in Adams et al. 2006). Apart from good emotions being a facilitator for generation of good ideas (Simonton 1977), a culturally creative outside environment being a facilitator of production of creative thoughts (Simonton 2000), a risk-taking attitude and having the right training and expertise are all crucial for someone to be creative (Simon 1986).

Innovation Diffusion: Identifying Barriers and Processes of Change

Innovations could differ from one another in what could be termed their technical, social and economic characteristics, and these factors affect their diffusion as well. Changing attitudes, clearing information bottlenecks the pre-innovation/new product development stage. The extent and pace of the diffusion process depends on the personality characteristics of the potential adopters, as well as the efficiency with which the network channels can function (Agarwal 1983). One of the obstacles is focalization, which is thought to be a-creative thought that pays too much attention to doing something, though in an asymmetric appraisal of information (Legrenzi 2005; West 2002; Rank et al. 2004); the recent national election results in India reflect this bias in processing information: while one party was aggressively rejected, the selection of the prime minister was done without having sufficient understanding. Legrenzi (2005) gives examples of the 3 Bs (bed, bus and bath) as a way to defocalize. Another factor is fixation, which presents a challenge in the 'openness to experience' and receiving information without inherent biases. While focalization prevents us from selecting useful and important information, fixation is a block in receiving new information. Fixations are emotional, cognitive, social rigidities, narrowness within us that blurs information and depletes our ability to innovate. Yet another barrier is a cross-fertilization of the two, which Legenzi (2005) calls quasi-creativity, dealing with scenarios that require restructuring without the need for external problem-solving, but with a definite need for internal problem-solving – dealing with fixations, obstacles and other mental math that complicate the picture. This requires bringing in defocalization, as well as working through one's fixations.

One of the adhesives that binds creativity to innovation, as a social phenomenon, is empathy. West (2002), Legrenzi (2005), Rank et al. (2004) allude to innovation being diffused creativity that needs cooperation, conflict, group think and, more importantly, empathy that underpins the process of innovating. Empathy is the ability to be in the shoes of someone else, and creativity is also a process of decentralizing or deconstructing an idea or process in the minds of other people. Legrenzi (2005) talks about 3 Ts, technology, talent and tolerance, as the cornerstone of innovation and innovation diffusion. Empathy (tolerance) provokes diffused creativity, and then intuition, skill and resources make innovation possible. 'Creating the conditions for innovations is equivalent to creating as many variants as possible' (Legrenzi 2005, p. 55).

Technological innovations are marked by patents and trademarks. In the earlier sections, we discussed how these erase the 'creator' and provide a categorization and a symbol to the innovation. While technological innovations can usurp individual creativity, it is very important to keep individual creativity alive in the process of innovation diffusion.[7] The 3Bs and 3Ts help combat cognitive bottlenecks in creativity and innovation implementation. The finished product of innovation, though an independent product, cannot be cut off from its journey that began with the creator's idea and continued through the various processes of transformation.

Laws governing different types of intellectual property (such as patents in the case of industrial technological inventions, copyrightable artistic works or trademarks on brand names) require the innovator (creator) to overcome a stipulated threshold of innovativeness (creativity). Fromer (2010), in her investigation into the sources of divergence between patent laws and copyright laws in terms of their respective protectability standards,[8] finds that the distinctions between the two intellectual property laws essentially relate to psychological findings on creativity. Fromer states that it is important to acknowledge the psychological underpinnings of creativity in analysing intellectual property (Fromer 2010, p. 3).

Given that one of the goals of intellectual property law is to give incentives to the creator of the innovative work, it is important to turn towards studies in psychology that deconstruct the entire process by which creators (scientists or artists) create a piece of work and individual users appreciate it.[9] Rebecca Tushnet (2009, p. 51) concurs with this view and goes on to say that "psychological and sociological concepts can do more to explain creative impulses than classical economics. As a result, a copyright law that treats creativity as a product of economic incentives can miss the mark and harm what it aims to promote."

---

[7] It must be noted that, in certain legal systems (particularly in the US), the inventor is a legally accepted and recognized person who is believed to have the intellectual dominion over the entire inventive process.

[8] In the context of granting legal protection through a legal system to the creator of an intellectual property. For instance, to get patent protection for an invention, a law demands that the invention be novel, it must be an inventive step in the field and it must have an industrial applicability.

[9] See Dreyfuss (1987) for more details.

Although a society may, in general, assimilate a high level of newness of an inventive product which results from scientific creativity, it typically prefers to have only a limited level of current products resulting from artistic creativity. This, according to Fromer, partly explains the difference in protectability standards across patent law and copyright law.[10]

Four kinds of knowledge are important in thinking of innovation diffusion: (1). New ideas that emanated from a creative process and contain a new piece of information, as well as having an identified creator. (2). Non-determinism: the creative process is non-determinisitic in that it is not ascribable to some mechanical procedural calculation. (3). Constraints: the process is characterized by some constraints and should have developed some obligated actions to address those. (4). Previous elements and experience: the creative process is not created from scratch, but has a history, and the context that drives it and the elements that triggered it need to be in synch (Legrenzi 2005, p. 67). Innovation diffusion remains a multistage process in which presence, kinds and dynamics of markets present the most challenging of barriers. For example, technological innovations can make direct benefits to the rural poor, but its efficacy depends on how well integrated these are with the markets (Berdegue and Escobar 2002). The real test of the innovation lies in working through the challenges associated with the 'social use' of it, as understood by the markets (for some suggestions on addressing exclusion, see Zohir 2013).

Some Self-introspective Questions: Poverty or Innovations?

The following questions are a primer for addressing psychosocial quandaries around innovations as a mechanism for change for the rural poor. Posed by a psychologist and an economist, there is indeed some idiosyncrasy to the questions raised. There is much that psychology economics and law can tell us about creativity, deprivation and, most importantly, the need for change.

---

[10] Refer to Raymond Loewy's MAYA principle – Most Advanced Yet Acceptable. Loewy's popular design heuristic can be helpful in relating novelty with consumer preferences broadly. "He believed that, the adult public's taste is not necessarily ready to accept the logical solutions to their requirements if the solution implies too vast a departure from what they have been conditioned into accepting as the norm" (Raymond Loewy Estate's website).

Where Is Novelty and Innovation in the Lives of the Poor and Why Is It Necessary?

Most times, it is when old solutions do not necessarily work or cease to be relevant that individuals seek novelty. There are other dimensions to seeking this change – the work environment, demands of the environment and external pressures, the competition around successful delivery of a work/project. Novelty, as we have read in previous sections, is about the 'restructuring' of ideas and ideas that are to be implemented in ways that are unusual.

Poverty is a reality – an intergenerational and multidimensional one, as we know by now – however, it is also a state of mind (Kumar 2012). Researchers worldwide have talked about nutritional, physical, emotional, and social deprivations emanating out of poverty conditions. Marginality is more multilayered with an even greater number of adversities, and one would wonder if there is any hope left amidst such deprivations for change! Novelty, amidst other needs, could be a driver enabling the marginalized to work their way out of poverty. The focus on change of conditions, practices, and hindrances is also about change in attitudes, openness and the desire to live well – what Maslow would call *'self-actualization'*. He would say that each individual is endowed with a potential, and in conditions under which this potential can be nurtured, the individual's self-worth and capabilities can be actualized. Sen (1999) talks of beings and doings- functionings – as a measure of capabilities. We know that in poverty and marginality, both individual potential and opportunities to realize it are thwarted, so novelty-seeking attitude and fervor is important. Poverty compromises hope and capabilities, however, this is also an intellectual bias that practitioners do not challenge. The public projection of poverty- that all abilities are compromised – also plays a part. This projection by others, as well as the self-defeatist feeling in the marginalized themselves, needs to be changed. So, novelty is needed and is also aspired to by the marginalized as a way towards changing their future.

Nandy (2002) argued that the conceptualization of poverty and who is poor or not is an intellectual process (or defense), as people don't necessarily always think in those terms. This may not be the same for the experience of marginality in which oppressions and invisibilities are very severe. Economists tend to reduce the poor to statistics (also see Sainath 1996), and discursive stories about how poverty might entail richness of experience and knowledge are seriously undermined. The poor reinvent, rediscover, recreate and re-innovate their limited resources, individual and collective strengths and time. A focus on 'frugal' innovations demands a change in the way others perceive the poor and marginalized so that their creativity and innovativeness are recognized. More often than not, we fail to recognize the novelty that is always there. Such innovations (mostly design) are simple, novel and have the capability to provide very affordable goods and services, particularly to the economically weaker sections of society. Radjou et al. (2012) define frugal innovation as the ability to solve technical/business problems – with an attitude of finding quick, creative, less costly, local market solutions in available resources that can be tested locally and then applied to other markets. Frugal innovations are necessary to develop appropriate, adaptable, affordable and accessible solutions, products and services (Radjou et al. 2012, p. 63; Basu et al. 2013). The real innovations are providing freedom and choices to the poor that institutional innovations can provide (Stirling 2009). The little good provided by education, health and social infrastructures goes a long way. We have several studies that validate this finding (Banerjee and Duflo 2011; von Braun and Gatzweiler 2014).

Do Poverty, Deprivation, and Adversity Mar Capacities That Drive Novelty-Seeking Behavior?

Sheth (1981) talks about people who resist innovations and suggests two factors that underpin their resistance: firstly, the stronger the inclination towards a particular behavior, the greater the resistance to change, and secondly, many possess an inherent uncertainty or experience aversive physical, social or economic reactions (among other perceived risks) towards innovation. In societies where creativity is not recognized, resistance to innovation could be an easy pattern to develop. We have looked at how cultures that engage in uncertainty avoidance might treat creativity and innovations differently. Uncertainty in conditions of poverty is a real challenge. Change and innovations can thrive or be severely impeded.

The sections above present a view in which the poor and the marginalized know that they need change and are capable of creating novelty and innovations. We have also looked at how their abilities get marred by the harsh circumstances of deprivation and dismal opportunities, but it is also the 'tag', a perceptual bias among those more privileged that the poor cannot innovate and need help, that is an equally problematic attitude. One of the arguments about redressing poverty emanating from marginality is to infuse a sense of novelty and drive towards change in the rung of the ladder that doesn't see opportunities, freedoms and choices in the same way. Adversities in the form of challenges or external demands in the context of poverty and marginalization, theoretically speaking, might provide direct insight into what would work better given a set of solutions. However, when these conditions become entrenched, people's capacities are overwhelmed and all energies are directed towards survival. In such instances, innovations and creativity have to be infused, introduced and harnessed to see the kinds of processes, products and initiatives that are needed take form. One of the issues about which developmental psychology can inform development and poverty studies is the circumstances under which an intervention is worthwhile. We know that the mother is the most important tool for an infant's survival. The infant waits for his mother eagerly and cries when she is not around; the delay lasting an $a + b + c$-minute is a learning exercise and bearable, but an $a + b + c + \ldots\ldots$ to z-minute delay may be unbearable and the infant could be psychically traumatized. Similarly interventions in poverty and marginality need to be made knowing that, as time elapses, the motivational and need structures change for the worse.

Open Access This chapter is distributed under the terms of the Creative Commons Attribution-Noncommercial 2.5 License (http://creativecommons.org/licenses/by-nc/2.5/) which permits any noncommercial use, distribution, and reproduction in any medium, provided the original author(s) and source are credited.

The images or other third party material in this chapter are included in the work's Creative Commons license, unless indicated otherwise in the credit line; if such material is not included in the work's Creative Commons license and the respective action is not permitted by statutory regulation, users will need to obtain permission from the license holder to duplicate, adapt or reproduce the material.

References

Adams R, Bessant J, Phelps R (2006) Innovation management measurement: a review. Int J Manag Rev 8(1):21–47. doi:10.1111/j.1468-2370.2006.00119.x

Agarwal B (1983) Diffusion of rural innovations: some analytical issues and case of wood-burning stoves. World Dev 11(4):359–376

Amabile TM (1996) Creativity and innovations in organizations. Harvard Business School Publishing, Boston

Banerjee AV, Duflo E (2011) Poor economics: a radical rethinking of the way to fight global poverty. Public Affairs, New York

Baregheh A, Rowley J, Sambrook S (2009) Towards a multidisciplinary definition of innovation. Manag Decis 47(8):1323–1339. doi:10.1108/00251740910984578

Basu RR, Banerjee PM, Sweeny EG (2013) Frugal innovation: core competencies to address global sustainability. J Manag Glob Sustain 1(2):63–82

Berdegué JA, Escobar G (2002) Rural diversity, agricultural innovation policies and poverty reduction. Network paper No 122. Agricultural Research and Extension Network, London

Bergland A (2013) Childhood creativity leads to innovation in adulthood. Psychology today. http://www.psychologytoday.com/blog/the-athletes-way/201310/childhood-creativity-leads-innovation-in-adulthood. Accessed 10 Aug 2014

Blijlevens J, Carbon C, Mugge R, Schoormans JPL (2012) Aesthetic appraisal of product designs: independent effects of typicality and arousal. Br J Psychol 103(1):44–57. doi:10.1111/j.2044-8295.2011.02038.x

Carbon C, Schoormans JPL (2012) Rigidity rather than age as a limiting factor to appreciate innovative design. Swiss J Psychol 71(2):51–58

Choi JN, Sung SY, Lee K, Cho D (2011) Balancing cognition and emotion: innovation implementation as a function of cognitive appraisal and emotional reactions toward innovation. J Organ Behav 32(1):107–124. doi:10.1002/job.684

Cohen LM (2011) Adaptation, adaptiveness and creativity. In: Runco MA, Pritzker SR (eds) Encyclopedia of creativity, vol 30, 2nd edn. Academic Press/Elsevier, Amsterdam/Boston

Costa PT Jr, McRae RR (1992) Revised NEO Personality Inventory (NEO-PI-R) and NEO Five-Factor Inventory (NEO-FFI) professional manual. Psychological Assessment Resources, Inc., Odessa

Csikszentmihalyi M (1996) Creativity: flow and the psychology of discovery and invention. Harper Collins, New York

Dreyfuss RC (1987) The creative employee and the Copyright Act of 1976. Univ Chicago Law Rev 54(2):590. doi:10.2307/1599800

Faerber SJ, Carbon C (2013) Jump on the innovator's train: cognitive principles for creating appreciation in innovative product designs. Res Eng Des 24(3):313–319. doi:10.1007/s00163-012-0148-7

Faerber SJ, Leder H, Gerger G, Carbon C (2010) Priming semantic concepts affects the dynamics of aesthetic appreciation. Acta Psychol (Amst) 135(2):191–200. doi:10.1016/j.actpsy.2010.06.006

Fromer J (2010) A psychology of intellectual property. Northwest Univ Law Rev 104(4):1441–1509

Gatzweiler FW, Baumüller H (2013) Marginality – a framework for analyzing causal complexities of poverty. In: von Braun J, Gatzweiler FW (eds) Marginality. Addressing the nexus of poverty, exclusion and ecology. Springer, New York, pp 27–40

Gruber HE, Wallace DW (1999) The case study method and evolving systems approach for understanding unique creative people at work. In: Sternberg RJ (ed) Handbook of creativity. Cambridge University Press, Cambridge, pp 93–115

Hekkert P, Snelders D, van Wieringen PCW (2003) Most advanced, yet acceptable: typicality and novelty as joint predictors of aesthetic preference in industrial design. Br J Psychol 94:111–124

Howells J (1995) A socio-cognitive approach to innovation. Res Policy 24(6):883–894. doi:10.1016/0048-7333(94)00804-3

Kaufmann G (2003) Expanding the mood-creativity equation. Creat Res J 15(2–3):131–135. doi:10.1080/10400419.2003.9651405

Kirton M (1994) Adapters and innovators: styles of creativity and problem solving. Routledge, New York

Kumar M (2012) The poverty in psychoanalysis: 'Poverty' of psychoanalysis? Psychol Dev Soc 24(1):1–34. doi:10.1177/097133361102400101

Leder H, Carbon C (2005) Dimensions in appreciation of car interior design. Appl Cogn Psychol 19(5):603–618. doi:10.1002/acp.1088

Legrenzi P (2005) Creativity and innovation. Icon 39(051):256011

McLaughlin N (2001) Optimal marginality: innovation and orthodoxy in Fromm's revision of psychoanalysis. Sociol Q 42(2):271–288. doi:10.1111/j.1533-8525.2001.tb00034.x

Nandy A (2002) The beautiful, expanding future of poverty: popular economics as a psychological defense. Int Stud Rev 4(2):107–121. doi:10.1111/1521-9488.00266

OECD/Eurostat (2005) The measurement of scientific and technological activities – Oslo manual: guidelines for collecting and interpreting innovation data, 3rd edn. OECD Publishing, Paris. doi:10.1787/9789264013100-en

Patterson MG, West MA, Shackleton VJ, Dawson JF, Lawthom R, Maitlis S, Robinson DL, Wallace AM (2005) Validating the organizational climate measure: links to managerial practices, productivity and innovation. J Organ Behav 26(4):379–408. doi:10.1002/job.312

Radjou N, Ahuja S, Prabhu J (2012) Jugaad innovation: think frugal, be flexible, generate breakthrough growth, 1st edn. Jossey-Bass, San Francisco

Rank J, Pace VL, Frese M (2004) Three avenues for future research on creativity, innovation, and initiative. Appl Psychol 53(4):518–528. doi:10.1111/j.1464-0597.2004.00185.x

Roseman IJ, Spindel MS, Jose PE (1990) Appraisals of emotion-eliciting events: testing a theory of discrete emotions. J Pers Soc Psychol 59(5):899–915. doi:10.1037//0022-3514.59.5.899

Sainath P (1996) Everybody loves a good drought. Penguin, New Delhi

Sawyer K (2006) Explaining creativity: the science of human innovation. Oxford University Press, Oxford. ISBN 9780199737574

Schumpeter J (1934) The theory of economic development. Harvard University Press, Cambridge, MA

Schweizer TS (2006) The psychology of novelty-seeking, creativity and innovation: neurocognitive aspects within a work-psychological perspective. Creat Inn Man 15(2):164–172. doi:10.1111/j.1467-8691.2006.00383.x

Sen A (1999) Development as freedom. Harvard University Press, Cambridge

Shavinina L, Seeratan K (2003) On the nature of individual innovation. In: Shavinina L (ed) The international handbook on innovation. Elsevier, Amsterdam. ISBN 978-0-08-044198-6

Sheth JD (1981) The psychology of innovation resistance: the less developed concept in diffusion
  research. Res Mark 4:273–282
Simon H (1986) How managers express their creativity. Across Board 23(3):11–17
Simonton DK (1977) Cross-sectional time-series experiments: some suggested statistical analyses.
  Psychol Bull 84(3):489–502. doi:10.1037//0033-2909.84.3.489
Simonton DK (2000) Creativity. Cognitive, personal, developmental, and social aspects. Am
  Psychol 55(1):151–158
Spektor E, Erez M, Naveh E (2011) The effect of conformist and attentive-to-detail members on
  team innovation: reconciling the innovation paradox. Acad Manage J 54(4):740–760. doi:10.
  5465/AMJ.2011.64870100
Stirling A (2009) Direction, distribution and diversity. Pluralising progress in innovation, sustain-
  ability and development, STEPS Working paper 32, STEPS Centre, Brighton, pp 1–43
Tushnet R (2009) Economies of desire: fair use and marketplace assumptions. William Mary Law
  Rev 51(2):513–546
Vandervert L (2003) The nature of innovation: the neurophysiological basis of innovation. In:
  Shavinina L (ed) The international handbook on innovation. Elsevier, Amsterdam
Ventura D-F, Cruz APM, Landeira-Fernandez J (2011) Psychology and innovation. Psychol
  Neurosci 4(3):297–298
von Braun J, Gatzweiler FW (2014) Marginality. Addressing the nexus of poverty, exclusion and
  ecology. Springer, Dordrecht/Heidelberg/New York/London
Weiner B (1986) An attributional theory of motivation and emotion. Springer, New York
West MA (2002) Ideas are ten a Penny: it's team implementation not idea generation that counts.
  Appl Psychol 51(3):411–424. doi:10.1111/1464-0597.01006
West MA, Farr JL (1990) Innovation and creativity at work: psychological and organizational
  strategies. Wiley, New York
Zohir S (2013) Exclusion and initiatives to 'include': revisiting basic economics to guide devel-
  opment practice. In: von Braun J, Gatzweiler FW (eds) Marginality. Addressing the nexus of
  poverty, exclusion and ecology. Springer, Dordrecht/Heidelberg/New York/London, pp 41–56

Chapter 5
An Optimization Model for Technology Adoption of Marginalized Smallholders

Deden Dinar Iskandar and Franz W. Gatzweiler

Abstract The rural poor are marginalized and restricted from access to markets, public services and information, mainly due to poor connections to transport and communication infrastructure. Despite these unfavorable conditions, agricultural technology investments are believed to unleash unused human and natural capital potentials and eleviate poverty through productivity growth in agriculture. Based on the concept of marginality, we develop a theoretical model which shows that these expectations for productivity growth are conditional on human and natural capital stocks and transaction costs. Policy recommendations for segment and location specific investments are provided. Theoretical findings indicate that adjusting rural infrastructure and institutions to reduce transaction costs is a more preferable investment strategy than adjusting agricultural technologies to marginalized production conditions.

Keywords Marginality • Infrastructure • Productivity growth • Human capital • Transaction costs

Background

This paper seeks to provide the theoretical support for interventions to increase the income-generating capacity of the rural farm households below the poverty line. In particular, we observe the impact of technology adoption and the transaction cost effects on the income generation capacity in specific segments of the rural poor. There is a role for agricultural technology innovations in influencing the poor directly by lifting constraints and increasing the output level of on-farm production (Irz et al. 2001). An empirical study from Mendola (2007) also emphasizes the potential role of technology in reducing poverty through the improvement of smallholders' production capacity.

In contrast to the economics of organization in which transaction costs are defined as costs which occur "... when a good or service is transferred across a

D.D. Iskandar (✉) • F.W. Gatzweiler
Center for Development Research (ZEF), University of Bonn, Bonn, Germany
e-mail: deden.dinar@gmail.com; gatzweiler@gmail.com

© The Author(s) 2016
F.W. Gatzweiler, J. von Braun (eds.), *Technological and Institutional Innovations for Marginalized Smallholders in Agricultural Development*,
DOI 10.1007/978-3-319-25718-1_5

technologically separable interface." (Williamson 1985, p. 1), this paper defines transaction costs as the costs that create barriers between rural households and input and output markets, and restrict market access, communication and interaction.

These costs mainly include transportation costs, due to the lack of well-maintained roads, long distances between the rural households and the market, and lack of affordable public transport facilities. Transaction costs also arise from the poor communication infrastructure for accessing and exchanging information regarding markets, products, and prices.

According to Reardon et al. (2001), insufficient access to public infrastructure raises entry barriers to more profitable labor markets. Renkow et al. (2004) examine the magnitude of fixed transaction costs that hamper the access to markets for subsistence farmers in Kenya. They predict that the impact of high transaction costs on the farmers' income is equal to a tax of 15 %. Therefore, the impact of infrastructure investment on farmers' welfare is equivalent to cutting a tax of identical size. A study by Stifel and Minten (2008) on transaction costs and poverty in Madagascar finds that the incidence of rural poverty increases with increasing remoteness, and the yields of major crops and the utilization of agricultural production inputs fall significantly with the distance to the market.

Our study categorizes the rural farm households below the poverty line into four segments (Fig. 5.1) according to labor and land endowments within the marginality framework of von Braun and Gatzweiler (2014). The households in the first segment are characterized by relatively higher labor capacity and land productivity. The households in the second segment feature higher land productivity, but lower labor capacity, while, contrastingly, the households in the fourth segment possess lower land productivity and higher labor capacity. The third segment represents the households under extreme poverty, with both low land productivity and low labor capacity. In this study, these extremely poor households will be referred to as the households under the survival line, since their main concern is to fulfill their basic needs for survival.

Theoretical Analysis

The Optimization Problem for Rural Households Under the Poverty Line

The income for a rural farm household is generated from the revenue of agricultural production (on-farm activities) and the revenue of renting out factor inputs, mainly labor, to off-farm activities. The rural farm household below the poverty line is assumed to depend on two primary inputs for agricultural production: land and labor. In addition to these main inputs, farm production also requires farming production input, such as farming equipment, fertilizer and seeds. We assume that the objective of the household is to maximize total household production

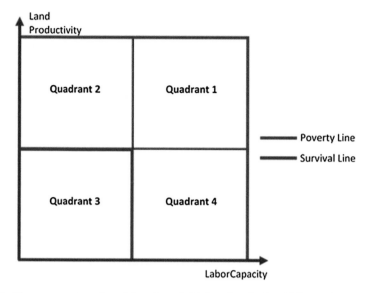

Fig. 5.1 The segmentation of rural farm households based on land and labor endowment

from on- and off-farm activities. After the introduction of technology, the objective function of a rural household is formulated in the following equation:

$$\underset{X,\hat{X},\widetilde{X},V}{Max} \left(\frac{1}{T} \left((p.Y) + \sum_{i=1}^{n} \left(\hat{w}_i \cdot \hat{X}_i \right) \right) - T \cdot \left(\sum_{i=1}^{n} \left(\widetilde{w}_i \cdot \widetilde{X}_i \right) + \sum_{j=1}^{m} \left(c_j \cdot V_j \right) \right) \right),$$

(5.1)

$$\text{S.t.} \sum_{i=1}^{n} \left(X_i + \hat{X}_i \right) \le \sum_{i=1}^{n} \left(\overline{X}_i + \widetilde{X}_i \right),$$

(5.2)

$$T \cdot \left(\sum_{i=1}^{n} \left(\widetilde{w}_i \cdot \widetilde{X}_i \right) + \sum_{j=1}^{m} \left(c_j \cdot V_j \right) \right) \le B_i,$$

(5.3)

where

$$Y = \prod_{i=1}^{n} \left(X_i^{\alpha_i} \right) \cdot \prod_{j=1}^{m} \left(V_j^{\beta_j} \right).$$

(5.4)

The income from on-farm activities is depicted by $(p..Y)$, where Y stands for the aggregate output of farm activities and p is the respective market price. Farm production is formulated as a Cobb-Douglas production function. Production output is determined by the production inputs (X_i), where i represents different types of input. Each input has a different elasticity, α_i, that represents the percentage of change in agricultural production output resulting from a 1 % change in the input i.

The technology adopted also contributes to farm production. The technology used in the production process is indicated by V_j, where j refers to different types of technology. The productivity output elasticity of each technology, β_i, indicates the percentage change of farm output arising from a 1 % change in the adoption of technology j. The production function is further characterized by $(\partial Y/\partial X^\alpha > 0)$ and $(\partial Y/\partial V^\beta > 0)$, meaning that production will increase with production input and adopted technology. The revenue generated from off-farm activities is formulated as $(\hat{w}_i \cdot \hat{X}_i)$, where $(\hat{w}_i)$ and $(\hat{X}_i)$ represent the price and the volume of input i used for other productive activities outside the farm household.

The total revenue of on-farm and off-farm activities should be adjusted by the transaction costs, T. The transaction costs occur because of spatial marginality and exclusion, specifically the difficulty of accessing the market because of the lack of public infrastructure and access to market information. The actual revenue will be discounted by transaction costs, since a certain proportion of household income needs to be spent to reach the market for selling farm output and buying the household's production input.

The costs of generating the household income can be divided into production costs and the costs of technology adoption. The production costs, $\left(\widetilde{w}_i \cdot \widetilde{X}_i\right)$, indicate the costs of production inputs that are not available in the household. $\widetilde{X}_i$ indicates the input i imported from outside the household, with $\widetilde{w}_i$ as its respective price. The cost of adopting technology is formulated as $(c_j \cdot V_j)$, where c_j is the price to adopt technology j. The presence of transaction costs (T) will increase the technology adoption costs, since the household has additional expenditures for reaching the input or technology market.

Equation (5.2) indicates the resource constraint faced by the household. The input i used for on-farm activities (X_i) and off-farm activities $\left(\hat{X}_i\right)$ is limited by the availability of the total input i, which is composed of the household's input endowment $\left(\overline{X}_i\right)$ and the input rented in from outside the household $\left(\widetilde{X}_i\right)$. Equation (5.3) is the budget constraint confronted by the household, which indicates that the total costs of employing additional inputs and adopting technology should not exceed the available production budget (B).

Given the input and budget constraints, the rural household maximizes the total income by deciding on the optimal amount of choice variables. Those variables include the amount of production input used for on-farm activities (X_i), the amount of input used for off-farm activities $\left(\hat{X}_i\right)$, the amount of additional input to be hired from outside the household $\left(\widetilde{X}_i\right)$, and the extent of adopted technology to be used for on-farm activities (V_j).

The following Lagrangean equation formulates the maximum income function for the household under the specified input and budget constraints:

$$L = \frac{1}{T}\left((p.Y) + \sum_{i=1}^{n}(\hat{w}_i \cdot \hat{X}_i) \right) - T \cdot \left(\sum_{i=1}^{n}(\tilde{w}_i \cdot \tilde{X}_i) + \sum_{j=1}^{m}(c_j \cdot V_j) \right)$$
$$+ \lambda_1\left(\left(\overline{X}_i + \tilde{X}_i \right) - (X_i + \hat{X}_i) \right) + \lambda_2\left(B_i - T \cdot \left(\sum_{i=1}^{n}(\tilde{w}_i \cdot \tilde{X}_i) + \sum_{j=1}^{m}(c_j \cdot V_j) \right) \right).$$

(5.5)

The Lagrange multipliers in Eq. (5.5), λ_1 and λ_2, measure the infinitesimal change in the generated income resulting from infinitesimal changes in the constraints. In the constrained optimization, λ_1 and λ_2 could be interpreted as marginal losses in the generated income due to the reduction in the availability of inputs and budget, respectively. These multipliers could also be interpreted differently as the marginal *income* of the *increase* in the available inputs and household budget.

Taking the first derivative of the Lagrangean equation will give the marginal income of each choice variable, i.e., change in the income generated by one unit change in the choice variable.

$$\frac{\partial L}{\partial X_i} = \frac{1}{T} \cdot \left(\frac{p.\alpha_i \cdot V_j^{\beta_j}}{X_i^{(1-\alpha_i)}} \right) - \lambda_1,$$

(5.6)

$$\frac{\partial L}{\partial \hat{X}_i} = \frac{1}{T} \cdot \hat{w}_i - \lambda_1,$$

(5.7)

$$\frac{\partial L}{\partial V_j} = \frac{1}{T} \cdot \left(\frac{p.X_i^{\alpha_i} \cdot \beta_j}{V_j^{(1-\beta_j)}} \right) \cdot -T \cdot c_j \cdot (1 + \lambda_2),$$

(5.8)

$$\frac{\partial L}{\partial \tilde{X}_i} = \lambda_1 - (1 + \lambda_2) \cdot T \cdot \tilde{w}_i.$$

(5.9)

Setting Eqs. (5.6), (5.7), (5.8) and (5.9) equal to zero will give the First Order Condition (FOC), the condition for the optimal level of each choice variable to maximize the income. Rearranging Eq. (5.6) equal to zero in terms of (X_i) will give the condition for the optimal level of input i as follows:

$$X_i = \left(\frac{p \cdot \alpha_i}{T \cdot \lambda_1} \cdot V_j^{\beta_j} \right)^{\frac{1}{(1-\alpha_i)}}.$$

(5.10)

To generate the maximal income, the level of utilized input i should be equal to the marginal income of the input and the extent of adopted technology, adjusted by the transaction cost and the marginal income loss by reducing the input availability i to the identical size as the employed input. The marginal income is determined by $(p \cdot \alpha_i)$, the product of output price and input i elasticity. This optimal condition implies that the utilization of input i in on-farm activities will increase with the output price and the input elasticity i, and decrease with the transaction costs and the marginal costs of losing the input availability to the same amount as the utilized input.

Equation (5.10) suggests that the optimal allocation of the inputs towards on-farm activities is determined by their elasticities. The higher the input elasticity, the more intensive the respective input could be used in production. Let Ld stand for land and Lb represent labor. The households in segment 2 with relatively higher land productivity, but lower labor capacity, $\alpha_{Ld} > \alpha_{Lb}$, will make use of land more intensively. On the other hand, the households in segment 4 with relatively lower land productivity, but higher labor capacity, $\alpha_{Ld} < \alpha_{Lb}$, will rely more on the utilization of labor to generate income from agricultural production. In segment 1, in which households have equally higher levels of land productivity and labor capacity, and in segment 3, in which the households suffer from equally low levels of land productivity and labor capacity, the contribution of labor and land utilization to the generated income is evenly balanced, $\alpha_{Ld} = \alpha_{Lb}$.

$$\frac{X_{Ld}}{(p \cdot \alpha_{Ld})^{\frac{1}{(1-\alpha_{Ld})}}} = \frac{X_{Lb}}{(p \cdot \alpha_{Lb})^{\frac{1}{(1-\alpha_{Lb})}}.} \tag{5.11}$$

The optimal condition for the allocation of those two inputs (Ld and Lb) on agricultural production is depicted in Eq. (5.11). When the households use two inputs, they will exhaustively use one particular input that gives the highest return (i.e., the input with higher elasticity) up to the point that the ratio of utilized input and the resulting marginal income between the two inputs is equal.

We can infer from Eqs. (5.10) and (5.11) that the presence of transaction costs, T, will reduce the optimal production input. Transaction costs discount the revenue from agricultural production. When the transaction costs increase, the optimal input for use in farm production will also decrease, since the actual revenue generated from the utilization of input is declining.

The condition for the optimal level of exported input for off-farm activities is given by the following equation:

$$\frac{1}{T} \cdot \hat{w}_i = \lambda_1. \tag{5.12}$$

The level of input used for off-farm activities will be optimal if the marginal revenue, which is the price of the input i adjusted by transaction cost, is equal to the marginal loss of generated income due to the reduction of input i availability. If the marginal revenue earned from off-farm activities is higher than the marginal loss, the optimal choice for farm households is to keep renting out the inputs. On the other hand, if the marginal loss is higher than the expected marginal revenue from renting out the inputs, then the rational household will keep the inputs for on-farm activities. In the presence of a transaction cost, the revenue from off-farm activities will be discounted, since the household will have additional costs to reach the input market.

Combining Eqs. (5.11) and (5.12) will link the decisions concerning allocation of the input between on-farm and off-farm activities.

$$X_i = \left(\frac{p \cdot \alpha_i \cdot V_j^{\beta_j}}{\hat{w}} \right)^{\frac{1}{(1-\alpha_i)}}. \tag{5.13}$$

Equation (5.13) indicates that increasing wages for off-farm work will decrease the allocation of input i to on-farm activities. Assuming that the transaction costs affects both the optimal input for use in on-farm and off-farm activities at the same scale, the transaction cost will not influence the decision.

The optimal level of technology adoption is then depicted by the following equation:

$$V_j = \left(\frac{p \cdot \beta_j}{T^2 \cdot c_j \cdot (1 + \lambda_2)} \cdot X_i^{\alpha_i} \right)^{\frac{1}{(1-\beta_j)}}. \tag{5.14}$$

Equation (5.14) says that the level of adopted technology j, (V_j), will be optimal if it is equalized to the marginal income of the adopted technology adjusted by the transaction cost and marginal income loss due to reducing the budget at hand. The marginal income is formulated as the product of output price and the elasticity of adopted technology on the generated income $(p \cdot \beta_j)$. The optimal level of adopted technology will increase with the output price and the elasticity of technology j, and decrease with the cost of obtaining the technology.

The contribution of technology to income generation does not work in isolation, but is a joint action in which the utilization of production inputs also takes part. Therefore, the condition for optimal adoption of technology is also influenced by the elasticity of input i, α_i. The optimal level of technology adoption and its contribution to income generation will increase (decrease) with a higher (lower) elasticity of the input production.

Equation (5.14) also indicates that the presence of transaction costs will reduce the optimal level of technology adoption at a multifold scale. Transaction costs hinder the adoption of technology in two ways: by discounting the actual revenue of production output and increasing the actual cost of acquiring the technology. Therefore, when the transaction costs and the price of technology are higher, it will be a rational option for rural households to decrease the adopted level of technology.

$$T \cdot \widetilde{w}_i = \frac{\lambda_1}{(1 + \lambda_2)}. \tag{5.15}$$

Equation (5.15) demonstrates the optimal condition for employing additional input production from outside households. In this equation, λ_1 represents the marginal income from increasing the available input. The optimal level of additional input i will depend on the costs of acquiring the input i, the marginal income of increasing the availability of input i, and the marginal costs of losing the available budget. The households will start buying additional input i when the marginal income from

increasing the input adjusted by the marginal cost of decreasing the current availability of the budget[1] (as a consequence of the payment made to buy the input) are
higher than the price of input i ($\widetilde{w}_i$) adjusted by the transaction cost.

The Optimization Problem for Rural Households Under the Survival Line

The extreme poor and marginalized rural households exist under worse conditions.
In our model, their capacities are constrained to fulfilling basic survival needs. The
budget constraint they live under restricts them from adopting agricultural technologies or buying additional inputs to increase production. Therefore, the constraints
in Eqs. (5.2) and (5.3) are changed into the following equations:

$$X_i + \hat{X}_i \leq \overline{X}_i, \qquad (5.16)$$

$$Bp \leq \dot{T}\left(\left(\widetilde{w}_i \cdot \widetilde{X}_i\right) + (c_i \cdot V_i)\right). \qquad (5.17)$$

$\dot{T}$ is the transaction cost confronted by the poorest households. We can expect
transaction costs to be higher for those households which are more marginalized,
therefore $\dot{T} > T$. Under the new constraints, the objective function for extremely
poor smallholders is the following:

$$\underset{X, \hat{X}}{Max} \sum_{i=1}^{n} \left(\frac{1}{\dot{T}}\left((p \cdot Y) + \left(\hat{w}_i \cdot \hat{X}_i\right)\right)\right). \qquad (5.18)$$

From the equation above, it is obvious that the one option available to the very
poorest households is limited to choosing the level of inputs to use in on-farm
activities and renting out their labor for off-farm activities. The problem of optimal
input allocation for generating maximal income is different from the problem of the
rural households in other segments, and is formulated by the following Lagrangean
equation:

$$L = \frac{1}{\dot{T}}\left((p \cdot Y_i) + \left(\hat{w}_i \cdot \hat{X}_i\right)\right) + \lambda_1\left(\left(\overline{X} - \hat{X}\right) - X\right). \qquad (5.19)$$

The first derivation of Eq. (5.19) results in the marginal income of each choice
variable,

---

[1] Spending the input for off-farm activities will reduce the available input at hand. This particular
concept of cost covers the possibility that this reduction will create cost for the farmer's income
generation.

$$\frac{\partial L}{\partial X_i} = \frac{1}{\dot{T}} \cdot \left(\frac{p \cdot \alpha_i}{X_i^{(1-\alpha_i)}} \cdot \right) - \lambda_1, \tag{5.20}$$

$$\frac{\partial L}{\partial \hat{X}_i} = \frac{1}{\dot{T}} \cdot \hat{w}_i - \lambda_1. \tag{5.21}$$

While the marginal income of renting out labor input for off-farm activities is theoretically identical (Eqs. (5.7) and (5.21) are exactly the same), it can be deduced from Eq. (5.20) that, as the extreme poor have fewer available inputs, the marginal income from input utilization for the poorest households is lower than that of the other households. That is reasonable, since the poorest households have not adopted (modern) technology in their agricultural production, whereas the production output is a joint result of all input utilization in interaction.

The conditions for the optimal level of input allocated for on-farm and off-farm activities are given in the following equations:

$$X_i = \left(\frac{p \cdot \alpha_i}{\dot{T} \cdot \lambda_1} \right)^{\frac{1}{(1-\alpha_i)}}, \tag{5.22}$$

$$\frac{1}{\dot{T}} \cdot \hat{w}_i = \lambda_1. \tag{5.23}$$

Equation (5.22) suggests that the optimal input used in on-farm activities will increase with output price and input elasticity, and decrease with transaction cost and the marginal cost of increasing input availability. This optimization behavior is equal to that of the less poor households, however, due to the absence of technological adoption, the optimal level of input utilization will be lower than the optimal level of the less poor households.

The decision to rent out the input for off-farm activities is identical to the other households (Eqs. (5.12) and (5.23) are identical). It depends on wages, transaction costs, and marginal income loss by reducing the current availability of input to spend outside the households. If the wage, after being adjusted by transaction costs, is higher than the marginal income loss for accessing labor markets, then the households will keep renting out the input up to the point where the wage and the marginal loss are equal.

Combining Eqs. (5.22) and (5.23) will link the decisions about allocating the input between on–farm and off-farm activities.

$$X_i = \left(\frac{p \cdot \alpha_i}{\hat{w}} \right)^{\frac{1}{(1-\alpha_i)}}. \tag{5.24}$$

Equation (5.24) indicates that the increase in revenue by renting out input i to off-farm activities ($\hat{w}$) will decrease the utilization of input i for agricultural production on-farm. The amount of inputs i that the poorest households intend to keep for on-farm activities is lower than that of the less poor households. Since the marginal income of the input i is lower, the poorest households are willing to rent out more input (labor) to generate household income.

Theoretical Support for Government Intervention

From the theoretical analysis, we can infer that the income generation capacity of rural households below the poverty line is determined by input elasticities, technology adoption level, and transaction costs. Particularly for the very poorest households, special attention should be given to increasing their available household budget so as to enable them to adopt productivity-increasing technologies to an extent that they would rather invest their labor on-farm than renting it out. Increasing the income of poor rural households requires the improvement of input elasticities and technology adoption, the reduction of transaction costs, and budget injection for extremely poor households. For many of the poorest households, an improvement in income elasticity and technology adoption could be a result of improving rural infrastructures, market access and land rights, which would also reduce transaction costs and improve proximity. However, if improving proximity by adjusting rural infrastructure and reducing marginality is perceived as too costly, a likely alternative for many rural poor will be to migrate to less marginal areas with better proximity and better access to markets.

Budget Injection for Extremely Poor Rural Households

The extremely poor households are suffering from a lack of budgetary capacity to support their production beyond survival levels. Therefore, one option for increasing their income-generating capacity is cash transfers from the government. Cash transfers have a direct increasing impact on the households' budget availability, moving the budget constraint from Eq. (5.3) to Eq. (5.17).

However, for the transfer to have a more permanent impact on sustainable income generation, cash transfers need to be large enough to cover the household's basic consumption needs, so that the rest of the cash transfer may be used for agricultural production, buying farming tools and seeds, and acquiring technology. An example can be found in the 2005 cash transfer program in Zambia, which shows that 29 % of the received cash transfer in the Kalomo district was invested in either livestock or agricultural inputs after the consumption of basic needs was satisfied (MCDSS and GTZ 2005).

The Improvement of Technology Adoption

We can see from Eq. (5.14) that the level of technology adoption is deterred by the availability and cost of obtaining the respective technology. The availability of technology and the cost of adoption could be defined as the function of government expenditure on research and development.

$$V_j = f(\dot{G}_j), \tag{5.25}$$

$$c_j = f(\dot{G}_j). \tag{5.26}$$

$\dot{G}_j$ is the government expenditure on research and development of agricultural technology j. V_j represents the technology j available for adoption by rural farm households, with c_j as its respective price. The availability of technology j is characterized by $\partial V_j / \partial \dot{G}_j > 0$, indicating that the availability of technology increases as the government increases spending on research and development. On the other hand, the cost of the adoption of technology j is featured with $\partial c_j / \partial \dot{G}_j < 0$, meaning that the cost decreases with government expenditure on research and development. If the government provides the subsidy for producing technology, the availability of technology will increase at a lower price and the level of adoption will increase.

However, the financial capacity of households to acquire the available technology will be different between poor households living adjacent to the poverty line and those who are extremely poor and living adjacent to or under the survival line. Therefore, besides cash transfers, cheaper technology needs to be made available to facilitate its adoption by the extremely poor households.

Improvement in Input Elasticities

Productivity improvements can also be achieved by improving input elasticity. Elasticity of input i is assumed to be the function of government investment I_i, which, in turn, is determined by government spending on that particular program, $\ddot{G}_i$.

$$\alpha_i = f(I_i), \tag{5.27}$$

$$I_i = g(\ddot{G}_i). \tag{5.28}$$

Equations (5.27) and (5.28) are characterized by $(\partial \alpha_i / \partial I_i) \cdot (\partial I_i / \partial \ddot{G}_i) > 0$, indicating that the elasticity of input i will increase with government spending . For instance, to increase the elasticity of land, the corresponding government program could be the provision of a better fertilizer funded by the government. Aside from directly providing the fertilizer, the government could also support a program to help the households make their own fertilizer. For example, the practical training program conducted by the United Nations Development Programme (UNDP) in Northern Nigeria provides practical guidance on how to make compost heaps and green manure for fertilizer (Onyemaobi 2012). The program now successfully yields better harvests for the rural households.

On the other hand, the increase in labor elasticity could be facilitated through the provision of training supported by the government. The role of the households' labor in agricultural production is not only limited to providing the work-force to cultivate the crops, but also acting as decision-makers and applying good agricultural practice. To succeed in farming, rural households need more training beyond basic literacy. They need training regarding the right crops to plant, the type and quantity of required inputs, and the methods for utilizing limited resources with greater efficiency. Better skill and knowledge will lead to higher return on labor employment in agricultural production.

Another example is a training program conducted by the UNDP in Northern Nigeria providing a practical demonstration on better farming techniques (Onyemaobi 2012). Other examples for increasing the farming skill of rural households are the farmer field schools. The season-long programs enable the farmers to meet regularly and learn new agricultural techniques. According to Davis et al. (2010), the farmer field schools have resulted in important improvements in farmer productivity. In particular, this approach is beneficial for poor farmers with a low level of primary education.

To produce effective results, the program should be targeted to address the right problems. For example, the households with lower land productivity call for provision of better and safer fertilizer. On the other hand, the households with relatively lower labor capacity require practical training to increase their skills and knowledge. These ways, the optimal condition, as indicated by Eq. (5.11), could be reached. Improving the education and skill of the farmer does not only enhance labor elasticity, but also increases technology absorption capacity. More educated and skilled smallholders will have a better capacity to adopt specific technologies and make use of them for accelerating income generation.

Reduction of Transaction Costs

Transaction costs result from the lack of public physical and institutional infrastructure and are a barrier to accessing the market. The difficulty in accessing the market impedes the opportunities to generate income. It reduces the market opportunity for agricultural products, decreases the returns to labor and land of on-farm activities, and increases the input costs, as well as the costs of adopting technology. It also reduces the opportunity of rural households to participate in labor markets for off-farm activities. Transaction cost, T, can be formulated as the function of government expenditure on public infrastructure, $\overline{G}$,

$$T = f\left(\overline{G}\right). \tag{5.29}$$

The equation above is characterized by $\partial T / \partial \overline{G} < 0$, meaning that transaction costs will decrease with increasing government spending on public infrastructure. In the case of the poor households under the survival line, Eq. (5.29) is slightly modified into $\dot{T} = f\left(\overline{G}\right)$, where $\dot{T} > T$.

Increasing the provision of public infrastructure will increase the actual revenue of agricultural production and off-farm activities, as well as lessen the cost of production, thus enhancing opportunities for generating income from agriculture. Public investment in transportation and communication infrastructure are particularly important as attempts to reduce transaction costs.

Access to public infrastructure leads to a reduction in those transaction costs, which the poor rural households have to carry when they access the output and input markets. Lower transaction costs could change the structure of relative prices for the poor farmer. This change will enable poor households to earn higher revenue from agricultural production and lower production cost, thus increasing their income. Lower transaction costs also allow the poor farm households to acquire the necessary additional inputs and technology; hence, they encourage the improvement in agricultural production that leads to higher agricultural output.

Finally, lower transaction costs may induce a change in the allocation of labor input between on-farm and off-farm activities. When rural households commit to more than one income-generating activity, the access to public infrastructure will influence the households' labor allocation decisions. The reduction in transaction cost due to the availability of public infrastructure will increase opportunities for poor rural households to participate in off-farm activities. On the other hand, lower transaction costs and improved public infrastructure, and the subsequent proximity and access to markets, may change labor allocation decisions to on-farm activities.

Conclusions

The theoretical optimization model for decision-making of marginalized smallholders on which we have elaborated assumes rational decision-makers. The likelihood of poor and extremely poor smallholders making decisions as elaborated in this optimization model correlate directly to the extent that these smallholders act rationally. This study provides theoretical evidence for increasing the income generation capacity of rural farm households below the poverty line by means of government interventions linked to the agricultural production process, like conditional cash transfers and improvements in institutional infrastructure. Particular concern should be given to the reduction of transaction costs, since high transaction costs reduce the revenue from on-farm as well as off-farm activities and increase the cost of using additional production inputs and adopting innovative technologies. Technology adoption, which has been advocated as one of the most promising ways to enhance the agricultural production capacity of the poor, is not as effective for productivity growth under the presence of high transaction costs. That is particularly relevant to marginalized smallholders. The provision of public infrastructure and improved institutions would lead to a reduction in transaction cost and increase income opportunities for poor rural households.

Segmentation of poor households provides differentiated recommendations for intervention strategies. For instance, the extremely poor households living under the survival line need more provision of infrastructure to overcome access barriers and cheaper technology than the poor households adjacent to the poverty line. Investments to increase input productivity also varies between different segments. Assuming that the households are rational, they will use those productive inputs which promise the highest return on income and thereby intensify production. Therefore, the government should invest to increase input productivity, so that their income generation capacity is increased.

Investments in technology in segments of rural society in which there is insufficient absorption capacity reduces the returns on technology investment, even if the technology is adjusted to the specific agro-ecological conditions. Productivity growth cannot be achieved in those segments, because the depreciation of human and social capital is larger than the investments in said capital. Improving the institutional infrastructure and reducing transaction costs by improving education and information and securing property rights would decrease societal depreciation, improve absorption capacities and make investments in technological innovations economically worthwhile.

From a broader agricultural development perspective, there is a trade-off between adjusting agricultural technologies to the marginalized production conditions of poor and extremely poor segments of rural society versus adjusting rural infrastructure and institutions to allow for the economically effective use of agricultural technologies. Theoretical findings indicate that adjusting rural infrastructure and institutions to reduce transaction cost is more preferable. However, it has become obvious that institutional and technological innovations need to go hand-in-hand. Therefore, both strategies need to be further informed by a spatially-specific approach.

Given the overall goal of productivity growth in agriculture, areas in which agricultural infrastructure is fragmented and marginalized will require investment in adjusting the technology to the locality. If these investments are not made, rural populations will most likely move to urban areas and both human and agro-ecological potentials will be lost. In areas in which agricultural infrastructure is less fragmented and marginalized, the use of agricultural technology which allows for the grasping of scale effects is economically advisable.

Open Access This chapter is distributed under the terms of the Creative Commons Attribution-Noncommercial 2.5 License (http://creativecommons.org/licenses/by-nc/2.5/) which permits any noncommercial use, distribution, and reproduction in any medium, provided the original author(s) and source are credited.

The images or other third party material in this chapter are included in the work's Creative Commons license, unless indicated otherwise in the credit line; if such material is not included in the work's Creative Commons license and the respective action is not permitted by statutory regulation, users will need to obtain permission from the license holder to duplicate, adapt or reproduce the material.

References

Davis K, Nkonya E, Kato E, Mekonnen DA, Odendo M, Miiro R, Nkuba J (2010) Impact of farmer field schools on agricultural productivity and poverty in East Africa. International Food Policy Research Institute, Washington, DC

Irz X, Thirtle C, Lin L, Wiggins S (2001) Agricultural productivity growth and poverty alleviation. Dev Policy Rev 19(4):449–466

MCDSS and GTZ (2005) External monitoring and evaluation report of the pilot social cash transfer scheme, Kalomo District, Zambia. Ministry of Community Development and Social Services and German Technical Cooperation, Lusaka

Mendola M (2007) Agricultural technology adoption and poverty reduction: a propensity-score matching analysis for rural Bangladesh. Food Policy 32(3):372–393. doi:10.1016/j.foodpol.2006.07.003

Onyemaobi K (2012) In Nigeria, farm training yields fuller harvests for rural women. Available from: http://www.undp.org/content/undp/en/home/ourwork/environmentandenergy/successstories/in-nigeria--farm-training-yields-fuller-harvests-for-rural-women/.   Accessed 2 Feb 2014

Reardon T, Berdegué J, Escobar G (2001) Rural nonfarm employment and incomes in Latin America: overview and policy implications. World Dev 29(3):395–409. doi:10.1016/S0305-750X(00)00112-1

Renkow M, Hallstrom DG, Karanja DD (2004) Rural infrastructure, transactions costs and market participation in Kenya. J Dev Econ 73(1):349–367. doi:10.1016/j.jdeveco.2003.02.003

Stifel D, Minten B (2008) Isolation and agricultural productivity. Agric Econ 39(1):1–15. doi:10.1111/j.1574-0862.2008.00310.x

von Braun J, Gatzweiler FW (2014) Marginality. Addressing the nexus of poverty, exclusion and ecology. Springer, Dordrecht/Heidelberg/New York/London

Williamson OE (1985) The economic institutions of capitalism. The Free Press, New York

Part II
Diversification of Agricultural Production and Income

Chapter 6
The BRAC Approach to Small Farmer Innovations

Md. Abdul Mazid, Mohammad Abdul Malek, and Mahabub Hossain

Abstract BRAC is a global leader in creating large-scale opportunities for the poor. This chapter describes how small farmer innovations are being developed by BRAC Agriculture and Food Security program. In collaboration with the Government and the International Agricultural Research Centers, the program aims to achieve food security and reduce hunger and malnutrition through increased environmentally sustainable agricultural production systems. The research focus is on cereal crops (rice and maize), vegetables and oilseeds. The program is currently implementing several innovative projects targeted at small farmers. BRAC is the largest market player, especially in hybrid seed (rice and maize) production and distribution in Bangladesh, and is gradually expanding to other countries, including Liberia, Sierra Leone, Uganda, Tanzania, South Sudan, Pakistan, Afghanistan, Myanmar, Nepal and Haiti.

Keywords Innovations • Smallholder farmers • Hybrid seeds • Community-based technology • Agro-credit • Gender

Introduction

BRAC is the world's largest development organization, with more than 115000 employees, roughly 70 per cent of whom are women, reaching an estimated 138 million people (BRAC 2013). Established by Sir Fazle Hasan Abed in 1972, following the independence of Bangladesh, BRAC is a developmental success story, spreading and implementing antipoverty solutions conceived in Bangladesh to 11 other developing

Md.A. Mazid (✉)
BRAC International, Dhaka, Bangladesh
e-mail: ma.mazid@brac.net

M.A. Malek
BRAC Research and Evaluation Division (RED), Dhaka, Bangladesh

University of Bonn-Center for Development Research (ZEF), Bonn, Germany
e-mail: malek.a@brac.net

M. Hossain
BRAC Research and Evaluation Division (RED), Dhaka, Bangladesh
e-mail: hossain.mahabub@brac.net

© The Author(s) 2016
F.W. Gatzweiler, J. von Braun (eds.), *Technological and Institutional Innovations for Marginalized Smallholders in Agricultural Development*,
DOI 10.1007/978-3-319-25718-1_6

countries in Asia, Africa and the Caribbean, making it a global leader in providing opportunities for the world's poor. BRAC's approach is holistic, utilizing a wide array of tools in the areas of education, health care, social and economic empowerment, finance and enterprise development, human rights and legal aid, agriculture and food security, as well as environmental sustainability and disaster preparedness. BRAC invests in respective community-owned human and material resources, catalyzing lasting changes and creating an ecosystem in which the poor have the chance to seize control of their own lives, making their interventions sustainable. Being the world's top development and humanitarian relief organization, BRAC is one of the few organizations which has originated in the global South. The organization is 70–80 per cent self-funded through a number of commercial enterprises that include a dairy and food project and a chain of retail handicraft stores called *Aarong*. BRAC maintains offices in 14 countries all over the world, including BRAC USA and BRAC UK.

BRAC AFSP is working in all agricultural sectors, while playing an important role in attaining self-sufficiency in the areas of food production in Bangladesh. The AFSP's goal is to contribute to achieving food security, as well as the reduction of hunger and malnutrition through increased environmentally-sustainable and diversified agricultural production. The program is working on research for development (R4D), technology validations and agricultural credit and marketing services, in line with its partnership with the Government and IARCs (International Agriculture Research Centers). This chapter focuses on small farmer innovations as a product of AFSP. The AFSP is pushing for R4D on cereal crops (rice and maize), vegetables and oilseeds (sunflower). To date, it has released ten hybrid rice, two hybrid maize, one quality protein maize (QPM) and nine vegetable varieties (hybrid and OPM). Other innovations from AFSP include short duration modern rice varieties that can be fitted into the rice cropping system, targeting four crops in a year, transforming traditional single-cropping systems to double/triple-cropping, etc.

The secondary sources used in writing this chapter have been gathered from the review of different documents, including the BRAC Annual Reports (various issues), the Program/Donor Submission Report, web-based information on the specific topic, the compilations of the BRAC AFSP, findings, and future potential works, etc., as well as in-person communication with relevant experts. The next section of this chapter describes the strategies and approaches that contribute to the BRAC Small Farmer innovations in Bangladesh and other partnering countries. The third section elaborates all existing innovations for SHs, and the final section will be a summary of the future of BRAC's innovations for small farmers.

BRAC Agriculture and Food Security Program: Partnership Is the Key

BRAC AFSP is working through research for development (AR4D), technology validations, and agricultural credit and marketing services. The main strength of the program is partnership with the Government, International Agriculture Research

Fig. 6.1 BRAC's approach to achieving food security

Centers IARCs, Private Companies and NGOs. AFSP began its activities in Bangladesh, and is now expanding into other partnering countries, because of a strong belief in the philosophy, "every country should have its own food and nutrition security" (Fig. 6.1). As of this writing, apart from Bangladesh, the organization has already initiated activities in Liberia, Sierra Leone, Uganda, Tanzania, South Sudan, Pakistan, Haiti and Afghanistan,

Sustainable development is a key issue for any institution, as well as for a country. BRAC has established partnerships with different stakeholders to ensure the following:

1. Funding support
2. Implementing different projects
3. Collaborating with partners
4. Ensuring more coverage both in country and out of country
5. Consideration of cost effectiveness and sustainability.

Partnerships with numerous local NGOs and INGOs gives BRAC an opportunity to operate cost effectively. About 120,000 employees of BRAC, for instance, can join with millions of employees of other NGOs to expand its program at reduced cost and without incurring any reduction in quality. More recently, an MoU (see next paragraph) has been signed between BRAC and African Rice, which allows BRAC to get seeds for varieties of stress-tolerant rice and an advanced breeding line for testing validation at the seed farm run by BRAC Liberia. African Rice provides BRAC Liberia with a rice variety that is tolerant of iron. Similarly, NaCRRI of Uganda provides vines of bio-fortified Orange Flesh Sweet Potato

(OFSP), which contains high beta carotene, a precursor of vitamin "A", for seed multiplication and dissemination to small farmers through the BRAC technology delivery model. The Bangladesh Rice Research Institute (BRRI) has developed stress-tolerant (flash flood, submergence, drought, salinity) rice varieties that have been disseminated to farmers through BRAC Bangladesh.

BRAC, through direct delivery, seeks partnership with donors belonging to international communities. Donors usually respond to calls for proposals. Then, a Memorandum of Understanding (MoU) or Letter of Agreement (LoA) is signed among partners through discussion of achieving a common goal. For instance, BRAC has an MoU with CIMMYT/Africa Rice/AVRDC. This organization also has some local partners, such as local NGOs, private sectors (PS), and corporate partners, to act as sharing parents of hybrid maize/rice/vegetables for validation and development at the BRAC seed farm.

Farmers' needs, preference, market value, nutritional value, food security and BRAC strategies are the main issues in prioritizing which crops need to be improved.

In Bangladesh, BRAC initiated a facility called Agricultural Research for Development (AR4D), with two research centres, one for rice and vegetables in Gazipur and one for maize in Sherpur, Bogra. Other initiatives include a one-plant tissue culture lab for potato, banana, and ornamental plants, one soil testing lab, nine seed production farms, two seed processing centres for rice, vegetables and maize, and eight seed storage facilities with a capacity of 2400 MT. BRAC's experience expanded overseas to countries which include Uganda, Liberia, Sierra Leone and South Sudan, establishing seed production farms and collective demonstration farms (CDF). Moreover, BRAC has established a one-plant tissue culture lab and a seed processing centre on a farm in Nakaseke, BRAC Uganda. BRAC has created linkages with different National Agricultural Research Systems (NARS) and Consultative Group on International Agricultural Research (CGIAR) organisations, and made partnership collaborations for R4D, technology validation and dissemination and agricultural credit and marketing. Such linkages have been established with the International Rice Research Institute (IRRI, Philippines), the International Maize and Wheat Improvement Center (CYMMIT, Mexico), the International Potato Center (CIP), the Asian Vegetables Research and Development Center (AVRDC, Taiwan), AfricaRice, the International Institute of Tropical Agriculture (IITA), the National Crop Resources and Research Institutes (NaCRRI) in Uganda, the Central Agricultural Research Institute (CARI) in Liberia, the Sierra Leone Agricultural Research Institute (SLARI) in Sierra Leone etc. BRAC's Agriculture program is also working with a number of multinational seed companies, having established agreements for sharing technology and marketing agroproducts. At present, partnerships with multinational seed companies include: the Yuan Long Ping High Tech Agriculture Co. Ltd. (China), the Pacific Seed Company (Australia), the Mahyco Seed Company (India), the Druk Seed Company (Bhutan), and the Seminis Vegetable Seed Company (India) Ltd. (India).

BRAC Agricultural Credit activities intervene through customized credit with improved agricultural technology and knowledge support. As a result, it is strongly collaborating with the public sector and donor bodies. BRAC began seed production in 1996 with the assistance of the Bangladesh Agriculture Development Corporation (BADC), under a project of the Ministry of Agriculture for rice, maize, potato, and vegetables carried out at its own farms in different Agro Ecological Zones (AEZ), and also through contact with farmers and markets through dealers.

While BRAC expands to other countries, it studies different scenarios which are prevalent in different countries on agricultural practice, seed systems, crop varieties, input situations, technology dissemination, extension services, agricultural tools, capacity building training for farmers, post-harvest loss, storing facilities, commodity marketing, private and public sector engagement in agriculture, etc. It also takes lessons concerning the major constraints for technology adoption which are listed below: low productivity, lack of stress-tolerant varieties (maize, rice, vegetables, cassava), seed admixture, soil acidity (upland), iron toxicity (lowland), imbalanced fertilizers and poor fertilizer management, water logging (maize), drought (rain fed rice), knowledge gaps (lack of modern cultivation practices, diseases, pests) and credit for inputs. The BRAC international agricultural program has also identified certain issues, for example, seed quality, seed storage at branch levels, preservation of vaccines due to poor electricity supply and inadequate transportation facilities, storage, low quality chicks, feed, and climate change, especially drought, excess rain which causes flash floods, soil erosion, water logging, poor growth of maize and vegetables, lack of communication, poor infrastructure, isolated transport facilities and inadequate market analysis, market system development and linkage with markets.

Based on these experiences, BRAC has been implementing relevant projects in those partnering countries to improve the above-mentioned agricultural constraints through the GPFA (Global Poverty Fund Association Project funded by DFID) in Tanzania, Liberia, and Sierra Leone, the JSDF (Japan Social Development Fund Project) through the World Bank in Uganda, LEAD (Livelihood Enhancement through Agricultural Development, funded by the UK Goverment) in Tanzania, the Agriculture & Livestock Extension Program funded by the Omidyar Foundation, the Agriculture & Livestock Extension Program funded by the Mastercard Foundation, the Livestock Project funded by EC in Liberia, the Demonstration Farm projects funded by OXFAM Novib and TUP (Targeting the Ultra Poor) in South Sudan, and Seed Production funded by AGRA in Liberia and Sierra Leone, among others. BRAC International has been improving agricultural situations through these programmes in such areas as availability of quality seed for farmers through the establishment of BRAC seed farms and seed processing centres, establishing tissue culture labs, seed distribution, identifying suitable crops and varieties, other input supports, organizing capacity building training for farmers, distributing agricultural tools, management practice, best suitable cropping patterns, technology dissemination through demonstration, meetings, farmer field days (FFD), providing extension services, post-harvest management training and linking farmers with markets.

BRAC Innovations for Small Farmer Agriculture

Technology Innovation

In the case of technological innovation processes for agriculture, BRAC proceeds in both ways: Agricultural Research for Development (AR4D) and Action research. For AR4D, this organization has its own agricultural research centers and a demonstration plot called the BRAC Agricultural Research Center (BARDC) located in Gazipur and Sherpur, Bogra in Bangladesh. In these centers, laboratory research and field trials are normally conducted with the help of an agronomist, plant breeder, horticulturist, etc. For action research, BRAC uses its contract farmers in different districts of Bangladesh. When BRAC wants to disseminate new technology, varieties or methods into any localities to verify their performance at the farm level, it first disposes its experiment to contract farmers and teaches them to apply those varieties or methods at the field level. Secondly, each farmer is directed to adopt it differently. Then, having used the same method but with different directions, each farmer is asked to identify any distinguishing features, for example, what time is best for particular rice varieties, in the output of their respective fields. Finally, the best process or varieties are identified based on agro products yield, growth duration, quality taste, appearance and market value etc.

Since 2001, BRAC has initiated hybrid rice research and development activities. BRAC introduced different exotic hybrid rice varieties from China and India and conducted adaptive trials in different regions of Bangladesh, registering seven exotic hybrid rice varieties with the National Seed Board (NSB). Subsequently, three hybrid rice have been developed by BRAC scientists and duly released by the NSB under the names Sakti, Sakti-2 and Sakti-3. Development of MV rice in a short duration was given priority against cold-tolerant and/or escape cold and/or escaped terminal drought conditions at Panicle Initiation (PI) to the flowering stage, so that it can be added to rice-based cropping systems and could be grown for three-four non-rice, rice and/or four rice crop systems during 2012–2013, obtaining 18 MT per ha per year with judicious fertilizer management.

Maize is now considered to be the third most important cereal crop after rice and wheat. It can be grown in winter (rabi or dry season) or alone after rice, and in the summer season (kharif I/pre-wet season) after potato. Maize was introduced into Bangladesh in 1975, having mostly composite varieties (OPV). Farmers did not accept maize widely because of low yield and lack of market facility. Since 1993, the maize crop's popularity has risen due to the introduction of hybrid varieties by BRAC. This was due to higher grain yield and increased market demand for maize as a poultry feed. BRAC initially popularized hybrid maize varieties through introduction of Pacific-11 in 1993, and later on, Pacific-984, Pacific-747, and Pacific-759. BRAC engaged farmers by providing quality seeds along with buy-back guarantees lasting 2–3 years to promote hybrid maize and the use of the maize grain as an ingredient in poultry feed. BRAC also introduced stress-tolerant crops and fish varieties to the cropping systems to combat the adverse effect of climate change due to flash flood submergence, drought, salt, and high and/or low temperatures. BRAC also accommodated high value non-rice crops like maize, sunflower and sesame in the rice-based cropping systems through the use of short maturing crop varieties (two-four crops/year), converting single cropping areas to double or triple cropping areas to enhance food security and land productivity. Other than that, BRAC introduced oil crops (sunflower) into the rabi (winter) season and sesame into the kharif I season in a saline environment and introduced vegetable cultivation through the pyramid technique in tidal wet lands.

BRAC also collects breeder and/or foundation rice seeds from Bangladesh Rice Research (BRRI) and the Bangladesh Institute of Nuclear Agriculture (BINA), maize and vegetables from the Bangladesh Agricultural Research Institute (BARI) and from international companies (hybrid and OP), testing and validating these at the BRAC Agricultural Development Centers (BARDC), and then multiplying, processing, and distributing the seeds through a market dealer. BRAC also validates and disseminates stress-tolerant varieties of rice for use against drought, salinity and submergence, and short duration rice varieties that could increase yield and land productivity, and, consequently, the income of farmers.

Thus, BRAC has been developing, introducing and promoting different technologies, which include practice, method, and crop varieties encompassing hybrid, open pollinated (OP), stress-tolerant, short duration, early/ late varieties, etc., all of which affect greater yield, production and price. These BRAC innovations are listed below:

Hybrid and OP seed varieties

1. Rice hybrid seed: HB-09, Jagoran, Alloran, Shakti, Shakti-2, Shakti-3 and Sathi
2. Maize hybrid seed: Pacific-984, Pacific-747, Pacific-759, Uttaran, Uttaran-2 & Uttaran-3
3. Vegetables seed:

   (i) Bitter gourd variety: Bulbuli (Hybrid)
   (ii) Ridge gourd: Green Star (Hybrid)
   (iii) Tomato variety: Tripti-1 and Tripti-2 (Hybrid)
   (iv) Sponge gourd: Green Star (Hybrid)
   (v) Sweet gourd: Beauty (Hybrid)
   (vi) Eggplant varieties: Super Singnath, Giant Green, Shruvi (Hybrid)
   (vii) Okra variety: Evergreen (Hybrid)
   (viii) Red Okra (hybrid)
   (ix) Cabbage variety: KzE-739 (Hybrid)
   (x) Cucumber variety: Shufalla-1(OP)
   (xi) Radish variety: Shufalla-40 (OP)
   (xii) Indian spinach variety: Shufalla Palang-1(OP)
   (xiii) Bottle gourd: Green Supper (OP)
   (xiv) Vitamin A rich varieties: Red Spinach (OP), Red LYB (OP) and orange flesh sweet potato (OFSP)

4. Seasonal change varieties developed targeting a high price in market

(a) Usha and Asha summer country bean varieties (country bean normally grown in the winter season in Bangladesh)

5. Stress-tolerant varieties

   Flood (50 per cent), drought (20 per cent) and salinity (30 per cent) are the main stress environment in Bangladesh, where rice frequently suffers from considerable shock in attempting to maintain its full yield potential. The nature and extent of these environments vary with season, topography and location.

(a) Saline-tolerant variety

30 per cent of the lands of Bangladesh are affected by varying degrees of salinity. About 1.02 million hectares in the coastal areas are affected by various degrees of salinity, varying with the season. In the dry season, soil and river water salinity increases, while during the monsoon season, it goes down. Therefore, land use has temporal and spatial variations with the season. BRAC introduced and promoted saline-tolerant rice varieties, such as BRRI dhan47, BRRI dhan60 BINA 8, and BINA10, in the rabi season (dry season), and BRRI dhan41, BRRI dhan53, and BRRI dhan54 in the aman or monsoon (wet season) along the coastal belt, allowing farmers to get crops from fallow land and affecting their food security.

(b) Drought-tolerant variety

Irrigation facility has not been equally available in and around Bangladesh due to high and low ground water. As a result, 20 per cent of the land suffers from drought. Drought occurs mainly due to uneven distribution of rainfall. The northwestern part of Bangladesh is treated as a drought-prone area due to poor rainfall. It is one of the major abiotic constraints for rice grown (5.7 m ha) under rain-fed conditions in Bangladesh and causes a substantial reduction of yield. In this case, BRAC introduced and disseminated a short duration (115–118-day) rice variety, e.g., BRRI dhan33, BRRI dhan39, and BINADHAN 7, which will escape the terminal drought if they are transplanted by July 15th in the drought prone area. Some of the Aman (wet season) varieties, like BRRI dhan56 (110 d), BRRI dhan57 (105 d), and BRRI dhan62 (100 d), being high zinc rice, can be grown within 100–110 days in order to avoid drought and provide room for a second crop, perhaps a non-rice variety, such as an early potato, vegetable, or mustard, in profitable rice-based cropping systems.

(c) Submergence-tolerant variety

50 per cent of the land in Bangladesh is affected by flash floods. Crop submergence due to flash floods is a significant risk to the agriculture sector of Bangladesh threatened by climate change. Bangladesh has a total area of 14.8 million hectares. Out of this area, 50 % is affected by different types of flash floods. In those cases, BRAC introduced and promoted a flash flood tolerant Swarna *Sub*1 (NSB released as BRRI dhan51), a BR11 *Sub*1(NSB released as

BRRI dhan52), a BINA rice variety such as IR64 *Sub*1 (NSB released as BINADHAN11) and a Samba masuri *Sub*1 (NSB released as BINADHAN12) in flash flood prone areas. The farmers in this case get crops from flash flood and/or drought-affected areas, which helps in establishing food security.

6. Short duration varieties

BRAC introduced and promoted the short duration rice varieties, such as BRRI dhan33, BINA7, BRRI dhan56, BRRI dhan57, and BRRI dhan62, into Bangladesh so that one crop (like the short duration mustard varities BARI sarisa 14, Bari sarisa15, and Bari sarisa 16) can be grown in a year in addition to the Boro and Aman cropping patterns, as a result of which total annual production is increased by increasing cropping intensity and improving crop diversification.

G-1: Market Share of BRAC Seed in Bangladesh

BRAC Seed and Market Size

(a) BRAC markets seeds through two different sales channels, which includes the following-appointing new dealers and using existing seed dealers: BRAC appoints new local dealers to market the products. Along with the new dealers, BRAC is also using its extensive seed dealer network to distribute in rural Bangladesh. BRAC is currently linked with over 4000 seed retailers across Bangladesh. The market share (%) in Bangladesh is shown in Fig. 6.2.

(b) BRAC is also working with seed markets through Community Agriculture Promoters (CAPs) in other countries.

Fig. 6.2 Market share of BRAC seed in Bangladesh

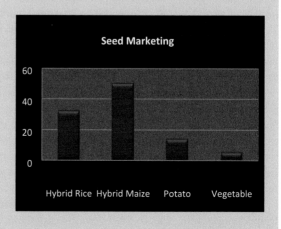

7. BRAC's innovation in cropping patterns changes and converts a single cropping area to a double or triple cropping area by promoting the short duration rice crops, depending on land types and profitable cropping patterns. These are:

   (a) Sunflower in Rice based cropping system Rice-Sunflower-Fallow
   (b) Rice-No rice: Rice-Potato-Mungbean/Jute, Rice-wheat-Mungbean/Jute
   (c) Rice-Maize-mungbean, Rice-Potato-Maize
   (d) Rice-Vegetables

8. Innovative potato storage for smallholder farmer households in Bangladesh

The construction of small "Ambient Type Potato Storage" used local materials such as bamboo, straw, and locally made concrete, essentially becoming a construction project carried out by local carpenters. BRAC provided technical support to farmers on to how prepare a special type of cold storage for potatoes at minimum cost. These ambient type spaces are used to store 6–8 t of table potato for 3–4 months, allowing farmers to get a roughly 50 per cent increase in the price of potatoes compared to the price during harvest.

9. Sunflower cultivation in the coastal/saline belt in Bangladesh

Sunflower cultivation is easier, cheaper and more profitable, and it is possible to cultivate in the salinity-induced soil of the coastal belt during the fallow season. It requires very little irrigation and small amounts of fertilizer and pesticide. One kg of sunflower seeds brings 500–600 ml of oil, a greater amount than that of any of the other oil seeds, and it is very good for human health

Technology with Financial Support

BRAC has been providing financial support along with technology in the field of farming through the Borgachasi/Sharecropper Union Programme (BCU) Project. Tenant farmers are very vulnerable in terms of not having easy access to formal financial institutions for credit or loans, aggravating their credit needs. Since BCUP farmers are tenant farmers, they don't have enough wealth which they can present to formal financial institutions as collateral. Therefore, to purchase inputs (seeds, pesticides, fertilizers, tillage implements, etc.), they need financial support. Sometimes these farmers require loans to meet immediate needs, or during the time of harvesting, marketing, and storing what they are producing. Newly innovative technologies and machineries also require financial support. Under the BCUP project scheme, BRAC provides credit to farmers as per the demand for purchasing inputs, tools and irrigation equipment, and continues to provide technical support to farmers and to address their ongoing production and post-harvest problems.

   BRAC also offers Agriculture Credit + (Borgachasi/Sharecropper Union Programme) with the NCDP Northwest Crop Diversification Programme (NCDP), the SCDP Second Crop Diversification Program (SCDP) and partial grants for quality inputs and tillage for AFSP. The Central Bank of Bangladesh

(Bangladesh Bank) offered BRAC BDT 5 billion ($75 million US) to provide loans to tenant farmers/sharecroppers. BRAC took this challenge and has been experimenting since October 2009, working in 210 Upazilas in 46 districts (twenty working regions in five working divisions) in 2013. BRAC's target is to reach 300,000 sharecropper/tenant farmers in the next 3 years with credit and proven agricultural technologies. BRAC started seed production in 1996 with the assistance of BADC, producing rice, maize, potato, and vegetables on their own nine farms in different AEZ (agro-ecological zones) and creating around 4000+ contract farmers.

Agriculture Commodities Marketing Support

Rural markets for farm produce are not generally developed on the basis of which farmers suffer from losses while they are selling their products in season. BRAC realizes this, and has worked to establish a value chain between farmers and BRAC social enterprises, such as maize which will be used as food for the farmers and feed for their livestock, giving scope and future direction to farming techniques. It provides marketing support to farmers through the following techniques and methods:

- The purchase of seed through contract farmers
- Two poultry feed mills which purchase maize through contract farmers
- Established community collection centres (CCC) so that farmers can easily bring their products to the CCC and sell them at a proper price
- Established links with farmers and value chain actors as a result of which farmers are ensured that their products will be sold.

Extension Innovation

The public sector extension program usually has very limited coverage in terms of access and providing information to the marginal and poor farmers. Thus, BRAC's approach is to disseminate agricultural technologies through large scale block demonstrations (D4D) in which farmers participate; this can create a larger impact amongst the participating farmers and lead to a spillover effect amongst the neighbouring farmers in the village. Therefore, BRAC, under its extension approach, will organize a group of 40–100 marginal farmers for block demonstration, giving a partial grant for quality inputs (quality seed, fertilizers, tillage) in order that they will be able to cultivate and use modern varieties of crops, fishes, production technologies and practices. The beneficiaries will also receive sectorwise adequate training and the latest information for getting better production from their fields. BRAC is currently targeting the disbursement of crops and/or sectorwise partial grants, providing technical support/follow up for crop production and integrating rice-fish-vegetable cultivation in dykes, while organizing farmer field days for the dissemination of learning, aiming to reach 60,000 farmers in 50 sub districts (Upazilla) belonging to 12 districts by the year 2015.

BRAC Innovations and the Future Outlook for Agriculture and Food Security Programs

Bangladesh

BRAC plans to introduce agro-consumer products to the market, including sunflower oil and spices. It will be marketing sunflower oil by pursuing BRAC's unique value-chain approach to procuring inputs for sunflower seeds from the salinity-affected southern parts of Bangladesh. This is to ensure increased cropping intensity in the seasonally fallow line plant in the coastal areas and guaranteed fair price for the farmers who cultivate this new crop. It also aims to adopt a unique door-to-door marketing approach with vegetable seeds to promote home-gardening amongst women from marginalized rural households. By 2015, the research and development wing of our agriculture programme is expected to release up to three new types of short duration maize varieties, an inbred rice variety, two hybrid rice varieties and two vegetable varieties. They are also currently in the advanced stage of developing a cold-tolerant shorter-maturity rice variety.

BRAC International

- Promotion of bio-fortified crop varieties such as high beta carotene (provitamin A) yellow cassava, orange fleshed sweet potato (OFSP), colorful red maize, red color okra, iron rich beans (IRB), enriched zinc rice etc., for reducing hidden hunger, meeting the micro nutrients and vitamins needs of the poor and improving food security
- Focus on climate change resilient agriculture adoption through location-specific technology, like raising pyramids in swampy land for pit method vegetable cultivation, or rice-fish cultivation with dyke vegetables in water logged conditions, iron toxicity tolerance rice, salt tolerant rice and flash flood submergence tolerance rice etc
- Commercial farming, from subsistent farming practices to selling of surplus in Africa through Agriculture value chain development, small scale mechanization, post harvest processing and storing
- Increasing cropping intensity through the use of short-duration varieties of rice and high value non-rice crops such as maize, sunflower, vegetables etc. in rice/non-rice cropping systems in farm fields

- Quality seed/seedling production of disease-free bananas, micronutrient-rich and provitamin A rich vines of OFSP, stem cuttings from yellow cassava through in vitro tissue culture
- Technology accompanied by financial access for farmers through Agri-financing/credit +
- Post-harvest management, especially the harvesting and threshing of rice and maize, drying, packaging, storing and travelling
- Market linkage between farmers and traders/private sectors for commodity marketing, especially getting yellow cassava to millers, maize to poultry feed producers, OFSP to value-adding biscuit and bread factories, and leaves and young vines for cultivating vegetables.

Conclusions

BRAC Agriculture and Food Security Programs have been achieving food security and reducing hunger and malnutrition through increased environmentally sustainable agricultural production systems which work through Agriculture research for development (AR4D), technology validation and dissemination of special quality seeds, bio-fortified crop introduction and promotion, extension services, irrigation, and the building of farmer capacity, agricultural credit and marketing services through partnership with development agencies, government, the private sector and International Agricultural Research Centers (IARCs). BRAC's approach is to disseminate agricultural technology through participatory demonstrations with farmers, large-scale block demonstrations, and group and sub-group approaches with lead farmers, as well as organizing farmer field day (FFD) workshops and assorted fairs. BRAC has introduced and promoted diversified crops in farm fields for improving cropping pattern and intensity. It has facilitated market linkage between farmers' agricultural commodities and the private sector in value chains and has introduced collection points and contract farming. BRAC encourages female farmers' involvement in the program. 70 per cent of female farmers in African countries have become involved.

Open Access This chapter is distributed under the terms of the Creative Commons Attribution-Noncommercial 2.5 License (http://creativecommons.org/licenses/by-nc/2.5/) which permits any noncommercial use, distribution, and reproduction in any medium, provided the original author(s) and source are credited.

The images or other third party material in this chapter are included in the work's Creative Commons license, unless indicated otherwise in the credit line; if such material is not included in the work's Creative Commons license and the respective action is not permitted by statutory regulation, users will need to obtain permission from the license holder to duplicate, adapt or reproduce the material.

References

BRAC (2012) Agriculture and Food Security Program (AFSP). BRAC, Dhaka
BRAC (2013) Brac annual report 2013. BRAC, Dhaka
BRRI annual internal review meeting 2012–2013 held in 2013. BRRI, Gazipur
Malek MA, Ahasan A, Hossain MA, Ahmed MS, Hossain M, Reza MH (2015) Impact assessment of credit programme for the tenant farmers in Bangladesh, 3ie Grantee Final Report. International Initiative for Impact Evaluation (3ie), New Delhi, http://www.3ieimpact.org/media/filer_public/2015/09/18/gfr-ow31222-credit_programme_for_the_tenant_farmers_in_bangladesh.pdf

Chapter 7
Agricultural Research and Extension Linkages in the Amhara Region, Ethiopia

Tilaye Teklewold Deneke and Daniel Gulti

Abstract Agricultural innovation systems require strong linkage between research and extension organizations in particular, and among the various actors engaged in the agricultural sector in general. In the context of Ethiopia and the Amhara regional state, the agricultural research and extension system is characterized by a large number of actors in a fragmented and underdeveloped innovation system, resulting in very low national and regional innovation capacities. Farmers are generally viewed as passive recipients of technology. As a result, research outputs do not reach farmers and remain shelved in research centers. Instead, research and extension need to take place within interlinked, overlapping and iterative processes. This chapter reviews past initiatives to bring about integration among these actors to identify areas for improvement.

Keywords Research • Extension services • Ethiopia • Innovation systems • Innovation capacities

Introduction

Agriculture in Ethiopia is the most important sector of the economy, contributing over 46.89 % of GDP, employing about 80 % of the labour force, and serving as the source of 60 % of export earning (MoFED 2011). Due to this dominance, poverty, food security and the performance of the economy at large depends heavily on the performance of the agricultural sector. Though there are positive trends in recent years (during the first 5 year growth and transformation plan), the agricultural sector has grown by 6.6 % per annum on average (MoFED 2015), the sector has largely remained underdeveloped, low input-low output and barely subsistent. The level of productivity of major crops, such as teff, sorghum, wheat and maize, has remained below 2 t/ha (see Fig. 7.1). The country has historically been chronically

T.T. Deneke (✉)
Amhara Agricultural Research Institute (ARARI), Bahir Dar, Ethiopia
e-mail: ttddeneke@yahoo.com

D. Gulti
Agricultural Transformation Agency (ATA), Addis Ababa, Ethiopia

© The Author(s) 2016   113
F.W. Gatzweiler, J. von Braun (eds.), *Technological and Institutional Innovations
for Marginalized Smallholders in Agricultural Development*,
DOI 10.1007/978-3-319-25718-1_7

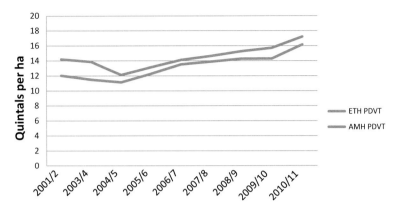

Fig. 7.1 Total grain productivity for the cropping years 2001/2–20010/11 for Ethiopia (ETH) and Amhara (AMH) (Source: Own analysis based on CSA data years 2001–2011)

food insecure and still depends on commercial imported food and food aid to some extent.

In order to avert this situation, the country followed a policy of using agricultural development as an engine of growth based on the justification that the sort of broad-based growth and rapid development that would lift the masses out of poverty in the shortest possible time lay in the agricultural sector. This was based on the fact that the vast majority of the population lives on agriculture and the country is endowed with abundant land and labour resources, although it is short in capital stocks. Over the last two decades, the GoE (Government of Ethiopia) has followed the pro-smallholder policy framework known as Agricultural Development Led Indus-trialization (ADLI). The policy is aimed at increasing the production and produc-tivity of smallholder agriculture to insure household level food security and generating capital from export earnings of agricultural products, as well as paving the way for industrial development through supplying adequate, high quality and cheap raw material for agro-processing sectors. This meant increased government investment in agricultural research, extension service, improvement of rural infra-structure and the provision of support services such as credit, external input supply, and agricultural commodity marketing.

Provision of such support in a coordinated, effective and efficient manner has, however, been a great challenge, as these support services are given by a number of separate government offices and nongovernmental organizations without laying down a viable and efficient mechanism for coordination and linkage. This has resulted in sluggish technological change that can be verified by the fact that the percentage of farmers using improved agricultural technologies, such as fertilizers and high yielding varieties, has continued to be low (in the Amhara region, only 15 % of the vast majority of agricultural technologies developed by the federal and regional agricultural research systems have been accessed and adopted by the farming community).

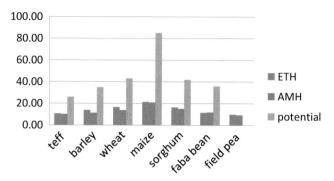

Fig. 7.2 Actual and Potential productivity of major crops in (quintals/ha) for Ethiopia (ETH) and Amhara (AMH) (Source: own analysis based on CSA data and data from the National Variety Register 2014)

Until 2014, over 960 high yielding varieties had been developed by the national agricultural research system, of which not less than 106 of them belong to ARARI. In addition to these, several improved agronomic practices, such as rate, method and time of fertilizer application, crop protection practices, seed rates and irrigation methods, have been recommended. Improved animal breeds, forage species, and feeding recommendations have also been developed. Soil and water conservation structures, and improved methods of managing problematic soils, such as acidic soils, saline soils and water logged heavy vertisols, have been tested and approved. Tree species suitable for farm forestry and agro forestry practices have been introduced, tested and adapted. Intermediate mechanization technologies and improved pre-harvest and post-harvest farm implements were also developed. Yet, average smallholder yields remain low (below 2 t/ha) for major crops and yield gaps between research plots and the farmers' fields remained high (see Figs. 7.1 and 7.2).

This dismal situation can, to a great extent, be attributed to the weak link and lack of coordination and alignment between agricultural research and extension systems at all levels of government. All the necessary organizations and support services related to the generation of knowledge and technology, input delivery, advisory service, and marketing and credit services are in place. But these actors all function in a largely uncoordinated and fragmented manner.

According to Havelock (1986), cited in Kassa (2008), coordination or linkage symbolizes two systems connected by messages to form a greater system. Agricultural research and extension are examples of two systems that can be linked by information flow and feedback (Munyua et al. 2002). Setting up the institutional linkage to foster proper information flow and effective collaboration is the most serious institutional problem in developing research and extension programmes.

Over recent decades, the Ethiopian government has made various efforts to build the capacity of the research and extension systems and strengthen their linkages to improve adoption and productivity. The linkage between research and extension systems plays a significant role in the generation and dissemination of appropriate technologies. Strengthening research and extension linkages must mean cultivating greater and more effective interaction among the stakeholders in the agricultural sector. To this end, several linkage initiatives have been tried out at different times with different levels of effectiveness. This chapter serves as an overview of past efforts and the current status of linkage between agricultural research and extension, as well as among the whole range of actors involved in agricultural commodity value chains in Ethiopia, with a focus on the Amhara region. The next section deals with the various models of research and extension linkage which will serve as theoretical underpinnings. The third section will present an overview of past linkage initiatives in terms of their strengths, weaknesses and challenges. The fourth section will analyze the current status of linkage among actors involved in agricultural research and development in Ethiopia. The last section draws conclusions and lessons from the foregoing analysis.

Research: Extension Linkage Models

At least three generic types of linkage model can be identified. These are the linear, top-down model, which is commonly known as the transfer of technology or "ToT" model, the farming systems research (FSR) model and the Innovation Systems Model. These three models have their own historical backgrounds, distinct features and resulting implications and mechanisms for linkage between the agricultural research and extension systems in particular, and linkage among the various actors in the agricultural commodity value chains in general.

The "ToT" model was the prominent model in the 1960s and 1970s. It is characterized by the separation of technology production and application. In this model, researchers are knowledge generators, extension agents are expected to transfer the knowledge to farmers, and farmers are passive recipients. It is a 'top-down' one-way communication model with information and technology flowing from researchers to end users via extension agents like a conveyer belt (Kassa 2008). Research, extension and adoption were viewed not as interlinked and iterative processes but as separate and compartmentalized processes that can be organized in different specialized structures.

This model has worked very well in the industrialized agriculture of developed western countries, as well as in the Asian green revolution, where the farming systems are not complex, diverse and risk-prone like those of Ethopia and the rest of Sub-Saharan Africa. The training and visit approach, commonly known as T&V, which has been instrumental in promoting the Asian green revolution technologies, was actually based on this model. The ToT model is still very much persistent in many countries, and people tend to think of the research and extension process as a separate and distinct process that can be linearly linked, despite the fact that it failed to deliver the intended rapid agricultural development through intensification in Sub-Saharan African counties that it had in the case of the Asian green revolution.

In the 1980s, the recognition that farmers are capable of actively adapting and coming up with new ideas and solutions to local problems, rather than being passive recipients, led to the birth of the holistic approach and Farming Systems Research (FSR). This approach is also mainly based on the idea that the reasons behind the low level of adoption of agricultural technologies is related to the fact that conditions for farmers were different from those of research stations, and the technologies developed were hardly suitable for their conditions. In the FSR model, the focus was generally on on-farm client-oriented research (OFCOR) and ways of linking farmers directly with researchers. The general approach in this model was a process in which a multidisciplinary team of researchers and extensionists identify and rank problems and opportunities, suggest alternative solutions, initiate joint on-farm experimentation and disseminate the results.

In the 1990s, the realization that previous approaches had failed to recognize institutional constraints and the usefulness of multiple actors led to a shift in focus from "technology" to "innovation". In this model, institutions are emphasized as being the main bottlenecks, not technology. It also recognized that complex problems require solutions that come out of interactions between many actors. Hence, along with the traditional actors in the Agricultural Knowledge and Information System (AKIS), those actors outside the AKIS who were parts of the value chain were also included.

History of Research and Extension Organizations and Linkage in Ethiopia

In Ethiopia, formal research and extension service was started in 1952 when the Agricultural and Technical School at Jima and the College of Agriculture and Mechanical Arts were established. They were modeled after the US land grant university system, which emphasized the integration of education, research and extension. The extension mandate was later transferred to the Ministry of Agriculture. Later on, research was also divorced from education when the Institute of Agricultural Research (IAR) was established in 1966. By compartmentalizing research and extension activities, the linear R-E-F model was adopted. At that time, there was no mechanism set for the coordination of research and extension, and this marked the formal divorce of research and extension that has existed ever since.

At present, the Ethiopian agricultural research system is characterized by a decentralized research structure in which there are federal and regional research institutes composed of a number of research centers spread across the various agro-ecologies of the country. There are 69 agricultural research centers under the federal, regional and university research institutes. Until recently, there was no umbrella organization coordinating research in Ethiopia (currently, the Ethiopian Agricultural Research Council is being established, modeled after the Indian

Council for Agricultural Research). The federal agricultural research institute known as the EIAR (Ethiopian Institute of Agricultural Research) itself is engaged in applied research not much different from what the regional institutes do and is not serving as a national agricultural research coordination or governing body. The relationship between the federal and region research institutes was largely competitive than collaborative. Structurally, the research institutes are under the Ministry or Bureaus of Agriculture, but they enjoy considerable autonomy and there is no direct reporting or accountability between the federal and region agricultural research institutes. However, there are, in fact, some collaborative research projects and voluntary research collaboration on some commodities.

The extension organization also has a decentralized structure that includes the federal ministry of agriculture, the regional bureau of agriculture, and the district (called Woreda in Amharic) offices of agriculture. There is such a huge army of extension personnel that there are three development agents (DAs) in every kebele (smallest administrative units). There is also one Farmers Training Center (FTC) in most of the kebeles.

Several attempts were made in the past to bring strong linkage, coordination, collaboration and efficient delivery of research and extension services to the farming community. The first attempt to address formal research, extension and farmer linkage was the establishment of the IAR/EPID (Institute of Agricultural Research Extension Project Implementation Department of MoA) on-farm research program in 1974. This joint program was mainly initiated for agricultural technology package testing and formulation of research recommendations for specific areas. The program was discontinued in 1977 due to budget problems. In 1980/81, the program was reinitiated with a new name, IAR/ADD (Agricultural Development Department). However, it too was not successful because it was poorly planned: there was no close monitoring of trial sites by researchers, extension personnel didn't know what the trials were about, and there was no feedback coming to the researchers. As a result, it was not possible to identify technologies suitable for immediate transfer to farmers.

Another linkage mechanism was designed in the early 1980s with the advent of Farming Systems Research (FSR). This focused on multidisciplinary surveys to identify production constraints, verify available technologies being used by farmers, and hand over those found to be superior. The main contributions of FSR research-extension linkage were providing feedback to researchers on the characteristics of technologies, providing researchers with information on farmers' problems, formulating recommendations appropriate to smallholder farmers, and generating useful recommendations for policymakers, but the program was expensive and time-consuming, and thus was phased out as project funds run out.

In 1985, following the lessons of FSR, and also mainly for the purpose of transferring technologies developed and shelved at IAR, IAR established the Research–Extension Division (RED). It was given the mandate of strengthening research-extension-farmer linkage by establishing and running the first formal linkage platform for research and extension, called the Research-Extension Liaison Committee (RELC). Since its establishment, the RED has played an important role in disseminating agricultural technologies, conducting demonstrations, providing training, and coordinating RELC linkage activities. Later on, in 2009, with the advent of Business Process Re-engineering (BPR), REDs were closed at most of the research institutes with the idea that they simply duplicated the work of extension departments of the Bureau of Agriculture (BoA). With the BPR, research and extension were viewed as separate processes which can be linked by a case team established in the BoA.

The Research-Extension Liason Committee was the first formal platform, established in 1986 at zonal and national levels to enhance vertical and horizontal integration of research and extension. RELC was primarily run by the REDs of IAR research centers. The primary role of zonal RELC was to review and approve research proposals and extension recommendations, identify training needs for SMSs, and oversee research-extension and farmer linkage in the respective zones of research centers. The national RELC was responsible for provision of overall policy direction and capacity building. Yet, RELC was largely ad hoc; meetings were not regular. It was also seriously affected by frequent changes in the organizational structure of MoA, resulting in discontinuities in many linkage arrangements and joint undertakings. Although the terms of reference of RELC stated that farmers were members of RELC, in practice, they were either passive participants or were not represented at all. It didn't have any legal status, which affected its decision-making power and institutionalizing accountability among members.

In the late 1990s, RELC was replaced by the Research-Extension-Farmers Linkage Advisory Council (REFAC). REFAC was organized at national, regional and research center levels. It was also run by the RED of the Ethiopian Agricultural Research Organization (EARO), funded by The World Bank's Agricultural Research and Training Project (ARTP). REFAC was responsible for the overall guidance of research and extension programs, and oversight of the linkage activities (FDRE 1999). The council had a chairperson and secretary. It had subcommittees or working groups, a women farmers group, representatives of farmers research groups, a resource management group and a farming systems group. Nevertheless, REFAC could not produce integration as expected due to the lack of commitment from stakeholders, the low involvement of certain key stakeholders and the fact that it was dominated by researchers. There was no frequent meeting of stakeholders and they met only once a year at all levels. The other serious problems in the REFAC was the lack of accountability of partners, i.e., failure to deliver on promises, and the fact that there was no clarity on the responsibility of actors.

Current Mechanisms for Research and Extension Linkage

In order to overcome the shortcomings of REFAC, a multi-stakeholder platform called the Agriculture and Rural Development Partners Linkage Advisory Council (ARDPLAC) was established in 2008. It is organized at national, regional, zonal

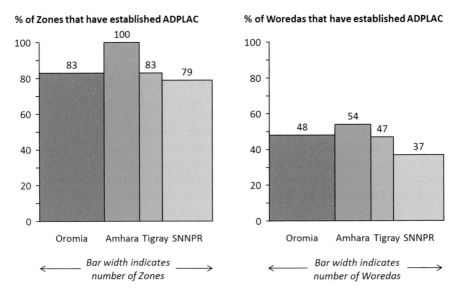

Fig. 7.3 Level of establishment of ADPLACs at Zonal and Woreda levels (Source: Communication with MoA- ADPLAC Expert 2012)

and district levels. In the Amhara Region, the council is institutionalized from the regional to the kebele levels, with its secretariat office being hosted at the Extension Process Service, under the Regional BoA. Unlike RELC and REFAC, ARDPLAC is coordinated by the Bureau of Agriculture or Ministry of Agriculture and, like its predecessors, is funded by World Bank projects. Later, the name was changed to the Agriculture Development Partners Linkage Advisory Council (ADPLAC) with the restructuring of the former Ministry of Agriculture and Rural Development into the Ministry of Agriculture, leaving all the rural development-related organizations, such as rural water service and rural roads authority, to stand on their own. Hence, these sectors are no longer part of the linkage platform, and only actors directly related to agriculture are included in the forum.

Structurally, ADPLACs are organized at national, regional, zonal, woreda and kebele levels that represent different levels of performance. Although ADPLACS are established in the majority of the zones, at the woreda level, there is still no ADPLAC in most of them (Fig. 7.3).

The current ADPLAC platforms, at all levels, face various systemic and operational challenges. First of all, like its predecessors, ADPLAC is not properly institutionalized at all levels. There are inadequate permanent staff and the platform is primarily supported by project money, rather than having permanent budget allocation. There is an ad-hoc nature to the planning and execution of linkage activities. And the platform lacks a long-term vision and strategy for systematically creating alignment across the various development actors for a greater impact.

Moreover, due to the lack of ownership and loose engagement, leadership, decision-making, communication, passion for a shared vision, enforced roles and responsibilities, and learning and sharing are all sub-optimal among the stakeholders. There are no agreed-upon roles and responsibilities and enforcement mechanisms to ensure that assigned roles are performed. This is partly due to weak stakeholder coordination and engagement capacity at all levels, which led to low awareness and shared vision among development partners about the functions and mechanisms of stakeholder platforms. Moreover, the frequent reshuffling of institutional representatives has also led to poor knowledge preservation and transition and loss of institutional memory. Whenever there is an urgent task at the offices, ADPLAC experts are relieved from their linkage duties and take up other roles.

To have a very successful engagement platform that responds to the needs of the participants and addresses their problems, an efficient monitoring, learning and evaluation (MLE) mechanism plays a crucial role. However, the current ADPLAC platforms are characterized by poor or, in many cases, non-existent MLE. There is no mechanism that gauges and documents stakeholder perception about benefits gained by individual stakeholders from participating in the platform and the overall learning process.

The other ADPLAC challenge is the issue of representation, in terms of both relevance and inclusiveness. Currently, membership is unmanageable and not really effective. It is important to include only relevant members to make ADPLACs effective. At present ADPLACs lack relevant and influential members, including those from the private sector, to make the platform inclusive and forward-looking. Additionally, all the stakeholders also lack commitment to the cause of ADPLAC.

Currently, the government is driving ADPLAC platforms at all levels and the participation of the private sector is very low. And apart from participating in regular meetings, there is no shared long-term vision of success among the stakeholders. Moreover, it is not very clear what do the ADPLACs aspire to achieve in the long-run? What will change as the result of the ADPLAC's activities? What are the core functions of the ADPLAC? What are the key performance indicators? How will the ADPLAC influence the overall modernization of agricultural advisory service systems?

In addition, role- and responsibility-sharing and tracking, as well as reporting among the members, are not attached to accountability. If all the stakeholders are convinced of the benefits of participating in the platform, then they can start to assume responsibility in assigning tasks to the right stakeholders and tracking their accomplishments and impact.

Although ADPLAC is more extensive and inclusive than its predecessors, its outreach to the woreda and kebele levels is rather limited. Awareness and effectiveness of ADPLAC decreases down the administrative ladder, starting at the regional level. There are three key reasons for this.

According to Demekech et al. (2010), the effectiveness of the ADPLAC challenged by poor abundance and weak strength of linkages, low awareness about what ADPLAC is among actors and poor facilitation for linkage creation. As can be seen

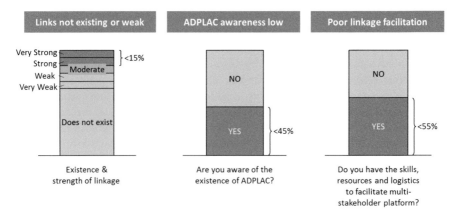

Fig. 7.4 Results from a survey among ADPLAC stakeholders in the Amhara region (Source: Demekech et al. 2010)

from Fig. 7.4 above, in the Amhara region, only about 15 % of the expected agricultural development actors' linkages are strong. And less than half of the key stakeholders know about ADPLAC. The majority of the stakeholders report that they have insufficient support for linkage creation (Demekech et al. 2010).

The ADPLAC lacks a long-term strategy for systematically creating alignment across the different development actors for greater impact (MoA RCBP 2012b). The limited linkage among ATVETs, extension systems, research and higher education means that ATVETs do not receive the necessary feedback to help them adjust and deliver training and technology transfer services that are up-to-date and relevant.

The conditions framing the effectiveness of ADPLAC at the federal level seriously affect linkage activities at the lower levels. Awareness of the partners at the woreda level about the purpose and benefit of the ADPLAC is limited. This is partly due to limited efforts at the regional level in terms of providing support and organizing events on how to develop and manage linkage activities. Woreda level ADPLACs also face limitations in resources for executing linkage activities.

Apart from the ADPLAC, to overcome the chronic linkage problem in Ethiopia, many projects have also tried their own version of linkage mechanisms. Most of the time they bring together stakeholders, whom they identify as important and influential for the success of their project objectives. The JICA funded Farmers Research Group project tried to strengthen the linkage among farmers, extension agents and researchers using on-farm, adaptive, participatory research. Following this experience other subsequent project used the same approach with the same aim (MoA-Rural Capacity Building Project 2012a). The USAID/AMAREW project tried research center-based district (woreda) level innovation platforms to scale up promising technologies. There were a number of woreda linkage committee established in the project districts in Amhara region. There was some degree of success in the dissemination of technology in the project districts, but the districts were too many to handle for the research centers. And they ceased to exist as the project phased out. The ACIAR (Australian Center for International Agricultural Research)-funded,

CIMMYT and Ethiopian counterpart-implemented conservation agriculture promotion project called SIMLESA has also come up with woreda-based, researcher-facilitated innovation platforms. There were about ten innovation platforms established in the project woredas, with two in the Amhara region.

There were operational and strategic innovation platforms in each project woreda. The operational platforms were established at the grassroots to facilitate coordinated action and joint problem-solving at local levels, while the strategic platforms focused on policy constraints and enabling environments. Establishment and operationalization of strategic platforms has been found to be difficult. However, there is some evidence that indicates that, although it is difficult to establish strategic platforms, once established, they are very effective, as they will bring top level politicians on board who are the real decision-makers in the highly hierarchical and top down Ethiopian National Agricultural Research and Extension System (NARES).

Conclusions

Generally, the Ethiopian NARES and that of the Amhara region agricultural research and extension system are characterized by large number of actors in a fragmented and underdeveloped innovation system, resulting in very low national and regional innovation capacity. The dominant thinking about the mechanism of integrating these actors, or simply about the linkage issue, continues to view research and extension as separate processes that should be organized in different organizations and can be linked linearly through periodic large assemblies and ad-hoc committees. In this view, farmers are also viewed as passive recipients of technology. As a result, research outputs do not reach farmers and remain shelved in research centers. Often, research and extension organizations compete for resources, mandates and influence rather than cooperating for a common purpose. In order to change this situation, this dominant paradigm needs to be replaced by the fact that research and extension are interlinked, overlapping and iterative processes. Hence, organization of research and extension institutes needs to take this into consideration. All past initiatives to bring about integration among these actors were project-funded, ad hoc and not institutionalized, creating discontinuity, as well as a lack of accountability. They also lack a strong M&E system, and are plagued by poor participation of other stakeholders, poor coordination and facilitation, poor institutional memory, and lack of clarity on the roles and responsibilities of participant stakeholders. Linkage brokering and management requires professional facilitation knowledge and skills. Such professionals are lacking and the profession not recognized. There are some successful project-based innovation platforms, but their successes were localized, not replicated and not sustainable as projects phased out. However, these experiences indicate that, in the context of a fragmented innovation system and extension systems characterized by top-down planning, strategic innovation platforms which bring together high level decision-makers could be hard to establish and manage sustainably, but once established and properly managed, they are ideal forms of linkage among the research and extension organizations and can easily promote region- or nationwide scaling up of proven technologies.

Open Access This chapter is distributed under the terms of the Creative Commons Attribution-Noncommercial 2.5 License (http://creativecommons.org/licenses/by-nc/2.5/) which permits any noncommercial use, distribution, and reproduction in any medium, provided the original author(s) and source are credited.

The images or other third party material in this chapter are included in the work's Creative Commons license, unless indicated otherwise in the credit line; if such material is not included in the work's Creative Commons license and the respective action is not permitted by statutory regulation, users will need to obtain permission from the license holder to duplicate, adapt or reproduce the material.

References

Central Statistical Agency (2003–2011) Ethiopian agricultural sample enumeration: statistical reports on report on land utilization: private peasant holdings from 2003 to 2011. Central Statistical Agency, Addis Ababa

Demekech G, Moges F, Zeleke G, Tesfaye K, Ayalew M (2010) Multi-stakeholder linkages in rural innovation processes in the Amhara Region, Ethiopia, Working document series 137. International Centre for Development-oriented Research in Agriculture, University of Bahir Dar, Amhara Region Agricultural Research Institute and Bureau of Agriculture and Rural Development, Wageningen/Bahir Dar

FDRE (The Federal Democratic Republic of Ethiopia) (1999) Agricultural research and training project: Ethiopian research—Extension-farmer linkages strategy, vol I. FDRE, Addis Ababa

Havelock RG (1986) Linkage: a key to understanding the knowledge system. In: Beal GM, Dissanayake W, Konoshima S (eds) Knowledge generation, exchange and utilisation. Westview Press, Boulder

Kassa B (2008) Agriculturalresearch and extension linkages in Ethiopia: a historical survey. Haromaya University, Dire Dawa

Ministry of Agriculture (2014) National Crop Variety Register (2014), Issue no. 17. Federal Democratic Republic of Ethiopia, Ministry of Agriculture, Addis Ababa

MoA Rural Capacity Building Project (2012a) The performance of FREGs: costs, benefits and intervention options for improved sustainability. Haromaya University, Dire Dawa

MoA Rural Capacity Building Project (2012b) Performance of agricultural development partners' linkage advisory councils. Haromaya University, Dire Dawa

MoFED (Ministry of Finance and Economic Development) (2011) Ethiopia: building on progress. A Plan for Accelerated and Sustained Development to End Poverty (PASDEP) (2005/06-2009/10) Volume I: Main Text. MoFED, Addis Ababa

MoFED (2015) Federal Democratic Republic of Ethiopia. Ministry of Finance and Economic Development, 2008 fiscal year budget (Amharic). MoFED, Addis Ababa

Munyua CN, Adams PF, Thomson JS (2002) Designing effective linkages for sustainable agricultural extension information systems among developing countries in Sub-Saharan Africa. In: Proceedings of the 18th annual conference of the association for international agricultural and extension education, Durban, pp 301–307

Chapter 8
Institutional Innovations for Encouraging Private Sector Investments: Reducing Transaction Costs on the Ethiopian Formal Seed Market

Christine Husmann

Abstract There is a considerable shortage of improved seed in Ethiopia. Despite good reasons to invest in this market, private sector investments are not occurring. Using an institutional economics theoretical framework, this chapter analyzes the formal Ethiopian seed system and identifies transaction costs to find potential starting points for institutional innovations. Analyzing data from more than 50 expert interviews conducted in Ethiopia, it appears that transaction costs are high along the whole seed value chain and mainly born by the government, as public organizations dominate the Ethiopian seed system, leaving little room for the private sector. However, recent direct marketing pilots are a signal of careful efforts towards market liberalization.

Keywords Ethiopia • Institutions • Private sector • Transaction costs • Improved seeds

Introduction

About 80 % of the Ethiopian population depend as smallholder farmers on agriculture for their livelihoods (CSA 2012a). These smallholders suffer from a very low productivity (see, e.g., Seyoum Taffesse et al. 2011). To increase productivity, improved inputs like seeds, fertilizer and better farming practices are crucial (see, e.g., von Braun et al. 1992; Conway 2012).

An earlier version of this chapter has previously been published as: Husmann C (2015) Transaction costs on the Ethiopian formal seed market and innovations for encouraging private sector investments. Quarterly Journal of International Agriculture Vol. 54(1):59–76

C. Husmann (✉)
Center for Development Research (ZEF), University of Bonn, Bonn, Germany
e-mail: husmann@uni-bonn.de

© The Author(s) 2016
F.W. Gatzweiler, J. von Braun (eds.), *Technological and Institutional Innovations for Marginalized Smallholders in Agricultural Development*,
DOI 10.1007/978-3-319-25718-1_8

Despite the presence of several seed companies, the agricultural input sector in Ethiopia is currently unable to satisfy the demand for improved seed in the country (Ministry of Agriculture (MoA) 2013). However, there are several reasons to invest in Ethiopian agricultural input markets: not only is the market large in terms of the number of people, high rates of economic growth and investments in infrastructure indicate huge potential, especially in the middle and long runs. Furthermore, in the last two decades, innovative business approaches have emerged that add social as well as financial returns to a company's bottom line, and thus augment the reasons for companies to invest in poor countries and poor people (Baumüller et al. 2013).

Empirical studies suggest that the current situation in Ethiopian agricultural input markets is not the efficient outcome of demand and supply meeting at a certain price, but that institutions drive up transaction costs, i.e., the "costs of coordinating resources through market arrangements" (Demsetz 1995, p. 4), leading to insufficient supply of and unmet demand for agricultural inputs (Alemu 2011, 2010; Bishaw et al. 2008; Louwaars 2010; Spielman et al. 2011). From a theoretical perspective of allocation, this situation is a market failure, since the lack of supply of agricultural inputs at current prices implies welfare outcomes below an achievable optimum. Thus, the current lack of inputs can be defined as a market failure in this sense (see Arrow 1969; Bator 1958 for detailed discussions of this argument).

Market failures are a result of high transaction costs. Transaction costs, however, are determined by the institutional structure of an economy (North 1989). North (1990) defines institutions as "the rules of the game in a society, or more formally, the humanly devised constraints that shape human interaction" (North 1990, p. 3).

The free market cannot serve as the fictive first best option whose approximation can guide the design of an institutional setting, since transaction costs drive a wedge between producer and consumer prices such that, even in theory, 'free markets' do not lead to Pareto efficient results when transaction costs are taken into account (Arrow 1969; Demsetz 1969). Thus, a comparative approach evaluating real alternative institutional arrangements based on the identification of the relevant transaction costs that determine economic performance is appropriate for studying transaction costs and the functioning of markets (Williamson 1980; Demsetz 1969; Acemoglu and Robinson 2012).

Against this background, an analysis of the institutional setting and the transaction costs arising for agricultural input markets is carried out to get a better understanding of the reasons for the observed market failure and to assess possible solutions for these frictions. Only if these costs are reduced is there a chance that the private sector can expand activities to make improved seed accessible to the poor as well.

Transaction cost economics has been applied to study many different problems of economic organization. Masten (2001) stresses the importance of transaction cost economics for the analysis of agricultural markets and policy, as well as vice versa, the potential that the analysis of agricultural markets has for refining transaction cost theory (see also Kherallah and Kirsten 2002).

Transaction costs are generally found to be high on agricultural markets in poor countries and have a considerable influence on farmers' marketing decisions.

Several studies show that transaction costs are closely related to distance and that distance from markets negatively influences market participation and thus incomes (Alene et al. 2008; de Bruyn et al. 2001; Holloway et al. 2000; Kyeyamwa et al. 2008; Maltsoglou and Tanyeri-Abur 2005; Ouma et al. 2010; Rujis et al. 2004; Somda et al. 2005; Staal et al. 1997; Stifel et al. 2003). More specifically, Staal et al. (1997) find that transaction costs rise in greater proportion than transportation costs due to factors such as increasing costs of information and risk of spoilage of agricultural products. Furthermore, costs of information and research are found to impact smallholders' marketing decisions (Gabre-Madhin 2001; Staal et al. 1997; de Bruyn et al. 2001; De Silva and Ratnadiwakara 2008; Kyeyamwa et al. 2008; Key et al. 2000; Maltsoglou and Tanyeri-Abur 2005).

However, Holloway (2000) and Staal et al. (1997) find a positive effect from organizations of collective action, such as cooperatives, in reducing transaction costs. These benefits accrue to both producers and buyers, as cooperatives reduce the cost of information for both sides and take advantage of economies of scales in collection and transport.

Less is known about transaction costs arising on the side of the private sector when companies try to market to poor smallholders. Recent studies engaged in initial analysis of constraints for companies entering agricultural markets in poor countries remain vague, but indicate that "(a) laws, policies or regulations that constrain business operations; (b) government capacity to respond quickly; and (c) access to capital" are the main hurdles named by the private sector to realizing investments in African agriculture (New Alliance for Food Security & Nutrition 2013, p. 6).

Against this background, the following chapter begins to fill this knowledge gap by analyzing the institutional setting and the resulting transaction costs that arise when selling improved seed to poor farmers in Ethiopia. The study uses primary data obtained through expert interviews that were conducted by the author in Ethiopia in 2011, 2012 and 2013. These interviews are analyzed concerning the importance of different types of transaction costs in providing incentives and disincentives to expand seed production. To ensure anonymity for the informants, only the stakeholder group of the informant is provided in the text in square brackets. Thus, if one or more experts from a stakeholder group provided information, this is indicated in the citation in the following way: [1] manager of an international seed company; [2] manager of a private Ethiopian seed company; [3] manager of a public Ethiopian seed company; [4] member of a farmer's organization; [5] government employee; [6] employee of a public research organization; [7] employee of another organization (banks, Agricultural Transformation Agency, etc.).

Results show that the formal Ethiopian seed system is largely controlled by the government and public organizations. Based on a *de facto* monopoly of breeder seed, the government forces seed companies to market all seed through one government-controlled distribution channel, at prices determined by the government. This limits profit margins and incentives to expand seed production. The only exception to this system are the international seed companies that operate in

Ethiopia, as these produce their own varieties and are thus not dependent on the breeder seed provided by the public research institutions. Thus, the government bears especially high transaction costs to sustain a system that does not lead to satisfactory outcomes. However, direct marketing pilots have been started that allow Ethiopian seed companies to market their seed directly to farmers for the first time, which may indicate a first step towards market liberalization.

Seed Production in Ethiopia

In the following, only the case of seeds of major crops is discussed. These major crops are 18 crops selected by the Ethiopian government: teff, barley, wheat, maize, sorghum, finger millet, rice, faba (fava) bean, field pea, haricot bean, chickpea, lentil, soybean, niger seed, linseed, groundnut, sesame and mustard. Institutions differ for other seeds, such as fruit and vegetable seeds, and for other agricultural inputs like fertilizer or agrochemicals. However, due to space limitations, only the circumstances concerning the 18 major crops are discussed below.

Only 2.9 % of the farmers in Ethiopia reported using improved seed in 2011 (CSA and MoFED 2011, p. 20). The contribution of the formal seed sector as a percentage of cultivated land was only 5.4 % in 2011, with considerable variability among different crops (Spielman et al. 2011). Low technology adoption rates can occur for many reasons (Degu et al. 2000; Feder and Umali 1993). In Ethiopia, one important reason is the substantial lack of improved seed (see MoA 2013).

In 2011/2012, seed supply covered only 51 % of stated demand for barley, 24 % for wheat, 16 % for rice, 30 % for millet and 60 % for faba bean (MoA 2013). The supply of maize, wheat and teff seeds has improved considerably over recent years. But still, only 20 % of the area cultivated with maize, 4 % of the wheat area and less than 1 % of the teff area are cultivated with seed from the formal sector (CSA 2012b).

In the Ethiopian case, it is important to distinguish between different types of seed companies. Generally, in this chapter, a private seed company is understood as a firm with a business and a seed producing license, producing seed on its own account and bearing the full risk of the business. Thus, cooperative unions or farmers employed as seed producers by seed companies, or other organizations such as non-governmental organizations (NGOs), do not fall into this category. However, this does not imply that seed companies produce all their seed themselves, companies can also hire farmers to produce the seed on their behalf.

For the following analysis, it is helpful to differentiate between public seed enterprises, private Ethiopian seed enterprises and international (private) seed enterprises. There are five public seed companies in Ethiopia: the Ethiopian Seed Enterprise (ESE), the Amhara Seed Enterprise (ASE), the Oromia Seed Enterprise (OSE), the South Seed Enterprise (SSE) and the Somali Seed Enterprise. The ESE was the only seed company in the country for several decades before some private seed companies entered the market. The regional public seed enterprises were established recently, starting with ASE and OSE in 2009. Their statutes foresee them producing different kind of seeds for Ethiopian farmers without profit-making being a primary goal (Amhara Regional State 2008).

The number of private Ethiopian seed enterprises is not clear. In 2004, 26 firms were licensed to produce seed but only eight firms were active in seed production (Byerlee et al. 2007). Other sources mention 33 seed producing companies but without specifying who they are (Atilaw and Korbu 2012). In 2011, 16 private seed enterprises were listed in the business directory, but it is not clear whether they were all operating at that time.

Two international seed enterprises are producing some of the selected major crops (as at July 2013), Hi-Bred Pioneer and Seed Co. Both focus on the production of hybrid maize, while one of them also produces smaller quantities of wheat, teff and beans ([1]).

Why Is There Not More Investment in Seed Production?

If the stated demand is much higher than seed production, the question arises as to what is preventing private seed companies from increasing investments in seed production to tap this market. The answer to this question lies in the institutional setting governing seed production and distribution in Ethiopia.

As illustrated in Fig. 8.1, the Ethiopian seed system is quite complex. The process of seed production starts with an assessment of seed demand, which is carried out by the Development Agents (DAs) on *kebele* (village) level. Information about seed demand is then passed up the governmental administrative ladder and collected by the Bureaus of Agriculture (BoA) and the MoA. On the basis of this information, the MoA orders production of quantities of various crops at the ESE, while the BoAs determine production portfolios for the regional public seed enterprises and the private seed companies in the area.

All Ethiopian seed companies – public and private – get their pre-basic and basic seed from public research institutes (see also Fig. 8.1). Only the two international seed companies operate with their own varieties. This is of great importance, because getting pre-basic seed from national research institutes comes with a contract entailing a clause that obliges the companies to sell all produced seed back to the government – at prices to be determined by the government and often announced on short notice.

The MoA determines the quantities of seed to be distributed to each region on the basis of the demand assessment; the BoAs define the quantities for each zone, and so forth. Seed distribution is usually managed by farmer cooperative unions, who bring the seed to the zones and the primary (multipurpose) cooperatives that pick the seed up in the zonal warehouses and bring it to the *woredas* (districts) and *kebeles*. Unions charge for transport, uploading and unloading, but they make only small profits from seed distribution, with profit margins being determined by the regional governments ([3]; [4]).

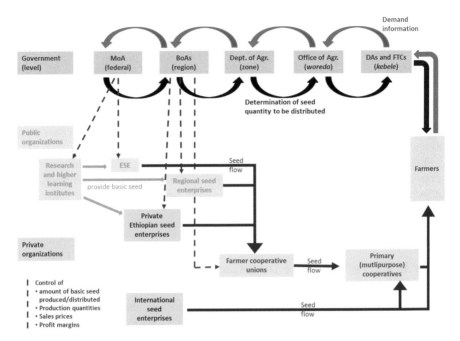

Fig. 8.1 The formal seed system in Ethiopia (Source: author)

An important implication of this seed system is the lack of agro-dealers, as seed distribution is organized in one government-controlled distribution channel. This also has important implications for other agro-dealers, as it makes it extremely difficult and expensive for them to market agricultural inputs outside larger agglomerations.

The private seed enterprises in Ethiopia mainly focus on hybrid maize production because it offers the highest profit margins. For this reason, supply of hybrid maize seeds has improved considerably since the regional seed enterprises started operation, from 88,000 quintals in 2006/2007 to 357,000 quintals (1 quintal = 100 kg) in 2010/2011 (Dalberg Global Development Advisors 2012). Private companies now produce about 40 % of the hybrid maize seed sold in the country (Alemu 2011).

Some companies also produce varieties of wheat, teff, beans, rice, soybean, sesame and sorghum. But all of these crops except hybrid maize are only produced in very small quantities, despite large cultivation areas. Thus, there are large untapped markets for these crops where demand is substantially higher than supply (see Table 8.1; MoA 2013). However, with the limited size of land for seed production, companies focus on the production of the seed with the highest profit margin as long as there is demand for that seed.

Table 8.1 Difference between supply and demand of improved seed of various crops

	Difference between demand and supply in quintals (2011/2012)	% of demand not met
Wheat	200,720	21 %
Teff	10,211	11 %
Maize	39,666	9 %
Barley	101,924	49 %
Sorghum	16,433	92 %
Rice	13,638	84 %
Millet	967	70 %
Faba bean	19,918	40 %
Field pea	47,769	84 %
Chick pea	11,035	63 %

Source: MoA (2013)

Institutions Preventing the Private Sector from Increasing Seed Production

Various institutions in the current seed system prevent private seed companies from increasing seed production and making it available to smallholders. Important constraints for the private Ethiopian seed enterprises result from the fact that none of them do their own breeding. Some managers express the intent of importing new parental lines for breeding to escape the strict government interference. However, breeding is a difficult business that requires additional land and high-skilled and experienced plant breeders, as well as technical facilities. Accordingly, seed producers need to get more land assigned by the government to start their own breeding, which takes a long time and is insecure. Additionally, it is difficult to hire experienced plant breeders in Ethiopia because, currently, plant breeders are government employees enjoying secure jobs and other privileges. Thus, it is difficult to attract them to private companies. This problem is aggravated by the fact that areas dedicated to plant breeding are always remote because breeding requires isolated land plots. These circumstances oblige companies to pay high salaries to plant breeders, since skilled people often do not want to live in remote areas ([2]). Moreover, the installation of the necessary technical facilities requires additional working capital, which is difficult to get.

On the other hand, several experts assume that some seed enterprises are quite content with the present form of the contracts, because they minimize risks as long as the government commits itself to buying all produced seed ([7]).

Another institution disadvantaging private Ethiopian seed companies is related to the distribution of seed. Farmers can select the varieties they want to purchase, but they are usually not given the choice to opt for one particular source. It even often happens that the farmer cooperative unions or the primary cooperatives mix seed or refill bags from one seed with another to make transportation easier, which confuses farmers about the quality of seed of different producers ([2]). Two problems arise as a result: first, this prevents companies from establishing a brand name, and second, it hampers response to complaints by farmers about seed quality because the producer of the seed is not clearly identifiable.

Price determination is another point posing major difficulties for the private Ethiopian seed companies. Compared to other Sub-Saharan African countries, the seed prices determined by the government are relatively low in Ethiopia. At first glance, this seems to be beneficial for the farmers but it also has considerable disadvantages concerning users' efficiency (Alemu 2010, p. 24) and can lead to a crowding out of the private sector ([2]). The prices of major crop seeds are negotiated by the BoAs, the board and the management of the public seed enterprises. These prices are then binding for maximum prices for the seed of all Ethiopian seed enterprises. Prices are based on estimations of farmers' willingness to pay for seed, but there is no systematic assessment of farmers' willingness or ability to pay (Alemu 2010). Prices vary considerably across regions and from year to year. In 2011, e.g., hybrid maize BH-540 was sold at 2000 Ethiopian Birr (ETB) per quintal in Oromia, while in Amhara, the price was 1500 ETB per quintal ([2]; [3]). In the 2010–2011 cropping season, Pioneer Hi-Bred sold its hybrid maize at 2784 ETB per quintal and sold all its stock ([1]).

What Is the Nature of Transaction Costs Arising in the Ethiopian Seed System?

Although it is not possible to quantify transaction costs resulting out of the presented institutions in the seed system, since neither the companies nor the government keep detailed records of their costs, the nature of the transaction costs involved and the distribution of these costs can be identified.

Costs for market entry have not been high in the past. Until now, it was not difficult for private companies to start a seed business. Business owners need (1) an investment license, (2) a competence license and (3) a business license if they produce the seed on their own land. If the company does not operate on its own land but hires farmers to produce the seed, it does not need the business license. Requirements to get the licenses are clear and the application procedure usually takes only a few weeks ([2]). However, private sector stakeholders fear that procedures will become more tiresome and lengthy, as the government might want to suppress additional competition for the regional public seed enterprises ([2]).

International seed enterprises that bring their own varieties face very high transaction costs for market entry ([1]). Bureaucratic procedures are unclear and lengthy. New varieties that are brought to the country need to get registered in a procedure that usually takes 3–4 years ([1]; Dalberg Global Development Advisors 2012).

Costs for market information and pricing are moderate, since demand is still very high for improved seed of all crops. For international seed companies marketing their own varieties, considerable costs arise for promotional activities, since it takes several years to gain the farmers' confidence in a new brand. Many field days, demonstration plots and gratis seed packages are needed to convince farmers of the benefits of improved seed ([1]).

For Ethiopian seed enterprises, pre-contractual activities are organized by the government. Although there is no law or regulation fixing it, a *de facto* monopoly of the public research institutes implies a monopsony for seed, as the government obliges the seed companies to sell all seed back to it. The government is then responsible for the marketing of the seed. In terms of transaction costs, this means that, for the companies, costs for searching for customers and costs for information about the market do not arise, because their product portfolio is largely determined by the government and they have to sell the produced seed to the government. This is changing with the direct seed marketing pilots (see next section), in which companies are responsible for demand assessment themselves.

Advertisement costs do not arise for Ethiopian seed companies, since marketing is done by the government with the help of farmers' cooperatives, and farmers cannot chose the source of their seed.

For the government, pre-contractual transaction costs are considerable. Government employees spend much time collecting data about seed demand and distributing seed. The typical time the head of extension in a *woreda* spends on collecting seed demand per season is 1 month, i.e., 2 months a year for both cropping seasons, and 45 days on distributing seed to the farmers ([5]). In the regional BoA of the Southern National, Nationalities and Peoples' Region (SNNP), for example, five full-time employees are charged with organizing seed supply and distribution ([5]). Additionally, employees in the zonal departments of agriculture and in the MoA are involved, but it is not clear how many people dedicate their working time to seed distribution there.

Contract formation (bargaining) is similarly simplified for companies, since the prices of major crop seed are negotiated by the BoA, the board and the management of the public seed enterprises. Since government regulations avoid direct contact and contracts between seed companies and farmers, there is no room for negotiations between customers and companies about prices or other parts of the contract ([2]).

The post-contractual transaction activities of contract execution, control, and enforcement are also minimized for seed companies by the actual government regulation. The theory of self-enforcing agreements (Furubotn and Richter 2005) ceases to be valid, since the seller of the seed is not the producer and complaints are usually not transferred back to the producer. The farmer cannot retaliate by ceasing to purchase the product if the product turns out to be of bad quality because, first, he cannot identify the producer, and secondly, because he cannot choose between different producers, such that the only alternative is not to buy improved seed at all.

As a result, it appears that, in the current situation, transaction costs are mainly born by the government. Governmental agencies assess demand and organize distribution of seed, and public banks finance the time elapsed between seed delivery by the seed enterprises and payments of the farmers. Promotional activities are done by the DAs, if at all. These transaction costs are very high and are not justified by satisfying outcomes in terms of quantities of seed produced, seed quality and timeliness of delivery.

Rather, the system has considerable disadvantages: cooperatives have to carry the burden of transporting seed, which keeps them away from other tasks, such as training for farmers or output marketing, on which they should actually focus ([7]). The current distribution network is also the reason for the lack of agro-dealers in the country, which is detrimental for the international seed companies and for other traders of agricultural inputs.

The Direct Seed Marketing Pilots

Increased pressure from private seed companies and other stakeholders led to the first trials in which Ethiopian seed companies could directly sell their seed to farmers. Starting in the Amhara region in 2011 and followed by Oromia and SNNP in 2012, Ethiopian seed companies were allowed for the first time to directly market their seed. These pilots have been scaled up in 2013. While in Amhara and Oromia, the direct seed marketing was restricted to hybrid maize, other varieties such as wheat could also be directly marketed in five *woredas* in SNNP.

Preliminary results of the Amhara pilot suggest that seed availability and timely delivery was better in project *woredas* than in non-project *woredas* (Astatike et al. 2012). The pilot also revealed that demand estimations for the pilot *woredas* were quite inaccurate. The project was not reiterated in Amhara in 2012, since the ASE was left with a lot of unsold seed that the government decided to sell preferentially in 2012 in the framework of the normal seed distribution system.

Concerning the 2012 direct seed marketing pilot in Oromia and SNNP, preliminary results indicate that all companies were able to sell almost all their seed. Participating companies in both regions even felt that they could have sold more seed if they would have had better demand information and fewer difficulties with transportation and storage in the *woredas* ([2]; Benson et al. 2014).

Still, in both *woredas*, more improved seed was sold than in any other year before, and more than was initially foreseen (ISSD 2013). This might have various reasons. First, the shops of agro-dealers were open 7 days a week and during the whole day, while the cooperatives distributing the seed usually only open for two afternoons a week due to the lack of full-time employees. Secondly, seed was available on time before planting and until planting was finished. Thus, previously well-known problems of late arrivals of seed were avoided. Third, agro-dealers are said to provide good technical advice to farmers. This, together with some promotion by the companies, might have increased awareness and trust in the seed. Finally, some farmers reported that they also bought seed for their relatives living in neighboring *woredas* who saw the benefits of early seed arrival and technical advice from the agro-dealers ([2]; [5]).

The main benefits of the pilots can be summarized as:

- Traceability of the seed, and thus increased accountability for seed quality, which increases farmers' trust;
- Time resources saved by DAs and Subject Matter Specialists, who were previously occupied with seed distribution and can now concentrate on training and advisory services for farmers;
- Farmers do not hold DAs responsible for seed failure, since seed distribution is now managed by agro-dealers, which considerably improves the relationship between DAs and farmers;
- Companies are rewarded for better quality, and thus have an incentive to improve on quality in the future;
- There is less seed fraud and storage damage as the value chain is much shorter.

The direct seed marketing trials can be seen as an important step towards market liberalization. However, the cessation of the pilot in Amhara shows how fragile such changes are. Improvements in the methodology and careful evaluations of the project will be needed to smooth the way towards market liberalization for companies, as well as for farmers.

Despite the generally very positive experience with the recent direct seed marketing pilots, some difficulties remain. An especially crucial point is the cost for transportation and agro-dealers. In 2012, sales prices were determined by the government and companies were not allowed to add up transportation costs and agro-dealer commissions, despite considerable expenses for long stretches of transport, which drove their profit margins towards zero or even below that. These problems led some companies to step out of the process. Other challenges are the lack of storage facilities in the *woredas* and a lack of trained agro-dealers.

Institutional Innovations to Improve Seed Supply and Access to Improved Seed

It can be doubted that the relationship between the sum of transaction costs and outcomes in terms of efficiency of seed production and distribution is optimal in the Ethiopian system. Of course, it is difficult to evaluate efficiency without a counterfactual. Yet, the analysis of the seed system reveals that institutions do not govern the seed market in an optimal manner: despite the high investments of time and other resources, inaccuracies in the demand assessments regularly lead to deficient outcomes that distort optimal seed production and distribution. High costs of capital and other burdens imposed by the government, e.g., concerning variety registration, prevent Ethiopian seed companies from investing in their own breeding, which could improve the availability of high-quality seed in the country. Incentives for optimizing seed quality are distorted, since farmers cannot identify the source of their seed and prices are independent from quality. Thus, access to pre-basic seed and/or support for their own breeding efforts, which would include the assignment of appropriate land plots and the availability of plant breeders and access to finance at reasonable costs combined with price incentives, could considerably improve incentives for private seed companies to increase production, and thus ameliorate the seed shortage in the country. The direct seed marketing pilots show that the government has recognized the need for change and may slowly deregulate the market.

To ensure supply of improved seed of all crops, contributions from the private sector will be needed. Even if the new regional seed enterprises expand and optimize their production in the near future, it is unlikely that they can satisfy the seed demand of all farmers in the country. This is also acknowledged by the government (World Economic Forum 2012).

In the current system, there is no strong incentive for many seed producers to begin making themselves more independent from the government. It is uncertain (for some even unlikely) that their profits would increase much, but business would become much riskier

To incentivize domestic as well as foreign investment, a well-designed and stepwise market liberalization is needed. Incremental institutional changes are required that provide incentives for the private sector to increase seed production and diversity in the product portfolio and to improve seed quality. Yet, the costs of such changes in terms of welfare losses of other stakeholders must be carefully evaluated. Some concrete innovations that are most likely to increase incentives for the private sector and result in better input supply for farmers in the middle- and long run are discussed below:

A central aspect for Ethiopian seed companies is that they need access to pre-basic seed of the varieties and in the quantities of their choice and the ability to market it in areas and at prices according to their firm strategy. Since public seed companies are not obliged to make profits according to their statutes, these enterprises can ensure that even the marginalized poor have access to improved seed in case the private companies develop strategies that focus on other market segments.

Microfinance institutions (MFIs) or farmer cooperatives need to provide credits to the farmers. Without a credit facility, a rise in seed production will hardly benefit the majority of the peasants. MFIs are already serving many farmers but are still far from being omnipresent. However, extended coverage is needed to back up the want for improved inputs with purchasing power. The extension of coverage, however, needs to be accompanied by lending methodologies that ensure repayments to avoid the high default rates that have eroded the credit system in the past.

Access to credit is a decisive factor not only for the farmers but also for the seed companies if they are to increase seed production and the diversity of varieties. Collateral requirements and costs for negotiations with the banks need to be lowered such that seed companies have a realistic chance of accessing finance at reasonable costs.

Another fundamental precondition for a more vibrant private sector is the assignment of more land for seed production and breeding efforts. Yet, more seed production and especially independent breeding efforts that would free Ethiopian companies from most governmental control along the value chain also require high-skilled plant breeders. The education of such people is a long-term task that needs to be taken care of by the government in the form of support for universities and higher learning institutes.

Additional to these 'enabling changes', it seems adequate to abolish the security for private seed enterprises that all produced seed is bought by the government. As long as seed companies do not need to use entrepreneurial spirit and design competitive firm strategies, many of them may remain in their cushy position in which no huge profits are made but the government organizes the marketing and covers much of the risks.

If seed markets are liberalized and the centralized distribution system is replaced by free market competition, access to seed for the poorest may be at stake. Thus, in a transition phase in which seed supply is not enough to meet demand and the private sector focuses on farmers who are better-off and easier to reach, the public seed enterprises can cater to the poorest, as, according to their statutes, they do not need to make profits. Alternatively, subsidies for the marginalized poor and investment incentives for companies may be temporary measures to ameliorate inequalities.

In the long term, however, private and public seed companies should compete for better quality and lower prices, both catering to the marginalized poor and to non-poor farmers. However, since the poor are the largest customer group in the country in terms of the number of people and, if they have access to credit, also in terms of the amount of inputs bought, it will be crucial for companies to cater to the poor if the largest part of the Ethiopian market is to be developed for the future.

How Can These Changes Be Brought About?

Having identified some potentially fruitful institutional innovations, the question arises as to how these changes could be brought about. It is unlikely that the private seed companies can establish a lobby group that bargains with one voice for market liberalization any time soon. Yet, to change institutions, a critical mass of agents is needed that together can reach a certain size in terms of market share or political importance and collectively work towards an institutional change (Acquaye-Baddoo et al. 2010). At the moment, the private seed companies do not seem to have this critical mass and it is not clear whether or how many of them really aim at changing the system.

Thus, while companies can still be expected to push for changes, the current situation and the self-conception of the Ethiopian government require the government to be in the lead. This then entails the question as to what could motivate the GoE to enact market liberalizing changes. Several factors may be important in this regard, most notably successful role models, support by other stakeholders and successes with investment incentive schemes in other sectors in the country.

Successful role models are certainly conducive for the government to enact changes. However, while neighbouring countries operate under similar initial conditions, so far, they have not successfully managed the transition to a liberalized market either (Ngugi 2002; Tripp and Rohrbach 2001). A more promising example would be China, which also comes closer to the development path aspired to by the Ethiopian government. China started from similar preconditions and was successful in increasing seed production by encouraging private sector investments and ensuring a division of labour between public and private enterprises for tasks like seed breeding and training of breeders (Cabral et al. 2006; Meng et al. 2006; Park 2008; Peoples Republic of China State Council 2013).

Not only the Chinese but also the German seed sector may be of some relevance for Ethiopia. The German seed industry is – similar to the Ethiopian case – built up of mainly medium-sized companies. About 130 plant breeding and seed trading companies operate in Germany, 60 of them with their own breeding programs. Most of the seed producing and trading companies are organized in regional associations and under a national umbrella association, the German Plant Breeders' Association. This umbrella association is the central part of a network that serves as a platform for pre-competitive joint research projects, patent issues, and public variety testing, as well as for safeguarding plant variety protection (http://www.bdp-online.de/en/). The German example is relevant to the Ethiopian case in as much as it shows that market consolidation can be avoided (Fernandez-Cornejo 2004).

Apart from learning from role models, cooperation between governments can help to facilitate the entry of private companies into the market. Furthermore, support can be expected from international organizations like AGRA (http://agra-alliance.org) and the Integrated Seed System Development Project (http://www. issdethiopia.org), or initiatives like the *New Vision for Agriculture*, the *New Alliance for Food Security and Nutrition* and Feed the Future (http://www. feedthefuture.gov) that are ready to support the government in market liberalization efforts.

In addition to these sources of support, cooperation and support from NGOs is needed as well. The distribution of free seed by relief organizations or even by public entities in the context of agricultural development programs has been identified as one of the most serious constraints to seed system development (Tripp and Rohrbach 2001).

And finally, successful experiences from other sectors of the economy, especially other subsectors of agriculture, may motivate the government to support private sector investments. Such positive experiences can be drawn from the flower sector, where investments have been attracted by the government. Thanks to these programs, flower production increased significantly in the last two decades (Ayele 2006; see also e.g. The Embassy of Ethiopia in China 2013).

Another very important factor is the change in informal institutions. In Ethiopia, many parts of society are still considerably shaped by the country's socialist past. Entrepreneurial spirit is not very common ([5]), and scepticism concerning business and the belief in a state-directed economy are still common among government employees. But if the private sector is meant to increase operations, support needs to come from all levels of government and other parts of society, from universities to banks to consumers.

Open Access   This chapter is distributed under the terms of the Creative Commons Attribution-Noncommercial 2.5 License (http://creativecommons.org/licenses/by-nc/2.5/) which permits any noncommercial use, distribution, and reproduction in any medium, provided the original author(s) and source are credited.

The images or other third party material in this chapter are included in the work's Creative Commons license, unless indicated otherwise in the credit line; if such material is not included in the work's Creative Commons license and the respective action is not permitted by statutory regulation, users will need to obtain permission from the license holder to duplicate, adapt or reproduce the material.

References

Acemoglu D, Robinson J (2012) Why nations fail: the origins of power, prosperity, and poverty, 1st edn. Crown Publishers, New York

Acquaye-Baddoo NA, Ekong J, Mwesige D, Neefjes R, Nass L, Ubels J, Visser P, Wangdi K, Were T, Brouwers J (2010) Multi-actor systems as entry points for capacity development. Capacity.org. A gateway for capacity development, pp. 4–5

Alemu D (2010) Seed system potential in Ethiopia. Constraints and opportunities for enhancing the seed sector. International Food Policy Research Institute, Addis Ababa/Washington, DC

Alemu D (2011) The political economy of Ethiopian cereal seed systems: state control, market liberalisation and decentralisation. IDS Bull 42(4):69–77. doi:10.1111/j.1759-5436.2011. 00237.x

Alene AD, Manyong VM, Omanya G, Mignouna HD, Bokanga M, Odhiambo G (2008) Small-holder market participation under transactions costs: maize supply and fertilizer demand in Kenya. Food Policy 33(4):318–328. doi:10.1016/j.foodpol.2007.12.001

Amhara Regional State (2008) Document for establishment of Amhara seed enterprise. Government of Ethiopia, Addis Ababa

Arrow KJ (1969) The organization of economic activity: issues pertinent to the choice of market versus nonmarket allocation. In: The analysis and evaluation of public expenditure: the PPB system, vol 1. US Joint Economic Committee of Congress, Washington, DC, pp 59–73

Astatike M, Yimam A, Tsegaye D, Kefale M, Mewa D, Desalegn T, Hassena M (2012) Experience of direct seed marketing in Amhara region (unpublished report). Local Seed Business, Addis Ababa

Arrow KJ (1969) The organization of economic activity: issues pertinent to the choice of market versus nonmarket allocation. In: The analysis and evaluation of public expenditure: the PPB system, vol 1. US Joint Economic Committee, Washington, DC, pp 59–73

Ayele S (2006) The industry and location impacts of investment incentives on SMEs start-up in Ethiopia. J Int Dev 18(1):1–13. doi:10.1002/jid.1186

Bator FM (1958) The anatomy of market failure. Q J Econ 72(3):351. doi:10.2307/1882231

Baumüller H, Husmann C, von Braun J (2013) Innovative business approaches for the reduction of extreme poverty and marginality? In: von Braun J, Gatzweiler FW (eds) Marginality. Addressing the nexus of poverty, exclusion and ecology. Springer, Dordrecht/ Heidelberg/New York/London, pp 331–352

Benson T, Spielman D, Kasa L (2014) Direct seed marketing program in Ethiopia in 2013. An operational evaluation to guide seed-sector reform. International Food Policy Research Institute, Addis Ababa

Bishaw Z, Yonas S, Belay S (2008) The status of the Ethiopian seed industry. In: Thijssen MH, Bishaw Z, Beshir A, de Boef WS (eds) Farmers, seeds and varieties: supporting informal seed supply in Ethiopia. Wageningen International, Wageningen, pp 23–31

Byerlee D, Spielman DJ, Alemu D, Gautam M (2007) Policies to promote cereal intensification in Ethiopia: a review of evidence and experience. IFPRI discussion paper 707, Washington, DC

Cabral L, Poulton C, Wiggins S, Xhang L (2006) Reforming agricultural policies: lessons from four countries. Futures agricultures futures agricultures working paper 002. Future Agricultures, Brighton

Conway G (2012) One billion hungry. Cornell University Press, New York

CSA (2012a) Household Consumption and Expenditure (HCE) survey 2010/11. Central Statistical Agency, Addis Ababa

CSA (2012b) Agriculture – 2011, national abstract statistics. Central Statistical Agency, Addis Ababa

CSA, MoFED (2011) Agricultural sample survey 2010–11. Study documentation. Central Statistical Agency and Ministry of Finance and Economic Development, Addis Ababa

Dalberg Global Development Advisors (2012) Amhara seed enterprise strategy refresh. Dalberg Global Development Advisors, unpublished document

De Bruyn P, de Bruyn JN, Vink N, Kirsten JF (2001) How transaction costs influence cattle marketing decisions in the northern communal areas of Namibia. RAGR 40(3):405–425. doi:10.1080/03031853.2001.9524961

De Silva H, Ratnadiwakara D (2008) Using ICT to reduce transaction costs in agriculture through better communication: a case study from Sri Lanka. LIRNEasia, Colombo

Degu G, Mwangi W, Verkuijl H, Wondimu A (2000) An assessment of the adoption of seed and fertilizer packages and the role of credit in smallholder maize production in Sidama and North Omo Zones, Ethiopia. International Maize and Wheat Improvement Center, Mexico DF

Demsetz H (1969) Information and efficiency: another viewpoint. J Law Econ 12(1):1–22. doi:10.1086/466657

Demsetz H (1995) The economics of the business firm: seven critical commentaries. Cambridge University Press, Cambridge

Feder G, Umali DL (1993) The adoption of agricultural innovations. Technol Forecast Soc Chang 43(3–4):215–239. doi:10.1016/0040-1625(93)90053-A

Fernandez-Cornejo J (2004) The seed industry in U.S. agriculture: an exploration of data and information on crop seed markets, regulation, industry structure, and research and development. Agric Inf Bull AIB-786:1–81

Furubotn EG, Richter R (2005) Institutions and economic theory: the contribution of the new institutional economics, 2nd edn. University of Michigan Press, Ann Arbor

Gabre-Madhin E (2001) Market institutions, transaction costs, and social capital in the Ethiopian grain market. International Food Policy Research Institute, Washington, DC

Holloway G, Nicholson C, Delgado C, Staal S, Ehui S (2000) Agroindustrialization through institutional innovation transaction costs, cooperatives and milk-market development in the east-African highlands. Agric Econ 23:279–288. doi:10.1111/j.1574-0862.2000.tb00279.x

ISSD (2013) Direct seed marketing: additional channel to enhance seed marketing and distribution system in Ethiopia. ISSD, Addis Ababa

Key N, Sadoulet E, Janvry A (2000) Transactions costs and agricultural household supply response. Am J Agric Econ 82(2):245–259

Kherallah M, Kirsten JF (2002) The new institutional economics: applications for agricultural policy research in developing countries. Agrekon 41(2):111–134

Kyeyamwa H, Speelman S, van Huylenbroeck G, Opuda-Asibo J, Verbeke W (2008) Raising offtake from cattle grazed on natural rangelands in sub-Saharan Africa: a transaction cost economics approach. Agric Econ 39(1):63–72. doi:10.1111/j.1574-0862.2008.00315.x

Louwaars N (2010) The formal seed sector in Ethiopia. A study to strengthen its performance and impact. Local Seed Bus Newsl 3:12–13

Maltsoglou I, Tanyeri-Abur A (2005) Transaction costs, institutions and smallholder market integration: potato producers in Peru, ESA working paper. FAO, Rome

Masten SE (2001) Transaction-cost economics and the organization of agricultural transactions. In: Baye R (ed) Industrial organization, vol 9, Advances in applied microeconomics. Emerald Group Publishing Limited, Bingley, pp 173–195. doi:10.1016/S0278-0984(00)09050-7

Meng ECH, Hu R, Shi X, Zhang S (2006) Maize in China. Production systems, constraints and research priorities. International Maize and Wheat Improvement Center, Mexico

Ministry of Agriculture (2013) Data on demand and supply of improved seed of different crops. MoA, Addis Ababa

New Alliance for Food Security & Nutrition (2013) New alliance for food security and nutrition 2013 progress report. DFID, AU Commission and WEF, London

Ngugi D (2002) Harmonization of seed policies and regulations in East Africa. In: Rohrbach D, Howard J (eds) Seed trade liberalization and agro-biodiversity in Sub-Saharan Africa. Presented at the workshop on impacts of seed trade liberalization on access to and exchange of agro-biodiversity, International Crop Research Institute for the Semi-Arid Tropics, Matopos Research Station, Bulawayo, pp. 101–120

North DC (1989) Institutions and economic growth: an historical introduction. World Dev 17(9):1319–1332. doi:10.1016/0305-750X(89)90075-2

North DC (1990) Institutions, institutional change and economic performance. Cambridge University Press, Cambridge

Ouma E, Jagwe J, Obare GA, Abele S (2010) Determinants of smallholder farmers' participation in banana markets in Central Africa: the role of transaction costs. Agric Econ 41(2):111–122. doi:10.1111/j.1574-0862.2009.00429.x

Park A (2008) Agricultural development in China: lessons for Ethiopia. Briefing note prepared for the DFID funded study "understanding the constraints to continued rapid growth in Ethiopia: the role of agriculture". DFID, London

Peoples Republic of China State Council (2013) National plan for development of the modern crop seed industry (2012–2020). Unofficial translation, Peoples Republic of China State Council, Beijing

Rujis A, Schweigman C, Lutz C (2004) The impact of transport-and transaction-cost reductions on food markets in developing countries: evidence for tempered expectations for Burkina Faso. Agric Econ 31(2–3):219–228. doi:10.1111/j.1574-0862.2004.tb00259.x

Seyoum Taffesse A, Dorosh P, Asrat S (2011) Crop production in Ethiopia: regional patterns and trends, ESSP working paper 0016. IFPRI, Addis Ababa

Somda J, Tollens E, Kamuanga M (2005) Transaction costs and the marketable surplus of milk in smallholder farming systems of the Gambia. Ooa 34(3):189–195. doi:10.5367/000000005774378784

Spielman DJ, Kelemwork D, Alemu D (2011) Seed, fertilizer, and agricultural extension in Ethiopia. Ethiopia Strategy Support Program II (ESSP II) 20. IFPRI, Addis Ababa

Staal S, Delgado C, Nicholson C (1997) Smallholder dairying under transactions costs in East Africa. World Dev 25(5):779–794. doi:10.1016/S0305-750X(96)00138-6

Stifel DC, Minten B, Dorosh P (2003) Transactions costs and agricultural productivity: implications of isolation for rural poverty in Madagascar, IFPRI MSSD discussion paper 56. IFPRI, Addis Ababa

The Embassy of Ethiopia in China (2013) Horticulture and floriculture industry: Ethiopia's comparative advantage [WWW Document]. http://www.ethiopiaemb.org.cn/bulletin/05-1/003.htm. Accessed 26 Feb 2014

Tripp R, Rohrbach D (2001) Policies for African seed enterprise development. Food Policy 26(2):147–161. doi:10.1016/S0306-9192(00)00042-7

von Braun J, Bouis H, Kumar S, Pandya-Lorch R (1992) Improving food security of the poor: concept, policy, and programs. International Food Policy Research Institute, Washington, DC

Williamson OE (1980) The organization of work a comparative institutional assessment. J Econ Behav Organ 1(1):5–38. doi:10.1016/0167-2681(80)90050-5

World Economic Forum (2012) Accelerating infrastructure investments – world economic forum on Africa 2012. http://www.weforum.org/videos/accelerating-infrastructure-investments-world-economic-forum-africa-2012. Accessed 16 Aug 2012

Chapter 9
Agricultural Service Delivery Through Mobile Phones: Local Innovation and Technological Opportunities in Kenya

Heike Baumüller

Abstract The rapid spread of mobile phones across the developing world offers opportunities to improve service delivery for smallscale farmers. International and local companies have already started to capitalize on these opportunities although many mobile phone-enabled services are still at an early stage. Kenya has emerged as a leader in m-service development in Sub-Saharan Africa. This chapter assesses the key factors that have helped the local innovation scene to emerge and reviews existing agricultural m-services that provide Kenyan farmers with access to information and learning, financial services, and input and output markets. The potential impact of m-services is illustrated with the example of the price and marketing service M-Farm. Finally, the chapter assesses current mobile technology trends to offer an outlook on potential future applications.

Keywords Mobile phone • Kenya • M-service • Innovation • Smallholder farmer

Introduction

Kenya has emerged as a frontrunner in information and communication technologies (ICT) in Sub-Saharan Africa. The government has been actively supporting the ICT sector as one of the key drivers of economic growth. In addition to large international firms that are setting up offices in Nairobi, such as Nokia, IBM and Google, local start-ups have also been expanding rapidly. Kenyan entrepreneurs have greatly benefited from the growth of the local innovation environment in recent years, including the establishment of several innovation hubs, a growing pool of human resources, and access to finance from private investors. An increasingly well-connected customer base and improving infrastructure are also helping entrepreneurs to market their services.

H. Baumüller (✉)
Center for Development Research (ZEF), University of Bonn, Bonn, Germany
e-mail: hbaumueller@uni-bonn.de

© The Author(s) 2016
F.W. Gatzweiler, J. von Braun (eds.), *Technological and Institutional Innovations for Marginalized Smallholders in Agricultural Development*,
DOI 10.1007/978-3-319-25718-1_9

As a result, Kenyans have access to a growing number of services through their mobile phones (m-services). The mobile payment system *M-Pesa* is one of the most successful mobile payment systems in the developing world. M-services are also available in other sectors, such as education, health and entertainment. In the area of agriculture, mobile phones could be particularly helpful in extending the reach of services to rural populations by facilitating communication that is not restricted by distance, volume, medium and time (von Braun and Torero 2006). Several m-services are already offered to Kenyan farmers, including information, insurance and marketing services. Many of these are provided by local companies, although most remain at a small scale.

This chapter outlines the key factors that have supported the growth of the Kenyan m-services sector. It reviews the agricultural m-services currently available and presents a case study of one such service, *M-Farm*, which offers price information and marketing services to Kenyan farmers. The chapter concludes with a brief assessment of current mobile technology trends to provide an outlook on potential future applications in the agriculture sector.

Kenya's ICT Ecosystem for Local Entrepreneurs[1]

Opportunities...

> Kenya is rising fast as a technology powerhouse on the African continent and more so in Sub-Saharan Africa. (Afrinnovator 2012, p. 1)

Network Infrastructure

Kenya's growing ICT ecosystem is making the country an attractive place for local entrepreneurs to develop and deploy m-services. The first sea cable to link Kenya internationally came online in July 2009, thereby offering a faster and cheaper alternative to satellite connections (McCarthy 2009). Since then, three additional sea cables have been connected to landing points in Mombasa (Fig. 9.1) and another three cables are planned (Mbuvi 2013). Terrestrial fiber optic cables are starting to reach into all parts of Kenya and are expected to expand further following an agreement in June 2012 between the Chinese and Kenyan governments to provide financing for the National Optic Fiber Backbone Infrastructure (Wahito 2012). The government has supported this infrastructure expansion through various regulatory measures and financing (see below).

---

[1] Unless otherwise stated, this section draws on interviews with key informants in Nairobi carried out in April–June 2012.

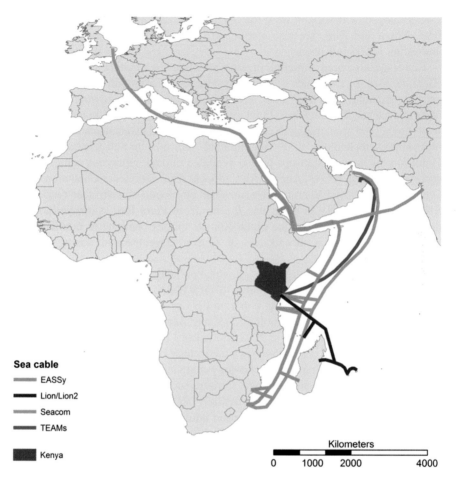

Fig. 9.1 Map of sea cables to Kenya (Data source: UbuntuNet Alliance (as of November 2012)) Cartography: Heike Baumüller

A Supportive Innovation Environment

One of the key factors driving the expansion of Kenyan technology start-ups is the innovation environment, which has grown in particular over the past 6–7 years. Several innovation hubs have been set up, led by the iHub and followed by others, such as the m:lab, the Nailab, the 88mph Garage or @iBizAfrica, which offer a space and infrastructure for developers, mentorship from more experienced entrepreneurs, and opportunities to interact with investors, fellow developers and business partners. The hubs have also helped to strengthen the connectedness of the local tech community, which Eric Hersman, co-founder of the iHub, believes has given Kenya a crucial competitive advantage over other countries (Hersman 2012). These innovation spaces were mainly driven by visionary entrepreneurs and tech developers with support from foreign investors or donors. Companies, such as Intel, Nokia and IBM, are also starting to link up with or invest in their own innovation spaces in Kenya.

Kenya has also been attracting investor attention "as a hub for ICT innovation" (Deloitte 2012, p. 17). Much of the interest has come from non-Kenyan investors and, in particular, so-called 'angel investors' who are willing to support start-up ideas and talents. The Savannah Fund, for instance, was launched in mid-2012 as a seed capital fund specializing in $25,000–$500,000 (US) investments in early stage high growth technology (web and mobile) start-ups in Sub-Saharan Africa.[2] Financing for Kenyan start-ups is also available through numerous competitions, such as Pivot East, IPO48, Apps4Africa, Google Apps Developer Challenge or the Orange African Social Venture Prize, in which developers can win seed funding of $10,000–$25,000 (US). In particular, the results of Pivot East, a competition for developers from East Africa in which Kenyan entries continue to dominate the winners' list, exemplify the success of Kenyan developers in raising start-up funding (Sato 2013).

The ICT sector can also draw on a growing pool of human resources and a young generation that is increasingly willing to take the risk of setting up their own technology companies. Training opportunities are expanding, notably through eMobilis, the first Mobile Technology Training Academy in Sub-Saharan Africa. The academy was established in 2008 and teaches both IT and business skills to enable young people to set up their own technology businesses. The graduates are highly motivated by seeing other technology companies succeed, such as Facebook and Instagram internationally and local start-ups such as *Ushahidi*,[3] *Kopo Kopo*[4] or *M-Farm*. The private sector is also increasingly tapping this potential, such as in the case of Safaricom, which, in collaboration with the @iLabAfrica of Strathmore University and Vodafone, has set up the Safaricom Academy where students can earn a Master of Science in Telecommunication Innovation and Development.

Government Policy

The development of the ICT sector has also been promoted by the Kenyan government. The sector has emerged as a key driver of economic growth, showing an annual growth rate of around 20 % and adding 0.9 % to annual GDP growth between 2000 and 2010 (World Bank 2010). To support the sector, the government adopted a national ICT policy in 2006 and set up an ICT Board in 2007. While the focus was initially on marketing Kenya as a hub for outsourcing ICT-related

---

[2] www.savannah.vc.

[3] A crowd-sourcing technology for collecting, visualizing and interactively mapping information (http://ushahidi.com).

[4] A platform to enable small and medium businesses to accept mobile payments and build relationships with their customers (http://kopokopo.com).

business, the government is also stepping up efforts to support local technology entrepreneurs. For instance, the ICT Board has launched the Tandaa grant which promotes the creation and distribution of locally relevant digital content and offers seed funding for local enterprises.

A number of regulatory steps have also supported ICT development in Kenya (Schumann and Kende 2013). In 2008, the government established a unified licensing regime which allowed any company to bid for a license with only a few requirements[5] and without restrictions on the number of operators allowed to build and operate ICT infrastructure. Other measures included investments in submarine and terrestrial fiber optic cables, the removal of a value-added tax for mobile handsets, support for the development of the internet exchange point in Nairobi, sharing of the state-owned electricity company's infrastructure and reduction in the cost of calling between different mobile networks. These measures have played an important role in attracting private sector investment, increasing competition, improving the quality of the network and reducing the cost of mobile access.

In an effort to further strengthen the sector, the government is developing Konza Technology City near Nairobi, which is being marketed as 'Africa's Silicon Valley'. Konza City is an integral part of the government's National ICT Master Plan 'Connected Kenya 2017', which was launched in February 2013 with the overall goal of becoming Africa's most globally respected knowledge economy by 2017 (Kenya ICT Board 2012). Specifically, the plan aims at developing 500 new ICT companies, 20 global innovations and 50,000 jobs. The government also adopted a National Broadband Strategy to establish faster and more reliable broadband connections around the country (Okutoyi 2012).

M-Pesa

M-services developers have also benefited from the success of Safaricom's mobile banking service *M-Pesa*. Since its launch in 2007, *M-Pesa* had expanded to almost 16 million active customers with over 90,000 agent outlets across the country (Safaricom 2016). While other mobile operators have also started to offer m-payment services, M-Pesa continues to dominate the market, accounting for 77 % of mobile money customers (as of June 2015) (CA 2015).

Through its widespread adoption, *M-Pesa* has helped to prepare the ground for m-services in Kenya, familiarizing many Kenyans with the use of their mobile phone for non-call related services. For instance, *M-Pesa* has been credited for the

---

[5] I.e., to have a Kenyan-registered entity with permanent premises, provide evidence of tax compliance, and, if foreign-owned, divest 20 % of ownership to Kenyans within 3 years of receiving the license.

relatively widespread use of SMS in Kenya (Boyera 2012) where 89 % of mobile users are sending SMS compared to 50 % in South Africa, 26 % in Nigeria and 20 % in Ghana (World Bank 2012). *M-Pesa* (and other m-payment systems) also provides supporting services for other companies offering m-services that require monetary transactions. Moreover, the agent network can be used to market other technologies, such as the first Intel-powered smartphone which is being sold exclusively through Safaricom to take advantage of the widely available and highly frequented Safaricom outlets (Macharia 2013).

A Growing Customer Base

The customer base for mobile phone-enabled services is growing rapidly, not least driven by Kenya's young and increasingly educated population. Almost 40 % of the economically active population was estimated to be below the age of 30 in 2012 (ILO 2011). School enrollment rates have been improving. By 2009, 50 % of children in their age group were enrolled in secondary school, up from a third in 2000.[6] The youth are tech-savvy and interested, exemplified by the fact that Kenyans are the second most prolific tweeters in Africa after South Africa.[7] According to the Kenya Technology, Innovation & Startup Report 2012, "[n]ever before has the digital consciousness of the Kenyan people been as alive as it is today" (Afrinnovator 2012, p. 2). This trend is also reflected in the rapid expansion of small ICT sellers, repairers and service providers in Nairobi who are servicing the low-income market in particular (Foster 2012).

Access to mobile phones is relatively high and improving. The majority of the population is covered by mobile services (85 % in 2008/2009[8]) thanks to a growing network of fiber optic cables. 3G networks are available (though do not always perform well) and plans to roll out LTE are also in place. By December 2013, mobile phone subscription rates were 77 per 100 people, up from 0.41 per 100 in 2000.[9] In 2010, the number of mobile phone subscribers for the first time exceeded the number of people above the age of 15 (Fig. 9.2). These rates compare well to the regional average of 75 per 100 across Africa and 95 per 100 in developing countries in 2013.[10]

Subscription rates only provide a general indication of mobile phone access in a country. The GSMA believes unique subscription rates in Kenya to be considerably lower than total subscription rates, at around 37 % (Makau 2012). Nevertheless, access

---

[6] World Bank, data.worldbank.org. Accessed 25 Jan 2012.

[7] According to a survey carried out in the last quarter of 2011 (Portland Communications 2012).

[8] Waema et al. (2010).

[9] CCK statistics, ca.go.ke/index.php/statistics. Accessed 9 Sept 2014.

[10] ITU statistics, www.itu.int/en/ITU-D/Statistics. Accessed 9 Sept 2014.

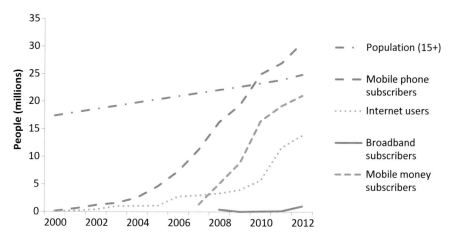

Fig. 9.2 Mobile phone, mobile money and internet penetration in Kenya. Note: The number of internet users was calculated by multiplying the share of the population using the internet (ITU) with the population (World Bank) (Data sources: ITU (mobile phone subscribers, share of population using the internet), World Bank (population), CBK (mobile money subscribers), accessed July 15, 2013)

to mobile phones is common in Kenya through the sharing of phones. One nationally representative survey found that 85 % of respondents used a mobile phone, although only 44 % owned a phone in 2009 (Wesolowski et al. 2012). Phone sharing was particularly prevalent among low income groups (Fig. 9.3) and in rural areas (even among higher income groups). Similarly, a survey of Kenyan farmers found that only around a third owned a mobile phone, but 84 % had used one (Okello et al. 2010).

The expanding mobile network also plays a critical role in facilitating access to the internet among Kenyan users. The vast majority of Kenyan internet subscribers (99 %) are accessing the web through mobile devices, including internet-enabled mobile phones and PCs with cellular modems (CA 2015). Internet usage began increasing significantly in 2010 (Fig. 9.2). While only around a third of Kenyans is estimated to use the internet, this share is almost three times higher than the African average (12 % in 2012) and one of the highest on the continent.[11] Internet uptake is particularly high by Sub-Saharan African standards if seen as a function of GDP, in part due to the low cost of internet compared to other countries in the region (Schumann and Kende 2013). Average download speeds from a local server are also considerably higher than in most Sub-Saharan African countries, with the exception of Rwanda and Ghana (in 2012) (ibid).

---

[11] ITU statistics, www.itu.int/en/ITU-D/Statistics. Accessed 15 July 2013.

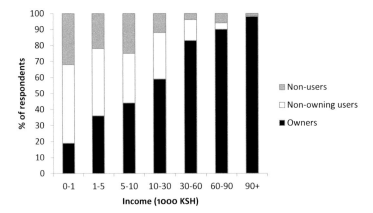

Fig. 9.3 Phone ownership and usage by income groups in Kenya (Source: Compiled by the author using data from Wesolowski et al. (2012))

. . . and Challenges

While Kenya's ICT ecosystem has come a long way in recent years, it is still maturing, and Kenyan entrepreneurs continue to face significant hurdles. Many start-ups struggle to move from initial idea to scale. The companies often do not involve enough marketing and business people due to a lack of funding, although these skills are particularly important as they seek to scale their businesses (Kieti 2012). Also, more mentorship and work experience in larger companies is needed to close the gap between a junior developer and the more senior established developers. Foreign companies could help start-ups graduate from small to medium-sized companies by outsourcing certain activities to local developers. However, lack of awareness of the local talent pool and difficulties in weeding out the good from the bad start-ups have so far prevented them from doing so.

There are also shortcomings in available training opportunities. While some universities are recognizing the importance of integrating ICTs into their curricula, there are no interdisciplinary courses that focus on building both sectoral expertise and practical software development skills. Moreover, university curricula are often insufficiently adapted to industry requirements. As Michael Macharia, CEO of Seven Seas Technologies in Kenya, observes, "there' s an urgent need to incorporate industry needs in university curricula across all our universities to ensure industry relevance" (cited in Mutua 2012).

A better understanding of the needs of the customers and the context in which the m-service is provided would also be beneficial. Companies rarely involve sectoral experts, such as health, education or agricultural specialists, to develop a product that meets specific needs. Also, too many m-services are developed with limited background research or interaction with potential customers. At times, developers appear too focused on building the next big idea or on pitching the idea at one of the numerous competitions. Much hope is pinned on earning big money by developing apps and selling them through the app stores, even though the revenue-generating potential is rather uncertain.[12] As a result, m-services risk turning into technology solutions, rather than solutions that address a particular demand. This problem is not restricted to Kenyan developers, however. As Ken Banks, the founder of *FrontlineSMS*, points out, in the ICT4D (ICT for development) community, "Mobile is still largely seen as a solution, not a tool" (Banks 2013).

Access, in particular to mid-level funding that would allow start-ups to scale, continues to pose a challenge. "There remains a gaping hole in the market where venture capital activity should be [...] there are few venture capital funds dedicated to funding [IT and mobile] entrepreneurs in East Africa" (Deloitte 2012, p. 19). Some investors are reluctant to engage with Kenyan start-ups because of limited exit opportunities, such as selling their interests to a larger investor. Investors are also often not aware of investment opportunities. In particular, Kenyan investors have so far not shown much interest in local tech start-ups, preferring safer and often bigger investments that bring high returns.[13] At the same time, "many of the nascent entrepreneurs are probably not yet ready for venture capital" (ibid). Indeed, start-ups sometimes hesitate to seek investors because they do not want to give up control of the company too early.[14]

Moreover, while the IT infrastructure is fairly advanced by regional standards, it still faces problems. Overall, the share of the population using the internet is still low at less than a third in 2012 (Fig. 9.2) and only 11 % of internet subscribers had access to broadband (CCK 2013). Access to the mobile network and internet has at times been disrupted by damage to the sea cables (Okuttah 2012), and power cuts continue even in Nairobi. In addition, while the liberalization of the licensing regime has helped to attract investors, critics complain that it has encouraged higher investments in profitable areas, such as the deployment of multiple fiber optic cables in wealthy neighborhoods (Schumann and Kende 2013).

Rural areas continue to lag far behind in terms of the reach and quality of networks and related services. The cost of supplying telecommunication services to as yet underserved areas has been estimated at KSh74 billion (ca. $825 million) (Mumo 2013). The government's Universal Service Fund, which aims to collect a share of industry revenues to finance the expansion of mobile services, has been slow to get off the ground and is expected to fall short of the KSH one billion target in its first year 2013/2014 (ibid). The main challenges include high operational costs due to limited access to electricity, roads and infrastructure security, low population densities and high license and spectrum fees coupled with unclear

---

[12] A survey of over 1500 developers from around the world found that around a third cannot rely on apps as their only source of income, even if they sell several apps. Only 14 % will earn between $500 and $1000 and 13 % between $1001 and $5000 per app per month, while 25 % will not generate any income at all (VisionMobile 2012).

[13] Paul Kukubo, Chief Executive Officer, Kenya ICT Board @ Pivot East, 5 June 2012.

[14] Benjamin Matranga, Investment Officer, Soros Economic Development Fund @ Pivot East, 5 June 2012.

spectrum policies in these areas (Apoyo Consultoria 2011). In addition to
network availability, download speeds also differ considerably within the country
and will continue to do so even with the government's new broadband strategy
(Okutoyi 2012).

M-Services for Kenyan Farmers

Kenya's agriculture sector is dominated by semi-subsistence, low-input and
low-productivity farmers (Jayne et al. 2003). Agricultural holdings tend to be
small at 2.4 acres on average (KNBS 2006).[15] Maize is the most widely grown
crop in Kenya. The staple food is produced by 90 % of rural households and
accounts for over 20 % of agricultural production (Bernard et al. 2010). Almost
two thirds of maize production is generated by small-scale farmers (ibid). The
second most widely grown crop is beans. Other important crops (i.e., with a
harvested area of more than 100,000 ha in 2011–2013) include sorghum, tea, cow
peas, coffee, wheat, pigeon peas, potatoes and millet.[16] Sugarcane is the main crop
in terms of production volume, followed by maize, potatoes and bananas.

Various m-services are already offered to Kenyan farmers (see Table 9.1,
excluding financial services). In most cases, assessing the reach and impacts of
these services is difficult in the absence of publicly available data on users and
impact assessments. Most of the services are offered by the private sector, including
Kenyan companies (*M-Farm, KACE, mFarmer, kuza doctor, Agrimanagr, iCow*,
radio stations), at times in collaboration with international companies (*M-Kilimo,
ACRE*). Only a few services are led by government departments (*National
Farmers' Information Service, Maize Variety SMS Service*) or international organizations (*Sokopepe, E-Farming*, index-based livestock insurance), and these are
often also implemented in partnership with the private sector.

Agricultural m-services can be grouped into four categories: information and
learning, financial services, access to agricultural inputs, and access to output
markets. Most of the Kenyan services focus on information provision. Several
deliver production-related information for crops (*ArifuMkulima, Sokopepe, kuza
doctor, M-Kilimo, NAFIS, E-Farming, Maize Variety SMS Service*) or livestock
(*iCow*) via SMS, phone calls and/or websites. Several radio stations also offer
interactive programs to which farmers can send questions by SMS to the radio
station which are then answered on air, in some cases using the software
FrontlineSMS to manage the incoming SMS traffic. Several services also provide
information on crop prices (see below and Box 9.1). The impact of these services

[15] "An agricultural holding is defined as all the land operated by a household for crop farming
activities. [...] A holding may comprise one or more parcels." (KNBS 2006, p. 159).

[16] Data on production area and volumes: FAOStat, faostat.fao.org. Accessed 2 July 2015.

Table 9.1 Examples of m-services offered to Kenyan farmers (as of June 2013)

M-Farm mfarm.co.ke	Daily crop price information, selling of produce, purchasing of inputs (on hold), start date: October 2010
Kenya Agricultural Commodity Exchange www.kacekenya.co.ke	Weekly crop price information, *Soko Hewani* to sell produce through radio auctions, start date: 1997 (company)
SokoPepe www.sokopepe.co.ke	Agricultural information (e.g., climate changes, product prices, services for farmers, agricultural methods), selling of produce, start date: October 2010
SokoShambani www.mfarmerkenya.org	Mobile trading platform to link potato farmers and restaurants
ArifuMkulima www.mfarmerkenya.org	Agricultural information (e.g., weather, diseases, calendar alerts, farm inputs, financial advice, agrovets)
kuza Doctor www.backpackfarm.com	Agricultural production information for 10 crops (20 crops planned) in English & Swahili (Luganda planned), start date: August 2011
M-Kilimo www.m-kilimo.com	Agricultural information (e.g., land preparation, planting, pest management, harvesting, post-harvest and marketing), date: 2009–2011
National Farmers' Information Service (NAFIS) www.nafis.go.ke	Agricultural information (e.g., crops, livestock, market prices on inputs and outputs, other info), start date: April 2008
E-Farming[a]	Agricultural information (e.g., soils, fertiliser application, agronomy, markets or pesticide use), start date: 2012
Maize Variety SMS Service www.kephis.org	Information on the most suitable maize variety to grow in the division
iCow www.icow.co.ke	Livestock production information (e.g., info about local services, record keeping, best practice, cow calendar) and virtual livestock market, start date: June 2011
FrontlnieSMS Radio radio.frontlinesms.com, www.organicfarmermagazine.org	e.g. The Organic Farmer, *Pur Mariek* (farm wisely) on Radio Nam Lolwe, agricultural information on the radio in response to farmers' questions
FarmerVoice Radio www.farmervoice.org	Agricultural information, start date: July 2009
ACRE (formerly Kilimo Salama) kilimosalama.wordpress.com	Insurance to protect crops against drought or flood, start date: 2009
Index-based livestock insurance livestockinsurance.wordpress.com	Insurance against drought-related livestock mortality, start date: January 2010
Agrimanagr www.virtualcity.co.ke	Supply chain management, start date: 2010
farmforce www.farmforce.com	Supply chain management, start date: 2012

[a]Okoth (2013)

Box 9.1: The Case of *M-Farm*

M-Farm was launched in October 2010 by a small Kenyan start-up company as an m-service targeted at smallholder farmers in Kenya. *M-Farm* provides wholesale market price information for 42 crops (legumes, fruits and tubers, horticulture, cereals and eggs) from five markets in Kenya (Eldoret, Kisumu, Kitale, Mombasa, Nairobi). Farmers can access the information by sending an SMS to a short code to access a searchable database. The information is also available through the website and two apps. In addition, *M-Farm* assists smallholder farmers in collectively selling their produce to large buyers through contracts, and connects buyers and sellers via an internet- and mobile phone-enabled platform.

Price information and marketing services can help increase agricultural productivity in a number of ways. Access to price and demand information can encourage agricultural technology adoption by reducing uncertainties about the expected profitability of a technology (Abadi Ghadim and Pannell 1999). In Kenya, many small-scale farmers rely on a limited number of middlemen or traders to receive price information, given that search costs for finding information elsewhere are often high (Eggleston et al. 2002). Without this information (along with other uncertainties), farmers may not produce the most profitable mixture of crops or use efficient technologies (Eggleston et al. 2002).

Access to market information and linkages could also increase the prices that farmers are able to obtain for their produce. Due to limited access to price information, price signals in many rural areas are often "faint or absent" (Eggleston et al. 2002, p. 5). As a result, farmers are unable to find the most lucrative market to sell their produce and transactions tend to become localised (Stigler 1961). Moreover, in the absence of selling options, farmers tend to establish long-term trading relationships with a few traders – a process also referred to as 'clientelisation' (Geertz 1978). The consequent lack of competition combined with information asymmetries between traders and farmers worsens their bargaining position to negotiate prices for their crops (Svensson and Yanagizawa 2009).

The case study showed that farmers are using the price information from *M-Farm* to plan production processes, i.e., when deciding what to grow, when to harvest and who to sell to. While most price enquiries are sent at the sales stage, farmers also request price information at earlier stages of production. However, information about demand is often seen as more important for decision-making than price information. The price information has also encouraged farmers to change their cropping patterns by expanding certain crops, but was less influential in encouraging them to introduce new crops.

(continued)

Box 9.1 (continued)

Evidence as to whether the price information had helped farmers negotiate better prices is inconclusive. While farmers felt that they had been able to obtain better prices, an analysis of sweet potato prices in Rachuonyo does not show marked changes since farmers started using *M-Farm* (although the data are too limited to draw strong conclusions). Rather than price increases, perceived income gains may be attributable to changes in cropping patterns and harvesting times. The price information does not seem to have induced farmers to change traders on a large scale.

The survey data also suggests that the radio offers a viable alternative to *M-Farm* in disseminating price information. A third of the farmers still obtains price information from the radio since they started using *M-Farm* (compared to 42 % before) which they regard as comparable to *M-Farm* in quality. The radio is seen as a good source of information, particularly in the early stages of production, while *M-Farm* becomes more important closer to the selling stage.

Source: Field research carried out by the author

has not been assessed in any detail. A small survey of *iCow* users found that 82 % of farmers were still using the service 7 months later (iCow 2010). 42 % of farmers thought their income had increased, with just over half attributing income increases to increased milk yield.

In terms of financial services, access to transmission services is common even in rural areas, owing to the widespread availability of mobile payment services, such as *M-Pesa*, *airtel Money*, or *Iko Pesa* by Orange. However, while m-payments are widely available, usage of the service among Kenyan farmers for agricultural purposes appears to be limited. A study carried out in three districts of Kenya found that although almost all respondents had heard about m-payments (mainly M-Pesa), just over half (52 %) had used the service (Kirui et al. 2010). Only 13 % of the money sent was used to pay for farming-related items, such as inputs (7 %) and farmworkers (6 %).

With regard to other financial services, including credit, savings and insurance, only a few m-services are on offer. Mobile payment providers have also recently started collaborating with local banks to provide other banking services. *Orange Money* (Telkom Kenya and Equity Bank) and *M-Swhari* (Safaricom, Commercial Bank of Africa and Vodafone), for instance, offer micro-loans and savings accounts (including interest) to their users. Two mobile phone-assisted insurance plans are available in Kenya, both of which insure farmers against extreme weather events that might affect livestock (index-based livestock insurance) or crops (*ACRE*).

None of the m-services reviewed here focus on input provision. *M-Farm* initially offered a service for collective sourcing of fertilizer, but put the service on hold due to liquidity constraints among farmers. *NAFIS* and *ArifuMkulima* provide price information for inputs, but it is unclear to what extent this function is operational.

Finally, several types of m-services aim at facilitating access to output markets. A number of information services have been developed in recent years which disseminate price information to farmers via SMS (*M-Farm*, *Sokopepe*, *SokoShambani*), USSD (*KACE*) and websites. Kenyan farmers are also able to sell their produce through internet- and SMS-supported selling platforms (*Sokopepe* and *M-Farm*).

In addition, the Kenyan company Virtual City uses mobile phones as part of their supply chain management system (*Agrimanagr*), allowing clients to record and track produce from delivery to final destination. A review of *Agrimanagr*'s performance showed that the system had reduced the delay in payments to farmers to 31 days (an improvement of 89 days) due to a faster consolidation of reports, cut purchasing times from 3 min to 22 s, and increased the average produce weight per transaction by 9–13 % with the use of electronic weighing technologies (Virtual City 2009). In addition, the Syngenta Foundation has trialed its supply chain management system *Farmforce* in Kenya which uses mobile phones to track deliveries from smallholder farmers to buyers.

Mobile Technology Outlook

Many of the m-services currently available in developing countries (including Kenya) are barely scratching the surface of what is technologically possible. With smartphone penetration and 3G networks still limited in many rural areas, most mobile applications for agriculture in developing countries are designed for low-tech mobile phones and delivery technologies such as SMS or voice services (Hatt et al. 2013; Qiang et al. 2011). Technologies being applied in precision agriculture, which employs ICT tools to monitor intra-field variations and manage crop production accordingly, offer a glimpse of the potential of modern ICTs to boost agricultural productivity. To date, however, adoption rates of these technologies have not lived up to expectations, even in countries with more advanced agricultural sectors, let alone among small-scale farmers (McBratney et al. 2005).

Recent technological advances could help to increase the use of modern ICT tools in agriculture. Technologies such as smartphones, tablets and sensors are becoming cheaper and, thus, more affordable for lower income users in the developing world. Mobile networks are also improving. In Africa, for instance, close to $4 billion (US) have been invested in new submarine cables, almost doubling the data capacities in just 2 years (Schumann and Kende 2013). By 2012, 40 % of the Sub-Saharan African population lived within 25 km of an operational fiber node following a roll-out of terrestrial fiber optic cables across the continent (Hamilton Research 2012). While rural areas still lag behind urban areas in terms of network coverage and speed, the gap is slowly closing. Improving access to hardware and infrastructure could lay the foundation for exploiting new mobile technology trends in agriculture.

Diversity of Mobile Connected Devices

The diversity of devices for accessing mobile services has been increasing in recent years. While basic and feature phones are still most prevalent in the developing world, high demand for mobile internet and price declines are expected to drive smartphone adoption in the future. Tablets may also offer a viable alternative to PCs in developing countries, in particular, lower-cost tablets now being produced in emerging economies, such as India and China. The expansion of these devices will support the growth of cloud computing, which is changing the way that m-services are used on personal mobile devices. The underlying idea of cloud computing is to offer computing, storage and software 'as a service' rather than running them on local IT infrastructure (Voorsluys et al. 2011). The mobile device then functions simply as an interface to access the service, which is run elsewhere, thus requiring less processing power than that required to run it on the device.

Many more and increasingly sophisticated agricultural m-services can be envisaged that take advantage of the technological capacities of different mobile devices, the enhanced computing powers of devices that use cloud- and web-based services, and the ability to access a service from multiple devices. For instance, smartphones or tablets can convey larger amounts of information than can be sent through an SMS, e.g., on different farming techniques, input suppliers, potential buyers or market prices. Cloud- and web-based services allow users to run more complex applications, e.g., to analyze price trends or access detailed weather forecasts. Web-based banking services could also enable farmers to make m-payments and access their account through multiple mobile devices.

Internet of Things

A technology trend that is predicted to revolutionize the way people live and work is the Internet of Things (IoT). In the IoT, "sensors and actuators embedded in physical objects . . . are linked through wired and wireless networks, often using the same Internet Protocol (IP) that connects the Internet" (Chui et al. 2010, 1). The underlying idea is not necessarily new. As the OECD (2012, p. 8) notes: "From the earliest days, in the use of information technologies, computers have processed signals from external sources". What has changed is the sheer scale, enabled through the declining cost and size of the required technologies, the use of the Internet Protocol, ubiquitous networks and significant increases in storage and computing powers (including cloud computing) (Chui et al. 2010; OECD 2012). As a result, communication modules can now be installed in nearly any device, thus allowing the internet to expand into previously unreachable places (Evans 2011).

In agriculture, the IoT has found application in precision agriculture (even if the terminology of the IoT is not necessarily used, especially in the early days of precision agriculture). Through the use of ICTs (such as global positioning and information systems, remote sensing or sensors to monitor climatic conditions, soils or yield), farmers can detect temporal and spatial variability across their fields to selectively treat their crop, either manually or through technologies that adjust their behavior in response to the gathered data. Uptake of these technologies in developing countries has, to date, been limited (Arab 2012). However, the rapid spread of mobile phones and networks, as well as advances in the IoT and related technologies, could lead to technology applications that are better adapted to the needs and capacities of small-scale producers.

A few examples of existing (small-scale) applications in developing countries highlight the potential of these technologies. For instance, data collection applications for mobile phones, such as *EpiCollect*, *Magpi* and *ODKCollect*, employ geo-tagging (using the phone's GPS) to gather location-specific data. For example, Makerere University in Uganda is using ODK Collect to automatically diagnose and monitor the spread of cassava mosaic disease (Quinn et al. 2011). In Kenya, GPS tracking devices attached to one cow in the herd enable livestock owners to monitor the movement of their animals and recover stolen cattle (Africa Agriculture News 2013). IoT devices are also being deployed in supply chain management. Virtual City's *Agrimanagr* and *Distributr* systems use mobile phones to collect data when farmers deliver the produce, e.g., weight and location (through GPS), and track the produce throughout the chain to the processing plant.

Capitalizing on Big Networks

The ubiquity of cellular networks coupled with the expanding reach and diversity of mobile devices will offer significant opportunities to collect, disseminate and exchange data and knowledge. Mobile connected devices are already gathering large amounts of data, e.g., on the location of the caller or calling patterns. In addition to these incidentally collected data, the devices can also be valuable sources of specifically collected data, e.g., through data collection tools or obtained through various IoT technologies (e.g., sensors, images or GPS tracking devices). Cloud-based services will facilitate the storage and analysis of such data and mobile devices can then be used as channels to disseminate the analyzed data. In agriculture, data collection could be used, e.g., to monitor crop disease outbreaks, gather information about input suppliers and prices, or collect information about crop damage from severe weather events for insurance purposes.

ICTs are also facilitating social networking and learning. Several initiatives have begun emerging in the agricultural sector which use ICTs to support social learning among farmers. In India, for instance, *Lifelong Learning for Farmers* offers learning modules as recorded audio content delivered to female livestock producers through mobile phones (World Bank 2011). Also in India, *Digital Green* recruits farmers to record videos with testimonials and demonstrations of farming techniques, market linkages or government policies which are distributed via the website and shown in villages using battery-powered projectors.[17] Another example is *Sauti ya wakulima* (The Voice of the Farmers)[18] in Tanzania which was initiated by a small group of farmers who share two smartphones to publish images and voice recordings about their farming practices on the internet.

Conclusions

Kenya has firmly established itself as the main ICT hub of East Africa and one of the leading ICT centers on the continent. While the sector is not without its challenges, it should also be born in mind that it is still early days for Kenya's technology scene. Start-ups need time to grow into full scale businesses and investors need time to build sufficient trust in the viability of the sector. To what extent this development will assist or even transform Kenya's agriculture sector is still an open question. While a number of m-services are already available to farmers, most are still in the pilot phase and their effectiveness and reach have not been assessed.

Moreover, most of the available m-services use simple delivery technologies. Current technology trends offer numerous opportunities to develop more sophisticated m-services for farmers. However, many of the new technological opportunities have not yet been realized in practice – neither in industrialized or in developing countries. It will be important to understand which of these technologies can realistically be applied to promote agricultural development in developing countries and which are most relevant in the given context. M-service developers will also need to ensure that their services continue to cater to a broad range of users rather than focusing overly on technologies that may not be within the reach of less-resourced farmers.

It is also important to stress that m-services can only ever be part of a broader solution. Farmers in the developing world face a multitude of challenges, some of which can be addressed through m-services, but many others of which cannot. Therefore, m-service should be embedded in complementary support programs and infrastructure developments to tackle other production and marketing limitations. Such complementary measures do not necessarily need to be implemented by the m-service provider, but can be the responsibility of other actors, such as companies, non-governmental organizations or government departments.

---

[17] www.digitalgreen.org.

[18] sautiyawakulima.net.

Open Access This chapter is distributed under the terms of the Creative Commons Attribution-Noncommercial 2.5 License (http://creativecommons.org/licenses/by-nc/2.5/) which permits any noncommercial use, distribution, and reproduction in any medium, provided the original author(s) and source are credited.
The images or other third party material in this chapter are included in the work's Creative Commons license, unless indicated otherwise in the credit line; if such material is not included in the work's Creative Commons license and the respective action is not permitted by statutory regulation, users will need to obtain permission from the license holder to duplicate, adapt or reproduce the material.

References

Abadi Ghadim A (1999) A conceptual framework of adoption of an agricultural innovation. Agric Econ 21(2):145–154. doi:10.1016/S0169-5150(99)00023-7

Africa Agriculture News (2013) Kenya uses GPS to tackle war against cattle theft. Africa Agriculture News, 11 Apr 2013

Afrinnovator (2012) Kenya technology, innovation & startup report 2012: a preview. Afrinnovator, Nairobi

Apoyo Consultoria (2011) Study on ICT access gaps in Kenya. Communications Commission of Kenya, Nairobi

Arab F (2012) The other M2M opportunity: enhanced utility access in emerging markets. GSMA – Mobile for Development, 3 Oct 2012

Banks K (2013) The truth about disruptive development. Stanf Soc Innov Rev, 16 Jan 2013

Bernard M, Hellin J, Nyikal RA, Mburu JG (2010) Determinants for use of certified maize seed and the relative importance of transaction costs. Presented at the AAAE third conference/AEASA 48th conference, African Association of Agricultural Economists (AAAE) and Agricultural Economics Association of South Africa (AEASA), Cape Town, 19–23 Sept 2010

Boyera S (2012) Mobile ICT & small-scale farmers? Voice Web, 28 Dec 2012

CA (2015) Fourth Quarter Sector Statistics Report for the Financial Year 2014/2015 (April-June 2015). Communications Authority of Kenya, Nairobi.

CCK (2013) Quarterly sector statistics report – third quarter of the financial year 2012/13 (Jan–Mar 2013). Communications Commission of Kenya, Nairobi

Chui M, Löffler M, Roberts RP (2010) The internet of things. McKinsey Q. https://www.mckinseyquarterly.com/The_Internet_of_Things_2538. Accessed 27 July 2011

Deloitte (2012) 2011 East Africa private equity confidence survey: promising 2012. Deloitte Consulting, London

Eggleston K, Jensen R, Zeckhauser R (2002) Information and communication technologies, markets and economic development. Discussion papers series No 0203. Department of Economics, Tufts University, Medford

Evans D (2011) The internet of things: how the next evolution of the internet is changing everything. Cisco Internet Business Solutions Group, San Jose

Foster C (2012) The other ICT "hub" in Nairobi. iHub, 17 Dec 2012

Geertz C (1978) The bazaar economy: information and search in peasant marketing. Am Econ Rev 68(2):28–32. doi:10.2307/1816656

Hamilton Research (2012) Africa's fibre reach increases by 32 million, to 40 % of population. Africa Bandwidth Maps, 13 Aug 2012

Hatt T, Wills A, Harris M (2013) Scaling mobile for development: a developing world opportunity. GSM Association, London

Hersman E (2012) Community connectedness as a competitive advantage. WhiteAfrican, 29 Nov 2012 iCow (2010) iCow impact study results. iCow. http://icow.co.ke/blog/item/15-icow-impact-study-results.html. Accessed 17 Aug 2012

ILO (2011) Economically active population estimates and projections, 6th edn. International Labour Organization, Geneva

Jayne TS, Yamano T, Nyoro JK (2003) Interlinked credit and farm intensification: evidence from Kenya. Presented at the international association of agricultural economists 2003 annual meeting, International Association of Agricultural Economists, Durban, 16–22 Apr 2003

Kenya ICT Board (2012) Connected Kenya 2017 national ICT master plan. Kenya ICT Board, Nairobi

Kieti J (2012) Evolving thoughts on innovation, entrepreneurship and economic growth. gmeltdown, 7 May 2012

Kirui OK, Okello JJ, Nyikal RA (2010) Awareness and use of m-banking services in agriculture: the case of smallholder farmers in Kenya. Presented at the 2010 AAAE third conference/AEASA 48th conference, African Association of Agricultural Economists, Cape Town, 19–23 Sept 2010

KNBS (2006) Kenya Integrated Household Budget Survey (KIHBS) 2005/06 – basic report. Kenya National Bureau of Statistics, Nairobi

Macharia K (2013) Why Intel chose Nairobi as Africa launch for YOLO smart phone. Cap. Lifestyle, 25 Jan 2013

Makau T (2012) The bane of "consultants" on African Telcos fortunes. Tom Makau, 17 Dec 2012

Mbuvi D (2013) Kenya launches national ICT master plan 2017, aims to connect all. CIO East Africa, 14 Feb 2013

McBratney A, Whelan B, Ancev T, Bouma J (2005) Future directions of precision agriculture. Precis Agric 6(1):7–23. doi:10.1007/s11119-005-0681-8

McCarthy D (2009) Cable makes big promises for African Internet. CNN.com/technology, 27 July 2009

Mumo M (2013) Sh74 billion needed to bridge Kenya's yawning digital divide. Dly Nation, 28 May 2013

Mutua W (2012) Mending Africa's Tech skills gap & tapping into it's youthful population to power innovation in Tech & the African renaissance. Afrinnovator, 3 Feb 2012

OECD (2012) Machine-to-machine communications. OECD digital economy papers. Organisation for Economic Co-operation and Development, Paris

Okello JJ, Okello RM, Ofwona-Adera E (2010) Awareness and the use of mobile phones for market linkage by smallholder farmers in Kenya. In: Maumbe BM (ed) E-agriculture and E-government for global policy development. Information Science Reference, Hershey, pp 1–18

Okoth P (2013) Research-based advice for farmers. ICT update, http://ictupdate.cta.int/Feature-Articles/Research-based-advice-for-farmers/%2870%29/1359544522. Accessed 13 Feb 2013

Okutoyi E (2012) Kenya to launch strategic plan for faster broadband rollout. HumanIPO, 14 Jan 2012

Okuttah M (2012) Cable repair work set to slow down Internet speeds. Bus Dly, 31 Dec 2012

Portland Communications (2012) New research reveals how Africa Tweets. Portland Communications, Portland

Qiang CZ-W, Kuek SC, Dymond A, Esselaar S (2011) Mobile applications for agriculture and rural development. World Bank, Washington, DC

Quinn JA, Leyton-Brown K, Mwebaze E (2011) Modeling and monitoring crop disease in developing countries. Presented at the twenty-fifth AAAI conference on artificial intelligence, San Francisco, 7–11 Aug 2011

Safaricom (2016) Safaricom Ltd. H1 FY16 Presentation. Safaricom, Nairobi

Sato N (2013) Kenyans dominate as PIVOT East 2013 winners are announced. HumanIPO, 27 June 2013

Schumann R, Kende M (2013) Lifting barriers to internet development in Africa: suggestions for improving connectivity. Analysys Mason Limited, London

Stigler GJ (1961) The economics of information. J Polit Econ 69(3):213–225. doi:10.1086/258464

Svensson J, Yanagizawa D (2009) Getting prices right: the impact of the market information service in Uganda. J Eur Econ Assoc 7(2–3):435–445. doi:10.1162/JEEA.2009.7.2-3.435

Virtual City (2009) Virtual city agrimanagr case study. Virtual City, Nairobi

VisionMobile (2012) Developer economics 2012: the new mobile app economy. VisionMobile, London

Von Braun J, Torero M (2006) Introduction and overview. In: Torero M, von Braun J (eds) Information and communication technologies for development and poverty reduction: the potential of telecommunications. Johns Hopkins University Press, Baltimore, pp 1–20

Voorsluys W, Broberg J, Buyya R (2011) Introduction to cloud computing. In: Buyya R, Broberg J, Goscinski A (eds) Cloud computing. Wiley, Hoboken, pp 1–41

Waema T, Adeya C, Ndung'u MN (2010) Kenya ICT sector performance review 2009/2010 (No 10), Policy paper series towards evidence–based ICT policy and regulation, vol 2. researchICTafrica.net, Johannesburg

Wahito M (2012) China to fund Kenya's fibre optic project. CaptialFM Bus, 28 June 2012

Wesolowski A, Eagle N, Noor AM, Snow RW, Buckee CO, Gómez S (2012) Heterogeneous mobile phone ownership and usage patterns in Kenya. PLoS One 7(4):e35319. doi:10.1371/journal.pone.0035319

World Bank (2010) Kenya economic update No 3. World Bank, Washington, DC

World Bank (2011) ICT in agriculture: connecting smallholders to knowledge, networks, and institutions. World Bank, Washington, DC

World Bank (2012) Information & communications technologies – IC4D 2012: maximizing mobile. World Bank, Washington, DC

Chapter 10
Identification and Acceleration of Farmer Innovativeness in Upper East Ghana

Tobias Wünscher and Justice A. Tambo

Abstract The generation of innovations has traditionally been attributed to research organizations and the farmer's own potential for the development of innovative solutions has largely been neglected. In this chapter, we explore the innovativeness of farmers in Upper East Ghana. To this end, we employ farmer innovation contests for the identification of local innovations. Awards such as motorcycles function as an incentive for farmers to share innovations and develop new practices. The impact of Farmer Field Fora is evaluated by matching non-participants to participants using propensity scores of observable characteristics. The results indicate that farmers do actively generate and test innovative practices to address prevalent problems. Moreover, this innovative behavior can be further stimulated by Farmer Field Fora, which were tested to significantly and positively affect innovation generation.

Keywords Innovation policy • Award • Contest • Upper East Ghana • Innovation behavior

Introduction

Global change forces farmers to adapt more rapidly to changing conditions than ever before. The generation of innovations that address these global challenges can be part of the adaptation portfolio. Earlier work of ours has established a robust causal relationship between farmer innovativeness and the resilience of farmers in terms of increased household income, consumption expenditure, food security, and reduction of the length of food shortages and the severity of hunger (Tambo and Wünscher 2014). While innovations are traditionally developed by research organizations for adoption by farmers, the farmer's own potential for the generation of innovative solutions has largely been neglected. Yet, farmer innovations have the advantage of having been developed within the environment in which the farmer operates. As such, they are likely to be adapted to local constraints and can be

T. Wünscher (✉) • J.A. Tambo
Center for Development Research (ZEF), University of Bonn, Bonn, Germany
e-mail: tobias.wuenscher@uni-bonn.de; tambojustice@yahoo.com

© The Author(s) 2016
F.W. Gatzweiler, J. von Braun (eds.), *Technological and Institutional Innovations for Marginalized Smallholders in Agricultural Development*,
DOI 10.1007/978-3-319-25718-1_10

expected to have good dissemination potential. Externally developed practices, on the other hand, i.e., those that were developed by non-farmer institutions such as universities, national and international research centers, often do not effectively address binding constraints among smallholders (Christensen and Cheryl 1994).

A local farmer innovation is defined here as a technology, practice or institution along the food chain which is different from common or traditional practice and which is developed primarily by a farmer or a group of farmers without external assistance, such as by extension agents, researchers or development workers. Likewise, a local farmer innovator is someone who has developed an innovation as defined above. In this, our definition is different from the one used by Rogers (2003) and the general adoption literature where an innovator is usually referred to as the farmer who is among the first to adopt a newly introduced technology.

Capital and formal knowledge constraints, as well as risk aversion and other factors, also set limits on what a farmer can do in terms of generating innovations. Farmer-based innovations are, therefore, to be seen as a complement and not a substitute to the traditional innovation system.

In this chapter, our objective is to assess whether farmers in one of the poorest regions of Ghana (Upper East Ghana) do, in fact, generate local innovations, and whether a newly introduced problem-solving instrument (Farmer Field Fora) can further stimulate innovative behavior, thereby increasing adaptation potentials for global change. Farmer Field Fora (FFF) are a platform for mutual learning and the development of technological and managerial solutions among agricultural stakeholders, particularly farmers, extension agents and researchers (Gbadugui and Coulibaly 2010). For the identification of local innovations, we employ a farmer innovation contest. Awards such as motorcycles function as an incentive for farmers to share innovations.

Awards are required under certain circumstances to overcome the secrecy of an innovation, or if innovations are simply not observable in the field. Reasons for secrecy include, for example, if the innovation gives the innovator a commercial advantage (Scotchmer 2004).

The paper continues with a section that outlines the implementation steps of the farmer contest and presents first results. Section "Impact Evaluation of Farmer Field Fora" presents the study details and results of the impact evaluation of Farmer Field Fora. We close in section "Conclusion".

Farmer Innovation Contest

The farmer innovation contest was implemented in Upper East Ghana. All farmers in Upper East Ghana were eligible and women were particularly encouraged to apply. Awards such as motorcycles, water pumps and roofing sheets served as incentives to share innovations with us. The contest was primarily announced through the local extension service of the Ministry of Food and Agriculture (MOFA). In workshops, extension agents were informed about the details of the contest. The extension agents' role was to spread the information within the study area, search for innovations, help farmers fill out the application form, and deliver the application forms to us. The extension agents were incentivized with a monetary award for each eligible application submitted. Received applications were scored by an independent selection committee. The selection committee consisted of eight members from four local stakeholder groups (farmers, MOFA, NGOs and research), each with two representatives. The selection committee members scored the applications on four criteria, namely innovativeness, economic potential, dissemination potential, and environmental and social sustainability. Scores ranged from zero to three. Zero represented no compliance (e.g., not innovative) and three represented highest compliance (e.g., highly innovative). If an application received zero for innovativeness, it was excluded from further consideration. Otherwise, the scores were added up and applications with the highest overall score were shortlisted for field visits. In the field, the selection committee members interviewed the applicant and, where appropriate, neighbors and other family members. The winners were then selected by the committee members in a final workshop. The awards were handed over in a ceremony on National Farmers' Day, which is organized by MOFA.

Between 2012 and 2013, we received 92 eligible applications (see Appendix 1). Table 10.1 shows the majority of applicants to have been male and, with a mean age of 47, mature and experienced farmers. Only three applications were received from farmer groups. We only received two applications with institutional innovations. All but two applications described innovations that were technical in nature. On average, the techniques were developed and implemented approximately a decade before the contest. This indicates that the innovations were not developed in response to the contest. The contest rather identified already existing innovations.

Most of the applications received addressed problems in animal husbandry, followed by post-harvest techniques for the storage of grain and seeds and the processing into higher level products (silage and yoghurt) (Table 10.2). Innovations in animal husbandry and crop management mostly addressed animal health and phytosanitation using local herbs. The effectiveness of these health- and phytosanitation-oriented innovations were generally difficult to assess within the context of the short field visits of the innovation contest because their functioning depended on often unknown ingredients of the herbs and their effectiveness could not intuitively be judged. All innovations required further evaluation in scientific

Table 10.1 Descriptive statistics of 92 applications (standard deviation in brackets, if applicable)

Variable	
Proportion of males (%)	79
Mean age of applicants (years)	47 (13)
Number of group applications	3
Proportion of technical innovations	98 %
Mean year of development	2001 (14)
Mean year of implementation	2003 (13)
Mean number of adopters	51 (109)

Table 10.2 Type of applications received in 2012 and 2013

Type of application	#	Comments
Animal husbandry		
Poultry	27	20 of these to treat sicknesses, 2 for feeding, 4 for breeding, poultry housing (1)
Livestock	12	of these, 12 for treatment of sicknesses
Fish	1	Fish feed formula
Other	1	Dog treatment
Subtotal	41	
Storage		
Treatment of seed/ grain	14	Use of different plants as treatment agent: Neem (4), Barakuk (1), Ash (1), Sheatree bark (1), Kwasuik plant (1), Dabokuka plant (1), Salt solution (3), Bicycle tubes (1), other (1)
Fermentation	1	Production of silage
Storage management	1	Cooling and ventilation of sweet potatoes
Subtotal	16	
Crop management		
Phytosanitary	15	Treatment of pests, termites, nematodes and weeds using Neem, Yookat, onion, tiger ants, Wacutik plant, diesel mixture, salt (9), Gloriosa fruit (1), other (2), prevention of pests applying onion seed inoculation (1), Neem leaves liquid spraying (1), prevention of worm infestation using millet seeds (1)
Production of planting material	1	Multiplication of sweet potato
Introduction of new crops	3	Introduction of crops from the south of Ghana (2), mushroom production (1)
Irrigation	1	Growing maize in dry season
Subtotal	20	
Water/soil conservation		
Water conservation	1	Recycling of fish pond water for irrigation and fertilization
Soil conservation	4	Use of innovative mixture of animal dung, liquid manure, zero tillage
Subtotal	5	
Processing & Marketing		
Adding value	1	Making yoghurt from cow's milk
Subtotal	1	
Trees and Forest	4	Forest management and conservation, afforestation, trees for control of microclimate
Other	5	Human health, farm products against Malaria, community-based extension agents, use of dogs for animal security, repellent for snakes
Total	92	

trials, but within the evaluation of the contest, scoring was based on intuition, observation, conviction and trustworthiness of the applicant. We also received a couple of innovations in soil and water conservation. For illustration, we present some of the innovations in more detail below.

Case 1: Using Fish Pond Water as Liquid Manure and Insecticide

Joseph Abarike Azumah, a fish farmer from Zuarungu, uses animal droppings such as cow, sheep and goat dung, as well as poultry manure, to supplement his locally-prepared fish feed. The fish then feed on the dung and add their own feces to the water. The water is recycled for gardening as natural manure. The innovation addresses the problem that nutrient rich water would normally be lost if it was released into the environment without further use. Its use as liquid fertilizer reduces the dependence on artificial fertilizer and also reduces the environmental impact. It is possible to combine this technique with the treatment of pests by soaking neem tree leaves in the pond in moderate quantities for it to be non-toxic to the fish. The water then acts as insecticide. The system was developed and implemented in 2008 and has been adopted by 25 farmers since.

Case 2: Use of 'Barakuk' to Store Seed

The Barakuk herb is harvested, dried and burned. The ash is then mixed with onion seed to prevent insects from attacking the seed. The process improves germination. Access to the herb was a problem during the development stage. However, the innovator, John Akugre Anyagre from Tilli, also experimented and succeeded in growing the herb on-farm, making the material readily available. One hundred and twenty farmers are known to have adopted the technique.

Case 3: Controlling Striga in Millet and Sorghum Fields Using Dried Onion Leaves

Striga is a common and severe problem in Africa. Abdul Rhaman Abieli from Missiga discovered that areas on his millet and sorghum fields where his family had dumped the leafy residues of onion production were free of striga in the following year. In order to scale up the application of onion leaves, they experimented with smaller quantities of onion and found the effect to persevere. Today, the onion leaves are pounded into powder and then mixed with the seed of millet or sorghum. Small amounts of water are sprinkled onto the powder to help it stick to the seed. One ball of dried onion leaves, the size of a fist, is enough to treat the seeds for one acre of onions. These small quantities rule out a fertilization effect. The innovation has been functional since 2001 and is known to have been adopted by approximately 50 farmers.

Impact Evaluation of Farmer Field Fora

As already indicated, this section addresses the impact of Farmer Field Fora on farmer innovativeness.

Farmer Field Fora

Farmer Field Fora (FFF) of the Root and Tuber Improvement and Marketing Programme (RTIMP) in Ghana are based on the successful implementation of the Root and Tuber Improvement Programme (RTIP) between 1999 and 2005. The RTIMP was initiated as a follow-up project, with major funding from the International Fund for Agricultural Development (IFAD). The RTIMP supports root and tuber crop production, increased commodity chain linkages and upgrading of technologies and skills within the value chain. The aim is to enhance income and food security to improve livelihoods of the rural poor and to build a market to ensure profitability at all levels of the value chain.

The RTIMP used the FFF as a platform for mutual learning among stakeholders in the root and tuber value chain, particularly farmers, extension agents and researchers. The main aim of FFF is to "build the capacities of farmers to become experts in the development of technologies and managerial practices to solve specific problems within the agro-ecological context of farming" (Gbadugui and Coulibaly 2010). It is a variant of the well-known Farmer Field School (FFS), a participatory extension model. The FFS approach was first introduced in Indonesia in the late 1980s by the FAO to help farmers deal with the pesticide-induced pest problems in irrigated rice, but has since spread to at least 78 countries and is highly promoted by many development agencies (Braun et al. 2006). Though it was mainly introduced to promote integrated pest management (IPM) practices in rice farming, its methods have been adapted to suit different farming activities and even non-farm topics in Africa (Braun et al. 2006; Davis et al. 2012). Unlike FFS, which gives little or no attention to farmer-developed innovations (Reij and Waters-Bayer 2001), FFF provides an opportunity for farmers to experiment with their own innovations, thereby strengthening their decision-making and innovation capacities.

The RTIMP-FFF in Ghana, which started in 2006, aims at improving farmer innovation and productivity of root and tuber crops in major production districts of the country. In each participation district, the FFF was developed for the most important root or tuber crop. This study is based on the sweet potato FFF in ten communities in three northern districts of Ghana. The main actors include researchers, extension agents, business advisors, farmers and processors, and they are all placed on an equal footing. During a participatory rural appraisal, the farmers determine the theme of the FFF, thereby ensuring that their priorities are addressed. The thematic areas normally selected by the farmers include improved crop varieties, integrated pest management (IPM), improved cultivation practices and integrated soil fertility management. There are also discussion sessions on non-farm topics. Each forum consists of a group of 30–40 farmers together with other key actors who meet regularly (usually weekly) in the field during a growing season. They engage in comparative experimentation using three plots: farmers practice (FP), integrated crop management (ICM) and participatory action research (PAR), with the assistance of a facilitator who stimulates critical thinking and discussions and ensures active participation. The participating farmers experiment with their own innovations or test new ideas on the PAR plots. Conventional practices and improved innovations are implemented on the FP and ICM plots, respectively.

There are many studies looking at the impact of farmer field schools (FFS) on outcome variables such as empowerment, technology adoption, household income and food security, but with inconclusive findings (for a review, see Braun et al. 2006; Davis et al. 2012, Table 10.1). Within this vast literature, however, there is little, if any, on the farmer innovation effects of FFS. This chapter provides empirical evidence on the potential of FFF, a variant of FFS, in stimulating innovation-generating behavior among farm households.

Empirical Method

We are interested in estimating the effect of FFF participation on farmer innovation. The challenge is that participation in FFF is voluntary; hence, farmers self-select to participate. Thus, participating farmers may differ systematically from non-participants in observed characteristics such as education, age and wealth, and unobserved characteristics such as entrepreneurship, risk behavior or motivation which might lead to biased estimates of the effect of FFF on innovation. Due to the self-selection bias, participants and non-participants are not directly comparable. To minimize this problem, we use propensity score matching (PSM), a non-parametric technique suggested by Rosenbaum and Rubin (1983). It involves matching FFF participants with non-participants who are similar in terms of observable characteristics (Caliendo and Kopeinig 2008). Though it only accounts for observables, it is less restrictive, as it does not impose any functional form assumption, which is a challenge with other estimation techniques, such as instrumental variable regression. We also try to minimize the bias stemming from unobserved heterogeneity by controlling for household risk preferences.

In the PSM, a probit regression was estimated using several covariates to obtain a household's propensity to participate in FFF. These covariates comprise household socio-demographic and economic variables (e.g., age, gender and education of the household head; household size and dependency ratio; access to services and the wealth position of the household). It also includes households' risk preferences.[1] We then use the propensity scores obtained in the first stage to match participants and non-participants in FFF. As a matching algorithm, we used kernel matching with a bandwidth of 0.3, but, for the robustness check, radius matching with a caliper of 0.05 and nearest-neighbor matching are also employed.[2] We conducted a matching quality test (Rosenbaum and Rubin 1983) to check if the balancing

---

[1] We measured households' subjective risk preferences using the Ordered Lottery Selection Design with real payoffs (Harrison and Rutström 2008).

[2] For a review of the different matching techniques, see Caliendo and Kopeinig (2008).

property is satisfied. Based on the kernel matching,[3] the test result (in Appendix 2) shows that, in contrast to the unmatched sample, there are no statistically significant differences in covariates between participants and non-participants in FFF after matching. Thus, the balancing requirement is satisfied. Using the PSM, we compute the average treatment effect on the treated (ATT):

$$ATT^{PSM} = E\big[Y(1) \mid FFF = 1, \, P(X)\big] - E\big[Y(0) \mid FFF = 0, \, P(X)\big] \qquad (10.1)$$

where $Y(1)$ and $Y(0)$ are the outcome variable (farmer innovativeness) for FFF participants and non-participants, respectively; FFF is a treatment indicator which is equal to 1 if the household is FFF participant and 0 otherwise; and $P(X)$ indicates the probability of FFF participation given characteristics X, which is obtained from the probit regression. The ATT measures the average difference in innovativeness between FFF participants and non-participants.

We use four different measures of the outcome variable, farmer innovativeness, to check if the results are sensitive to the indicator employed. The first (innovation_binary) is a binary variable which is equal to one if the household has, in the past 12 months, implemented any of the four categories of farmer innovation (i.e., invention of new practices or technologies, adaptation of exogenous ideas, modification of common or traditional practices and experimentation with new ideas), and 0 otherwise. The second (innovation_count) is a count variable that indicates the number of different innovation-generating activities implemented by a household in the past 12 months. In the third and fourth measure of FI, we consider the varied importance of each of the four categories of farmer innovation and constructed an innovation index using weights. In the third measure of FI (innovation index 1), we followed Filmer and Pritchett (2001) and used principal component analysis (PCA) to assign weights to each of the four innovation categories, and constructed a household innovation index. The final indicator (innovation index 2) also involves the construction of a household innovation index, but using weights obtained through expert judgements. A stakeholder workshop was organized, and 12 agricultural experts in the study region assigned weights to the four innovation categories based on an agreed level of importance for each category. They assigned weights of 0.4, 0.2, 0.3 and 0.1 for invention, adaptation of exogenous ideas, modification of traditional practices and experimentation, respectively.

---

[3] The other two matching estimators also yield similar results of matching quality, but are not reported, for brevity.

Data

The empirical analysis is based on data for the 2011–2012 agricultural season obtained from a household survey in the districts of Bongo, Kassena Nankana East and Kassena Nankana West in the Upper East Region, one of the poorest administrative regions of Ghana. The districts fall within the Sudan savanna agro-ecological zone. The area is characterized by a prolonged dry season and erratic rainfall. Agriculture is the main income source and a cereal-legume cropping system is predominant in the study region. The major crops are millet, sorghum, maize, cowpea, rice and groundnut. Most households also rear livestock.

The sample included FFF participants, non-participants from FFF communities and non-participants from control communities. We interviewed 409 households from 17 villages using a stratified random sampling. We first obtained from the district RTIMP project officers a list of all the 24 villages in the three districts where FFF had been implemented between 2008 and 2011. Then, we randomly selected ten participating villages across the three districts. We interviewed about 16–21 participants from each of these villages, resulting in a total of 185 FFF participants. We also obtained a list of all households in each of the FFF participating villages and randomly sampled and interviewed 99 non-participants across these villages. Since these non-participants are located in the FFF villages, they may potentially be exposed to some of the effects of FFF. To obtain a group of control farmers devoid of potential spillovers, we randomly selected seven villages (from the same three districts) that had similar infrastructural services and socio-economic conditions but not in close proximity to the FFF communities. Out of these, we randomly selected 125 farm households from a household list obtained from the District Agricultural Offices. Thus, our final sample consisted of 185 FFF participants and 224 non-participants, making a total of 409 sample farmers.

Data collection was conducted by experienced enumerators who were highly trained for this research. Interviews were conducted with the aid of pre-tested questionnaires and were supervised by the first author. The questionnaire captured data on household and plot characteristics, off-farm income earning activities, innovation-generating activities and access to infrastructural services, information and social interventions. The respondents were mainly FFF participants or household heads in the presence of other available household members.

Descriptive Statistics

In this section, we focus on four categories of farmer innovations. These are: developing new techniques or practices (hereafter, invention), adding value or modifying indigenous or traditional practices, modifying or adapting external techniques or practices to local conditions or farming systems and informal experimentation with original or external ideas. Thus, innovators are farm households

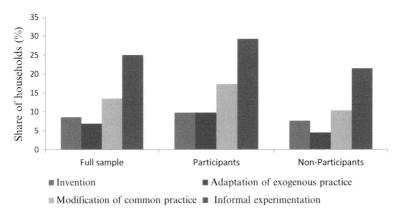

Fig. 10.1 Share of households that implemented innovation-generating activities

who have implemented any of these four categories of innovation-generating activities during the 12 months prior to the survey.

Figure 10.1 presents the share of households that implemented the four categories of innovation-generating activities and compares the results between participants and non-participants. Informal experimentation, which was implemented by 25 % of the sample households, constitutes the most practiced activity. A similar trend is observed when we compare the innovation activities of FFF participants and non-participants. This is expected, as experimentation is the first stage of most innovation processes. The figure also shows that, relative to non-participants, FFF participants implemented more innovation-generating activities in each of the the four categories, which seems to suggest that FFF participation enhances innovation capacity. Examples of innovations include: informal trials with or introduction of new crops or varieties in a community; testing and modification of planting distance and cropping pattern; using plant extracts as insecticide; new formulations of animal feed and new herbal remedies in the treatment of livestock diseases (ethnoveterinary practices); developing and using new farming tools; storage of farm products using local grasses; and new methods of compost preparation.

Table 10.3 outlines the description and mean values of the outcome indicators and variables used in estimating the propensity scores. The table shows that about 42 % of the sample households conducted at least one innovation-generating activity in the past 12 months.

Probability of FFF Participation

As mentioned, the first step in the PSM technique is the probit estimation of the propensity to participate in FFF, and the result is presented in Table 10.4. The result shows that FFF participation is influenced by household characteristics such as age,

Table 10.3 Description and summary statistics of variables

Variable	Description	Mean	SD
Outcomes			
Innovation_binary	Household has conducted innovation-generating activities (Binary)	0.42	0.41
Innovation _count	Number of innovation activities conducted by household (Count)	0.59	0.79
Innovation index 1	Household innovation index based on weights obtained through PCA	0.00	1.00
Innovation index 2	Household innovation index based on weights assigned by experts	0.13	0.21
Covariates			
Age	Age of household head	49.42	14.88
Gender	Gender of household head (dummy, 1 = male)	0.86	0.35
Household size	Number of household members	6.64	2.59
Dependency ratio	Ratio of members aged below 15 and above 64 to those aged 15–64	0.89	0.79
Education	Education of household head (years)	1.67	1.10
Land holding	Total land owned by household in acres	4.56	4.15
Livestock holding	Total livestock holding of household in Tropical Livestock Units (TLU)	2.92	3.41
Assets	Total value of non-land productive assets in 100 GH¢[a]	4.54	6.92
Off-farm activities	Household has access to off-farm income earning activities	0.76	0.43
Credit	Household has access to credit	0.26	0.43
Road distance	Distance to nearest all-weather road in km	0.54	0.84
Extremely risk averse	Household is extremely risk averse	0.40	0.49
Severely risk averse	Household is severely risk averse	0.22	0.42
Intermediately risk averse	Household is intermediately risk averse	0.14	0.34
Moderately risk averse	Household is moderately risk averse	0.04	0.20
Slightly to neutral risk averse	Household is slightly risk averse to risk neutral	0.11	0.32
Neutral to risk preferring	Household is risk neutral to preferring	0.09	0.30

[a]The exchange rate at the time of the survey was $1 (US) = GH¢ 1.90

gender of household head and household size. Participants are likely to be younger, and come from male-headed households of large size. Membership in a social group and credit accessibility also positively influence FFF participation. The negative and significant effect of road distance indicates that households living close to all-weather roads have a higher probability of participating in FFF. It is interesting to note that all the wealth-related covariates (i.e., land holding, productive assets, livestock holding and off-farm income) are not statistically significant. This seems to suggest that participation in FFF is inclusive of both resource-rich and resource-poor households. Finally, the result shows that a household's risk preferences do not affect FFF participation.

Table 10.4 Probit estimation of the propensity score

	Coefficient	Standard error
Age	−0.013***	0.01
Gender	0.381*	0.20
Household size	0.056**	0.03
Dependency ratio	0.057	0.08
Education	−0.013	0.02
Land holding	−0.019	0.02
Social group	0.368***	0.14
Livestock holding	0.019	0.02
Productive assets	0.000	0.00
Off-farm income	−0.136	0.16
Credit access	0.404***	0.15
Road distance	−0.221***	0.08
Severely risk averse	0.145	0.17
Intermediately risk averse	0.19	0.21
Moderately risk averse	0.237	0.35
Slightly to neutral risk averse	0.226	0.22
Neutral to risk preferring	0.343	0.24
Constant	−0.274	0.38
No. of observations	409	
LR chi^2(17)	46.75	
Prob > chi^2	0.000	
Pseudo R^2	0.083	

Effect of FFF Participation on Farmer Innovation

The estimated ATT is presented in Table 10.5. We find positive and significant effect of FFF participation on farmer innovation irrespective of the matching algorithm or how the outcome variable is measured. Using the kernel matching approach, for instance, the results show that the rate of innovation generation by FFF participants is 13.4 percentage points higher relative to matched non-participants. Furthermore, FFF participants are more likely to implement between 0.24 and 0.31 more innovations than non-participants, depending on the matching technique. Overall, the results suggest that FFF participation consistently and robustly enhances innovativeness in farm households.

We also conducted tests on the sensitivity of estimates to unobservable factors (Rosenbaum 2002). Running mhbounds for binary outcome variables (Becker and Caliendo 2007), for example, we obtained a critical value of gamma, $\Gamma = 1.40$ for kernel matching (model 1) which indicates that the ATT of 0.134 would be questionable only if matched pairs differ in their odds of FFF participation by a factor of 40 %.

Table 10.5 PSM estimation of the effect of FFF participation on farmer innovation

Matching algorithm[a]	Outcome	ATT	SE
Kernel matching	Innovation _binary	0.134***	0.051
	Innovation _count	0.239***	0.083
	Innovation index 1	0.268***	0.104
	Innovation index 2	0.054**	0.022
Radius matching	Innovation _binary	0.123**	0.055
	Innovation _count	0.235***	0.088
	Innovation index 1	0.255**	0.111
	Innovation index 2	0.054**	0.023
Nearest neighbour	Innovation _binary	0.178***	0.055
	Innovation _count	0.308***	0.089
	Innovation index 1	0.357***	0.112
	Innovation index 2	0.071***	0.024

***, **, * represent 1 %, 5 %, and 10 % significance level, respectively
[a]ATT estimates of kernel matching and radius matching were obtained by implementing 'psmatch2' command in Stata. ATT estimates of nearest neighbour matching were obtained using the 'teffects nnmatch' command with bias adjustment option in Stata 13

Conclusions

In this chapter, we explored the innovativeness of farmers in Upper East Ghana and evaluated whether farmer innovativeness can be stimulated by Farmer Field Fora, a platform for mutual learning and development of technologies and managerial skills. Using a farmer innovation contest with awards for the most innovative practices, we received 92 applications describing innovative and mostly technological approaches to farming. The results, therefore, indicate that farmers do actively develop innovations to address prevalent problems. Applying a propensity score matching approach, Farmer Field Fora were found to significantly and positively affect innovation generation. Overall, our results suggest good news with respect to the innovation capacity of farmers, and with respect to the ability of policy makers to foster this capacity. In light of global challenges such as climate change, fostering farmer innovation through Farmer Field Fora can therefore potentially act as a policy to enable autonomous adaptation.

Acknowledgements We thank the German Federal Ministry of Education and Research for funding within the WASCAL project (www.wascal.org). We are grateful for the support from the Ministry of Food and Agriculture (MOFA) in Ghana, as well as from the local farmers.

Open Access This chapter is distributed under the terms of the Creative Commons Attribution-Noncommercial 2.5 License (http://creativecommons.org/licenses/by-nc/2.5/) which permits any noncommercial use, distribution, and reproduction in any medium, provided the original author(s) and source are credited.
The images or other third party material in this chapter are included in the work's Creative Commons license, unless indicated otherwise in the credit line; if such material is not included in the work's Creative Commons license and the respective action is not permitted by statutory regulation, users will need to obtain permission from the license holder to duplicate, adapt or reproduce the material.

Appendices

Appendix 1: List of Applications Received in Innovation Contest Rounds 2012 and 2013, as Well as Additional 20 Innovations Identified in Surveys

Location	ID	Name/Brief description of innovation
Bolgatanga Municipal	1	Pisika (Cida acuta)
Bolgatanga Municipal	2	Brooder house for poultry (local fowls & guinea keets)
Bolgatanga Municipal	3	Treatment of animal eyes using 'yaae' roots or bark
Bolgatanga Municipal	4	Treatment of Alopecia using 'Sa-ire'
Bolgatanga Municipal	5	Treatment of livestock using periga, kuka, anriga trees
Bolgatanga Municipal	6	Organic manure farming
Bolgatanga Municipal	7	Using fishpond water as liquid manure and insecticide
Bolgatanga Municipal	8	Formulation of local fish feed
Talensi Nabdam	9	Production of yogurt from milk obtained from cattle
Talensi Nabdam	10	Extraction of neem oil from neem seed for the spray of crops to control pests
Talensi Nabdam	11	Preparation of silage for feeding livestock
Talensi Nabdam	12	Livestock feed formulation
Talensi Nabdam	13	Neem extracts from neem seed
Bawku West (Zebilla)	14	Reducing guinea keet mortality by using sorghum malt plus pepper water as dewormer
Bawku West (Zebilla)	15	Barakuk – a herb for treating livestock wounds
Bawku West (Zebilla)	16	Use of 'Yookat' herb to prevent and control termites
Bawku West (Zebilla)	17	Use of 'Barakuk' to store seed
Bawku West (Zebilla)	18	Combination of three herbs to treat fowl pox: 'Baker', 'Gbangdang' and 'Morag Kombri'
Bawku Municipal	19	Honey with mahogany for treatment of intestinal works in guinea fowl
Bawku Municipal	20	Use of dry onion leaves to control striga weed in millet & sorghum fields
Garu Tempane	21	Using neem seed oil for storage of crop seed
Garu Tempane	22	Herbal treatment for newly hatched chickens using 'Gbenatun' & Mango tree bark
Garu Tempane	23	Cowpea storage using wood ashes

(continued)

Location	ID	Name/Brief description of innovation
Kassena Nankana	24	Improving hatchability of guinea fowl eggs
Kassena Nankana	25	Improving survival rates of puppies
Kassena Nankana	26	Anti-snake weed or plant
Builsa	27	Sweet potato vine multiplication in artificial shade
Builsa	28	Traditional means of seed preservation
Bolgatanga Municipal	29	Introducing Southern crops to Bolgatanga municipality
Kassena Nankana	30	Predator control for poultry
Bongo	31	Planting and eating of Dawadawa fruit against the traditional belief of dying
Bongo	32	Using salt solution as a seed dresser
Bongo	33	Storing Bambara beans using solution from boiled shea tree bark
Bongo	34	Kuka (mahogany) bark for the treatment of chicken diseases
Bongo	35	Bicycle tube pieces with ku-enka for storage of seed and grain
Kassena Nankana East	36	Salt for controlling striga weed
Kassena Nankana East	37	Kenaf seed for hatching eggs
Kassena Nankana West	38	Salt to control termite in rice field
Kassena Nankana West	39	Peels of ebony and mahogany to control poultry diseases
Kassena Nankana West	40	Onion to control poultry disease
Kassena Nankana West	41	Compost preparation using a mixture of animal droppings and farm residue in a unique way
Kassena Nankana East	42	Control of nematodes using a mixture of salt solution and fungicide (rodimil plus)
Kassena Nankana East	43	Control of stubborn weed (Digitaria spp.) using a mixture of diesel, water and weedicide
Kassena Nankana West	44	Hatching of guinea fowl eggs using cotton and rag
Kassena Nankana East	45	Bark of Goa tree to treat Newcastle disease in poultry
Kassena Nankana West	46	Neem leaves to spray pepper and tomato against pests and diseases
Kassena Nankana West	47	Semi-intensive type of guinea fowl production, i.e., using mud to construct walls, and coops for laying with trees inside to provide shading
Kassena Nankana East	48	Spraying Gloriosa fruit solution to treat vegetable pests
Kassena Nankana West	49	Mixing millet seeds with dry cell content before planting to prevent worms from destroying the seeds

(continued)

Location	ID	Name/Brief description of innovation
Bolgatanga Municipal	50	Control of Newcastle Disease in poultry using Dawadawa
Bolgatanga Municipal	51	Prevention of fowl pox in poultry using Gubgo grass
Bolgatanga Municipal	52	Mushroom production in dry environment
Kassena Nankana West	53	Tree forest management
Kassena Nankana West	54	Pest management in pepper
Kassena Nankana West	55	Using secret groves to conserve forest
Kassena Nankana West	56	Preparation and application of liquid organic manure
Bawku Municipal	57	Preservation of Bambara nut
Bawku Municipal	58	Raw ebony fruit solution for treatment of fowl pox
Bawku Municipal	59	Preventing swollen gums and bleeding in animals
Bawku Municipal	60	Controlling worms and ticks using barakuk plant ruminants
Bawku Municipal	61	Treatment of foot and mouth disease in cattle using the "Pelinga" tree
Pusiga	62	Treatment of boils and skin diseases in ruminants using the bark of mahogany
Pusiga	63	Treatment of snake bite using water drained from boiled dawadawa seeds
Bawku Municipal	64	Destroying termites during storage
Pusiga	65	Controlling worms in dogs
Pusiga	66	Controlling rickets in chicks
Pusiga	67	Controlling worm infestation in guinea fowls using "Gberige" roots
Bawku Municipal	68	Treatment of chicken pox in poultry and fowl pox in poultry using henna paste solution
Binduri	69	Using millet ash solutions and salt petre solution to treat fowl pox
Binduri	70	Mahogany and neem extracts as water medications for poultry diseases
Binduri	71	Preservation and sweet potatoes
Binduri	72	Neem tree leaves to store maize
Builsa South	73	Treatment of guinea keets with kornamunig
Builsa North	74	"Kwasuik" plant for storage of seeds
Builsa North	75	Using striga plant as mosquito killer in rooms and surrounding
Builsa North	76	Deworming ruminants with "kpalik "plant
Garu Tempane	77	Growing maize in the dry season using residual rainfall and white Volta breeze
Garu Tempane	78	Community-based extension agents

(continued)

Location	ID	Name/Brief description of innovation
Garu Tempane	79	Onion seed and seedlings resistant to excessive rainfall and diseases
Garu Tempane	80	Zero tillage and fertilizer in water melon production in the dry season
Garu Tempane	81	Training dogs to watch tethered animals
Garu Tempane	82	Preventing termite attack on roots of seedlings (Mango, accasia) using earth worm
Garu Tempane	83	Maggot production for feeding chicks
Garu Tempane	84	Cyclical brooding fowl (increased brooding cycle)
Garu Tempane	85	Biological control of termites on young seedlings using tiger ants
Garu Tempane	86	Using artificial methods other than incubators to hatch eggs
Nabdam	87	All crop protection for storage using dabokuka plant
Nabdam	88	Deterring termites, especially on maize fields, using the "Wacutik" plant
Nabdam	89	Jetropher life fencing as snake repellent
Nabdam	90	Effect of micro climate in cocoe plant to fruit
Nabdam	91	Provision meeting ground (place) using afforestation
Nabdam	92	Livestock bones for poultry/pig feed formulation

Appendix 2: Test of Matching Quality (Kernel Matching)

	Unmatched			Matched		
	Participants	Non-participants	t-test	Participants	Non-participants	t-test
Age	47.03	51.81	3.20***	47.11	48.82	0.39
Gender	0.89	0.82	2.12**	0.89	0.88	−0.15
Household size	6.90	6.38	2.05**	6.86	6.61	−0.16
Dependency ratio	0.92	0.86	0.89	0.92	0.90	−0.05
Education	2.77	2.39	0.91	2.78	2.65	0.10
Land holding	4.51	4.60	−0.21	4.50	4.39	0.05
Social group	0.46	0.34	2.50**	0.46	0.41	−0.46
Livestock holding	3.02	2.56	1.37	3.03	2.63	−0.03
Assets	4.67	4.41	0.36	4.67	4.69	−0.12
Off-farm activities	0.76	0.75	0.05	0.76	0.77	−0.21
Credit access	0.32	0.19	3.10***	0.33	0.25	−0.13
Road distance	0.42	0.64	−2.55**	0.43	0.46	−0.04
Extremely risk averse (RA)	0.36	0.44	−1.66*	0.36	0.39	0.70
Severely RA	0.22	0.22	−0.17	0.22	0.23	−0.18
Intermediately RA	0.14	0.13	0.17	0.14	0.14	0.06

(continued)

	Unmatched			Matched		
	Participants	Non-participants	t-test	Participants	Non-participants	t-test
Moderately RA	0.05	0.03	0.90	0.05	0.04	−0.36
Slight to risk neutral	0.13	0.10	0.85	0.13	0.12	0.00
Neutral to risk preferring	0.11	0.08	1.30	0.11	0.08	−0.26
Median bias		9.10			3.10	
Pseudo R-squared		0.08			0.00	
p-value of LR		0.00			1.00	

References

Becker S, Caliendo M (2007) Sensitivity analysis for average treatment effects. Stata J 7(1):71–83

Braun A, Jiggins J, Röling N, van den Berg H, Snijders P (2006) A global survey and review of farmer field school experiences. International Livestock Research Institute, Nairobi/Endelea

Caliendo M, Kopeinig S (2008) Some practical guidance for the implementation of propensity score matching. J Econ Surv 22(1):31–72. doi:10.1111/j.1467-6419.2007.00527.x

Christensen C (1994) Agricultural research in Africa: a review of USAID strategies and experience, SD publication series, Technical paper No 3. Bureau for Africa, USAID Office of Sustainable Development, Washington, DC

Davis K, Nkonya E, Kato E, Mekonnen DA, Odendo M, Miiro R, Nkuba J (2012) Impact of farmer field schools on agricultural productivity and poverty in East Africa. World Dev 40 (2):402–413. doi:10.1016/j.worlddev.2011.05.019

Filmer D, Pritchett LH (2001) Estimating wealth effects without expenditure data-or tears: an application to educational enrollments in states of India. Demography 38(1):115–132. doi:10. 2307/3088292

Gbadugui BJ, Coulibaly O (2010) PRONAF's Farmer Field Fora (FFF). International Institute of Tropical Agriculture, Cotonou

Harrison GW, Rutström EE (2008) Risk aversion in the laboratory. Res Exp Econ 12:41–196

Reij C, Waters-Bayer A (eds) (2001) Farmer innovation in Africa. A source of inspiration for agricultural development. Earthscan, London

Rogers EM (2003) Diffusion of innovations, 5th edn. Free Press, New York

Rosenbaum PR (2002) Observational studies. Springer, New York

Rosenbaum PR, Rubin DB (1983) The central role of the propensity score in observational studies for causal effects. Biometrika 70(1):41–55. doi:10.1093/biomet/70.1.41

Scotchmer Z (2004) Innovation and incentives. The MIT Press, Cambridge, MA/London

Tambo JA, Wünscher T (2014) More than adopters: the welfare impacts of farmer innovation in rural Ghana. Paper presented at the 2014 Agricultural & Applied Economics Association's (AAEA) annual meeting, Minneapolis, 27–29 July 2014

Chapter 11
Gender, Social Equity and Innovations in Smallholder Farming Systems: Pitfalls and Pathways

Tina D. Beuchelt

Abstract Development processes, economic growth and agricultural moderniza-tion affect women and men in different ways and have not been gender neutral. Women are highly involved in agriculture, but their contribution tends to be undervalued and overseen. Sustainable agricultural innovations may include trade-offs and negative side-effects for women and men, or different social groups, depending on the intervention type and local context. Promising solutions are often technology-focused and not necessarily developed with consideration of gender and social disparity aspects. This paper presents cases of gender and social equity trade-offs related to the promotion and diffusion of improved technologies for agricul-tural development.The analysis is followed by a discussion of opportunities and pathways for mitigating potential trade-offs.

Keywords Gender • Marginality • Social disparity • Agricultural technologies • Women farmers

Introduction

Threats to future food security include climate change, overexploitation of natural resources, soil degradation and a change in demand structure for non-food uses of biomass. At the same time, the world is marked by enormous inequities in contem-porary living conditions (Anand and Sen 2000). Sustainable development and human development therefore need to go hand in hand given that *"sustainable development can only be achieved when both men and women have the opportuni-ties to achieve the life they choose"* (IISD 2013). However, development processes and economic growth have not been gender neutral; men and women are affected in different ways (Momsen 2010). The modernization of agriculture has changed the division of labour between women and men, often increasing women's dependent

T.D. Beuchelt (✉)
Center for Development Research (ZEF), University of Bonn, Bonn, Germany
e-mail: beuchelt@uni-bonn.de

© The Author(s) 2016   181
F.W. Gatzweiler, J. von Braun (eds.), *Technological and Institutional Innovations for Marginalized Smallholders in Agricultural Development*,
DOI 10.1007/978-3-319-25718-1_11

status as well as workload. It has displaced women from their traditional productive functions, and diminished the income, power, and status they previously had (Momsen 2010; Moser 1993).

In many Sub-Saharan Africa and South Asian countries, the agricultural sector is underperforming. According to the FAO (2011), one of the key reasons is large gender inequalities in access to and control over resources and opportunities which undermine sustainable and inclusive agricultural development. The inequalities relate to many assets, inputs and services, such as land, livestock, labour, education, extension, financing and technology. This imposes actual costs on the agriculture sector, limits its efficiency, and includes costs for the broader economy and society (FAO 2011). A change in the distribution of inputs and/or control over resources between female and male farmers can not only significantly increase productivity, food and nutrition security, but also positively affect education outcomes (Alderman et al. 1995; Meinzen-Dick et al. 2011; Quisumbing and Maluccio 2000; World Bank 2009). The FAO (2011) estimates that if women had the same access to productive resources as men, total agricultural output could be raised in developing countries, which, in turn, could reduce the number of hungry people in the world by 12–17 %.

The socio-economic and institutional context in which innovations are introduced is key for their adoption (Bayard et al. 2007; Shaw 1987; Umali et al. 1993). Important aspects are the socio-economic status of the household, access to and control over resources and services, and intra-household dynamics (Haque et al. 2010). Gender aspects are often central for the success of agricultural interventions and development because of the specific roles and responsibilities of women and men in the agricultural systems and value chains (Beuchelt and Badstue 2013; Carr 2008). However, solutions to low agricultural productivity often focus on technological innovations, but do not necessarily consider social and gender disparities. Evidence grows that innovations in agriculture can affect women and men differently within households and communities due to differences in power, roles and access rights (Doss 2001). Still, relatively little is known about how agricultural development programs can most effectively deliver outcomes of well-being and higher incomes in ways that acknowledge the differential access to and control over assets and that lead to more equitable outcomes (Meinzen-Dick et al. 2011). Therefore, this chapter aims first to identify differential impacts of technological innovations on women and men and the related reasons. Second, it looks at opportunities and pathways to increase gender and social equity when designing and fostering innovations for sustainable agriculture intensification.

This chapter is based on a comprehensive literature review. While the focus of the chapter is on gender, we also address social equity aspects, since they are often interlinked. The next section introduces concepts around gender and the adoption of agricultural innovations, as well as analytical categories in which trade-offs and opportunities for innovation may occur. The third section addresses trade-offs in technological innovations from gender and social equity perspectives, using the case of conservation agriculture and decentralized bioenergy production. The fourth section identifies opportunities and pathways to enhance gender and social equity with sustainable intensification processes; the last section concludes.

Gender and the Adoption of Agricultural Innovations

Concepts

Many research projects and development programs around technological and institutional innovations for sustainable agricultural intensification are built on the assumption that by targeting the "household", all members will (equally) benefit from the intervention. Typically, households are perceived as quite homogenous in terms of family structure, with the man as the household head who adequately represents the needs and preferences of all household members (Moser 1993). Empirical evidence, however, shows that households do not have a joint utility function or practice joint decision-making; unequal exchange, power imbalances and inequality exist within households and between husbands and wives (Quisumbing 2003). In smallholder and marginalized farming systems, limited resources are typically allocated according to the priority of the household and/or to the most powerful household member, who is usually a man (Ponniah et al. 2008).

Gender[1] is a determining factor in defining who does which activity, who owns a good or resource, who decides, and who has power (UNICEF 2011). Gender aspects relate directly to men's and women's roles and responsibilities in the farming household and to decisions about allocating resources or adopting technologies in farming systems. For example, in Africa, wide gender disparities exist over ownership and management of land, trees and other resources; certain crops, trees, or parts of them, or necessary management activities, are often specifically attributed to or used by either women or men (Carr 2008; Doss 2002; Kiptot and Franzel 2011; Schroeder 1993).

In Africa, female farmers, compared to male farmers, often show lower adoption rates of sustainable intensification practices such as high-yielding varieties and improved management systems (Doss 2001; Ragasa 2012). Ndiritu et al. (2014) find for Kenya that women have similar adoption rates of intensification practices such as soil and water conservation measures, improved seeds, chemical fertilizers, and maize-legume intercropping, but are less likely to adopt minimum tillage and animal manure for crops. They relate the observed adoption differences to gender differences in access to these technological innovations and to required inputs, resources or information, as well as other socio-economic inequalities and barriers for women.

Successful interventions are usually transformative, whether through creating opportunities, new commodities and services or through changing the ways people

---

[1] Following Beuchelt and Badstue (2013), the term gender is used to refer to the socially constructed roles, rights, and responsibilities of women and men and the relations between them. Men and women, and their relations, are defined in different ways in different societies and, influenced by historical, religious, economic, and cultural realities, the roles and relations between women and men change over time (Doss 2001).

do things (Meinzen-Dick et al. 2011). Also, the gender roles and responsibilities are dynamic and get renegotiated, reflecting the changes in socio-economic circumstances; it is thus difficult to predict a priori what the adoption effects will be within households and communities (Doss 2001). For example, after innovating, women farmers face difficulties in maintaining profitable market niches and risk losing control over resources such as land, as men often take over production and marketing when it becomes financially lucrative (Momsen 2010; World Bank 2009). Since there have been several detailed literature reviews on gendered constraints and opportunities in relation to the adoption of new agricultural practices and technological innovations (Doss 2003, 2001; Peterman et al. 2010; Ragasa 2012; World Bank 2009), this section will not deal with it further, but will concentrate on how to categorize opportunities and trade-offs in agricultural innovations.

Analytical Categories for Identifying Opportunities and Trade-Offs in Innovations

From a gender and social equity perspective, opportunities and trade-offs in innovations around sustainable agricultural intensification typically occur in several areas of the farming and food system (Fig. 11.1). These can be grouped into five analytical categories: food and nutrition security as well as diversity; resources and labour; information and technology; and income, marketing and value chains, as well as health aspects (Beuchelt and Badstue 2013). The identification of the categories is derived from a review of the literature on human rights-based approaches to development, particularly for agriculture, nutrition and women (Anderson 2008; Cornwall and Nyamu-Musembi 2004; Doss 2001; FAO 1998; Lemke and Bellows 2011; Rae 2008; Socorro Diokno 2013).

For each category, innovations can have different effects on women and men from different social groups which may also stretch out to other categories. The effects of technologies and interventions are likely to vary between individuals in a household or between different social groups, depending on the socio-cultural context, age, sex, skills, abilities, religion, social relations, including kinship ties, and economic status. It is important to ask who benefits, who loses and what the potential consequences are. There is an enormous heterogeneity and complexity among African and Asian households, including in regard to gender roles, therefore generalizations are not possible (Doss 2001).

Identifying potential gender or social equity trade-offs in itself may lead to opportunities to address them straight away in the research for and implementation of innovations. It also may lead to the discovery of complementary measures that can enhance the overall potential for positive human development impacts of the particular intervention.

Fig. 11.1 Analytical categories for identifying potential gender and social equity trade-offs or opportunities related to agricultural technologies (Source: Adapted from Beuchelt and Badstue (2013))

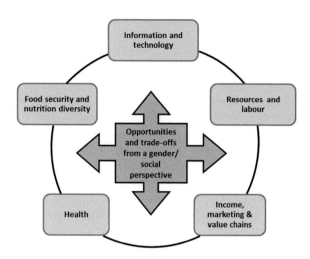

Technological Innovations in Agriculture from a Gender and Social Perspective

Several research studies have shown that women's labour burden can increase with new agricultural technologies and innovations. This happens when women take on additional tasks, or when their current tasks become more burdensome, for instance, when fertilizer application requires more weeding, or more output to be processed – both tasks often done by women (Doss 2001). Along similar lines, it is pointed out that *"an intervention that increases the amount of time women work in the field without considering childcare may improve food availability and diet, but hurt child welfare"* (Berti et al. 2004, p. 605).

A study by Paris and Pingali (1996) shows that the gender and equity impacts of a new labour-saving technology depend on the cultural and social characteristics of the local context. The introduction of a mechanical thresher in the Philippines reduced labour for both men and women, since threshing was much faster. Farmers were thus able to grow a second rice crop, which benefitted women, as it increased their employment opportunities in transplanting, weeding, and harvesting. The benefits outweighed the reduced labour demand for threshing. Contrarily, in Bangladesh, the introduction of a mechanical thresher affected poor and landless women negatively, because it replaced their work as a thresher. As cultural restrictions prevented these women from leaving their homestead, they could not look for alternative employment opportunities, and thus lost an important income source (Paris and Pingali 1996). Similar effects occurred in Vietnam, where new seeder technologies were promoted for rice production. As a consequence of adoption, more than half of the women from poor farming households, who previously worked as agricultural wage labourers in rice transplanting, lost this important income opportunity (Paris and Truong Thi Ngoc Chi 2005).

Palmer-Jones and Jackson (1997) report on treadle pumps and their gender effects in Bangladesh. The pumps were introduced as a pro-poor technology so that poor farmers could irrigate their fields. Though this often lead to an increase in production, food security and income, negative effects also occurred. Women living on small farms did the pumping in addition to their other household responsibilities, which not only raised their total labour burden but also its intensity. Although the treadling affected the women's ability to perform their other household tasks or reduced breast milk production, only in some cases did they receive support from their husbands. In other cases, poor women employed by better-off households had to use the treadle pumps as part of their employment activity. Women frequently suffered pain and exhaustion from the pump, even months after the work was finished. In general, more women than men used the treadle pump, especially those from poorer and female-headed households; however, the pump was clearly designed for the average weight and strength of a man and not of a woman (Palmer-Jones and Jackson 1997).

A positive example of unexpected gender effects is the case of the improved dual-purpose cowpea, which was developed to address problems of low productivity in northern Nigeria. As a result, productivity increased and with it the availability of food, fodder, and household income. Though cowpea production and sale is a male activity, additional income from the grain sales was also forwarded to the wives. The women saved the money, bought household goods or food, and invested in petty trading. Though it was not expected, the social and economic status of the wives from male adopters were improved (Tipilda et al. 2008).

Gender differences may also exist in regard to male and female farmers' crop preferences and varieties. Women and men rate maize characteristics differently and prefer different combinations of traits because of the intended maize consumption objectives, e.g., for markets, their own consumption, special dishes, feed (Bellon et al. 2003; Hellin et al. 2010). Whereas men often prefer high-yielding varieties to sell surplus production, women's reproductive roles often mean that they focus on food security and/or varieties that are palatable, nutritious and meet processing and storing requirements (Badstue 2006; Bellon et al. 2003). Improved maize varieties may also require longer cooking times, thus requiring more firewood and more female labour, and consequently are less preferable to women (Hellin et al. 2010). In Mexico and southern Africa, women's varietal preferences are also linked to their productive roles and income generation from the artisanal processing and sale of traditional maize products (Badstue 2006; Bellon et al. 2003; Doss 2001).

In the following section, two case studies are presented which describe in more detail the differential impacts of technological innovations regarding gender and social equity using the five analytical categories from section "Gender and the Adoption of Agricultural Innovations".

The Case of Conservation Agriculture

Conservation agriculture (CA) is globally promoted as a sustainable innovation for small and large farms (Derpsch et al. 2010; Hobbs 2007; Hobbs et al. 2008; Kassam et al. 2009; Valbuena et al. 2012). There are three key components to conservation agriculture: (i) Maintaining a permanent organic soil cover (through cover crops, intercrops and/or mulch); (ii) minimizing soil disturbance by tillage and other cultural operations; and (iii) diversifying crop rotations, sequences and associations (Kassam et al. 2009).

Evidence shows that CA can enhance soil quality and health, contribute to higher, more stable yields, and reduce production costs (Govaerts et al. 2005; Kassam et al. 2009). Depending on the local context, there are several constraints, including the cost of moving to and adapting CA practices for the specific farming system, the need to have access to inputs, markets, machinery, credit, and information, the availability of labour, the increases in weeds and pests (Baudron et al. 2007; Erenstein et al. 2012; Nyanga et al. 2012) and the competing uses of crop residues in smallholder systems for fuel, livestock fodder, and thatching (Hellin et al. 2013, Beuchelt et al. 2015).

Table 11.1 lists the effects of CA on women and men in smallholder agricultural systems. CA can imply diverse trade-offs from a gender and social perspective. The impacts of CA on certain social actors depend, among other things, on the specific context, the farming system, and local gender norms, and can be entirely different for other social actors or in a distinct context (Beuchelt and Badstue 2013).

The Case of Small-Scale Biomass Production for Decentralized Bio-energy

A lack of secure, sustainable and affordable energy is a big development constraint in developing countries (Amigun et al. 2011; Wiskerke et al. 2010) and has a disproportionate impact on rural women (Karlsson and Banda 2009). Though it bears a large potential for Africa, little research attention is directed to decentralized, local, small-to-medium-scale energy production based on local biomass (Ewing and Msangi 2009; Mangoyana et al. 2013). More research around the gender and equity impacts of decentralized energy schemes is certainly needed. The following analysis concentrates on smallholder biomass production for decentralized energy schemes, such as a small-to-medium-scale biodiesel plant located close to an agricultural area and oil mill (Amigun et al. 2011) or small-scale short rotation woodlots for fuelwood production (Wiskerke et al. 2010). The directions of effects depend highly on the local situation, for example, whether it is a biomass rich or dryland area, which technologies are used, and how large the dependence on local biomass is. Table 11.2 indicates potential effects of

Table 11.1 Potential effects of CA on women and men in smallholder agricultural systems

Categories	Potential equity effects of CA
Food security and nutrition diversity	+ Increased and more stable yields in 4–5 (10–12 at the outset) years and reduction of hunger period, benefits whole household + Crop rotation/intercropping increase nutrition diversity and food security when using food crops, helping women to fulfill their reproductive role − Herbicides and mulch layer may negatively affect traditional intercropping patterns and suppress use of wild vegetables typically managed by women → lower nutrition diversity and food security, esp. in hunger season; women often suffer disproportionately − Fewer residues available for feeding livestock – can also affect small livestock managed by women. If livestock is reduced, nutrition diversity is lowered or risks are increased − Food security may decrease if cash crops are used and income not spent (by men) on food
Health	+ Potential for better health, esp. of women, due to improved nutrition once higher yields appear amd due to rotation/intercropping, esp. with food crops and legumes (when not practiced before) + Reduced physical stress due to less land preparation (particularly benefiting men) and use of herbicides (particularly benefiting women) − Herbicides may contaminate ground water, wells, and ponds → risk for drinking water; women may have to walk further to get decent water − Herbicides may be a direct health hazard to household (HH) members, esp. to children, due to faulty application and storage of herbicides
Information & technology	+ Once technology is mastered, increased understanding of agricultural management − Highly knowledge intensive; may take women with lower education levels longer to learn, but projects/extension often do not account for this − Threat of male bias in decision-making when extension service/projects are gender blind and do not include women − Tendency to overlook womens' needs and constraints, especially when it comes to introducing machinery or working in mechanized systems − Mechanization may exclude women from its use (depending on gender norms)
Resources & labour	+ Mechanization reduces drudgery in land preparation and reduces land preparation in the long run, mainly benefits men but also women + Herbicides reduce work load, esp. for women, who usually do the weeding − Women and marginalized farmers often have insecure land title or rent land → land improvement due to CA can lead to risk of losing their plot − Crop rotation/intercropping may include putting "male" crops on female plots → women risk losing control over plot/harvest when growing "male" crops due to gender norms

(continued)

Table 11.1 (continued)

Categories	Potential equity effects of CA
	– Less residues/weeds available for livestock or fuel → increased labour burden for women to obtain alternative sources
	– Without herbicide use and/or using planting basins, increased labour burden to HH members, especially women and girls → mothers may neglect their children's welfare or nutrition to keep up with work
	– If herbicide applications are incompatible with intercrops, typically planted by women, gender disparities increase
	– Planting basins increase labour burden, esp. of women
	– Mechanization reduces need for hired, casual labour, eliminating an important income source for marginalized farmers, landless people
Income, marketing & value chains	+/– Higher potential income due to higher yields and lower production costs in mechanized systems, but when men alone make decisions about income, gender disparities can increase
	– In case of herbicide use, potential income loss when wild plants are sold for income
	– If herbicides/mechanization replaces rural workers, income loss of day labourers, especially women, due to limited employment opportunities in the rural sector
	– If crop residues become private property, poor landless shepherds/marginalized livestock owners may not be able to maintain their herds and lose their income source

Sources: Ackerman (2007), Berti et al. (2004), Beuchelt and Badstue (2013), Doss (2001), Giller et al. (2009), Govaerts et al. (2005), H. Nyanga (2012), Hellin et al. (2013), Kettles et al. (1997), Nyanga et al. (2012), Ramírez-López et al. (2013), Valbuena et al. (2012), World Bank (2009), Beuchelt et al. 2015

decentralized bioenergy production based on biomass which can serve as an analytical input for estimating impacts of new investments.

Opportunities and Pathways to Enhance Gender and Social Equity Through Sustainable Intensification

The ways and processes by which innovations are generated, adapted and disseminated are complex, given the many direct and indirect interactions between stakeholders (Aw-Hassan 2008). As the above-mentioned cases illustrate, it is not necessarily possible to predict how the introduction of new technologies may affect the patterns of labour, resource and land allocation between men and women, or how this, in turn, may influence who benefits and loses. Having highlighted potential trade-offs around agricultural innovations, the question remains as to how anticipated or emerging trade-offs can be converted into opportunities and pathways for making agricultural innovations more equitable and gender responsive, and thus, to expand the overall human development impact. In part, the

Table 11.2 Potential effects of smallholder biomass production for decentralized bio-energy on women and men

Categories	Potential equity effects of smallholder biomass production
Food security & nutrition diversity	+/− Food security should be little or unaffected, since not much feedstock is needed for a small bioenergy plant, but in case of competition, local food prices could somewhat increase → increased food insecurity could hit the most marginalized, esp. women
Health	+ In case of improved stoves, less burden to carry wood and less indoor air pollution, respiratory infections, and eye problems, benefiting mainly women and girls + (Maternal) health benefits from clinics with improved facilities from electrification
Information & technology	+ Potential for small-scale mechanization of laborious household tasks through decentralized energy – benefitting women, e.g., through mechanization of food processing (grinding), powering water pumps (no longer tiresome water fetching) + Access to electricity → e.g., improved school performance of children and enrollment of girls; less dependency of women on men, e.g., to recharge mobile phones + Women could be targeted in technology training, for supervision and plant management → increase in knowledge − Literacy constraints of smallholders, esp. of women; due to lack of technical know-how related to feedstock, its conversion may lead to their exclusion in decision-making and participation − Energy produced may be insufficient for whole village; only the better-off households might get energy or have money to pay for it → negligible benefits for marginalized HH − Tendency to overlook women's energy needs and constraints, esp. when it comes to introducing machinery → threat of male bias in projects
Resources & labour	+ Frees female labour/time when wood is no longer required or laborious household tasks, typically done by women, are mechanized, e.g., pumping water, grinding − Land scarcity / biomass scarcity → land previously given to women may be taken away → affecting their food security, income, status − Land insecurity may prevent investments in tree crops, esp. for women; having insecure land rights and land tenure system may inhibit access to biomass fuel for women − When using slow-growing crops like jatropha or oil palm, better-off farmers tend to benefit more than poorer farmers, and men more than women, due to land rights and available liquidity − Increased labour burden (for women) to procure biomass for the plant − Lack of willingness to pay for electricity/fuel generated by energy scheme – female labour is "for free" and women are often not very involved in decision-making in regard to energy
Income, marketing & value chains	+/−When feedstock sold to plant, income opportunities for both sexes; however, men tend to take over activities from women once profitable and invest less in food security + Women could be targeted to be involved in management of

(continued)

Table 11.2 (continued)

Categories	Potential equity effects of smallholder biomass production
	plants → employment/income opportunities + Electricity can provide business opportunities for women and men, e.g., many small retail businesses, such as phone charging or tailoring, are run by women − Commercial biomass activities such as charcoal and firewood trading are often male activities; income may not benefit women

Sources: FAO (2008), Farioli and Dafrallah (2012), Hunsberger et al. (2014), Karlsson and Banda (2009), UNDESA (2007)

response depends on the kind of impacts an innovation or development programme aims to have on women and men, as well as on social equity and whether explicit gender and equity goals were defined (Skutsch 2005).

Different dimensions of marginalized and smallholder lives, such as livelihood assets, institutions, food system activities, and food system outcomes, affect agricultural innovations and are affected by them, and all can imply equity issues. Equity issues in the 'livelihood assets' dimension relate to access and control over the natural, physical, financial, human and social capital (DFID 1999) needed for agricultural biomass production, processing and marketing. In the 'food system activities' dimension, they are connected to the food-related activities undertaken in the farming system and value chains by women and men of different social groups. In the 'institutions' component, they refer to formal institutions, such as legislative frameworks and policies, as well as informal institutions – social relations, values, and norms that shape beliefs and behaviours. All, but especially the norms, influence relationships between men and women. In the 'food system outcomes' dimension, equity issues arise due to differences in the actual situation and potential project or policy outcomes regarding food and nutrition security, health, poverty reduction and natural resource sustainability between women and men and different social strata. Mainstreaming equity issues entails the inclusion of a gender and social equity perspective for each dimension and requires strategies, as well as tactics, that take into account the power difference within and between female and male members of various groups, integrate advocacy to have open spaces for voices to be heard and enable people to recognize and use their agency (Cornwall 2003).

Though gender mainstreaming is commonly known and promoted, it is seldom fully practiced. To overcome trade-offs and use opportunities, gender and equity aspects need to be integrated into the project cycle, i.e., to be included in all stages of a project, programme or policy from the planning and design stage, during the implementation, in the progress monitoring and in the final evaluation (Arenas and Lentisco 2011; Aw-Hassan 2008).

At the planning and design stage, it is important to explicitly define whether an innovation/project (also) aims to improve women's welfare, increase the economic productivity of women and/or marginalized farmers or contribute to their empowerment (Skutsch 2005). These goals should ideally be defined together with the concerned stakeholders; however, this is often not feasible. Possible opportunities and trade-offs in agricultural innovations and interventions need to be carefully assessed for women and men of different social strata and age groups before the project starts. This implies a sound gender and social analysis of the specific intervention and the related target context, with particular focus on the analytical categories listed in Fig. 11.1. It is essential to know whether women or men are the direct users of a technology, who is considered responsible for different aspects of the innovative technology, who will make the investment and labour decisions and who will benefit from it, since this will have a bearing on who will be involved with and affected by the new technology. It is a good business practice to utilize a marketing survey in order to know the customers, their needs and priorities, before the project starts (Skutsch 2005). *"If it turns out that all such investment decisions are made by men, and if it is likely that this will result in decisions which are not in women's interests, then a strategy may have to be developed to counter this as far as possible"* (Skutsch 2005, p. 48). For real project success, it is key to do this as early as possible in the research and development process, and explicitly address the critical issues, ideally in a participatory process together with the relevant stakeholders of both sexes. A stakeholder analysis is very useful for understanding power issues and the impact of the innovation on the stakeholders, as well as the impact of the various stakeholders on the project or innovation. This provides opportunity to identify joint priorities, adjust research targeting and project design, and devise context specific alternatives or ways to mitigate negative trade-offs. It can include the combination of various technologies or approaches which, when used together, can compensate for trade-offs or enhance overall development impacts. The identification of non-traditional research and development partners with comparative advantages for addressing specific trade-offs can play an important role. For example, in a situation in which CA mechanization holds great promise for individual farm households to reduce labour input, but may happen at the cost of offsetting rural landless workers, collaboration with alternative partners with expertise in income-generating activities could be considered.

Before and during implementation, an analysis of the capacity of the implementing organizations to be aware of and handle gender and equity issues is helpful, as some organizations do not possess experience in this field, but are rather "technical focused" (Skutsch 2005). During implementation, a gender and social perspective can be incorporated into the activities through gender-responsive and gender transformative approaches, as well as a focus on empowerment of marginalized farmers.

Gender transformative approaches seek to address and eventually change gender norms, roles and imbalances of power when inequities are large and can easily be combined with agricultural extension. They raise awareness of gender roles and relations between women and men; foster – at a local pace – more gender-equitable relationships between both sexes while challenging the unequal distribution of resources and allocation of duties between men and women. They can also address the power relationships between different stakeholders and social actors (Consortium International Agropolis 2012; USAID and IGWG 2011). They, thus, are a complementing means to achieve agricultural intensification, improve livelihoods and gender equity, especially where current extension and technological approaches alone have had limited effect with regards to adoption of the promoted technologies, or an equitable benefit sharing between men and women. A successful example of gender transformative approaches regarding sustainable agricultural intensification is summarized by Beuchelt and Badstue (2013), based on experiences in Zambia (Bishop-Sambrook and Wonani 2008; Klos 2000).

In agricultural development, empowerment efforts are often viewed as an advanced form of participation that will improve project effectiveness through farmers making their own decisions, rather than only adopting recommendations. However, a large focus is still on an ex-ante decision of what is supposed to happen in the project and how rural people are supposed to live their lives, which is found in statements such as "30 % of farm households will use improved varieties" or "CA will be practiced on 20,000 hectares". Instead of controlling the development process, projects may go one step further and become entry points for empowerment through enhancing the means for and facilitating the process of intrinsic empowerment (Bartlett 2008). This requires a change in power relations not among the different social groups, but with the project planners and managers. Research and extension can support changes in knowledge, behaviour, and social relationships with the aim that poorer people are taking control of their lives, thus transforming the way they live their lives (Bartlett 2008).

Emerging trade-offs and negative effects, which were not anticipated in the planning stage, also need to be identified and addressed – a task of the monitoring process. Corrective measures may include new alliances with project partners who can help to mitigate trade-offs or embark on the opportunities. Sufficient time buffers for these unexpected events should be integrated into the planning phase. Reflection and joint learning processes, especially through participatory approaches, regarding effects on gender and social equity are crucial during monitoring, but also in the evaluation phase in which the project's success is assessed. This can be combined with disaggregated qualitative and quantitative data which also distinguishes for sex, age, economic and social strata, to describe and explain the observed changes and project effects among men and women in different groups of society. Gender separation during data-gathering phases in planning, implementation, monitoring and evaluation is suggested to obtain reliable information on gendered uses, constraints, opportunities and trade-offs around innovations (Skutsch 2005).

Conclusions

In summary, research and development for sustainable intensification face the challenges of (i) enhancing the food and nutrition security of poor men and women of all age groups; (ii) increasing gender and social equity and decreasing poverty and (iii) being environmentally and socio-economically sustainable. Development processes and agricultural modernization have affected men and women in different ways and have often increased gender and social disparities. Given the complexity of gender and social dynamics and their embeddedness in the agricultural and socio-economic contexts, innovations for sustainable intensification need to address these in order to reach desired development impacts.

There are many positive characteristics of agricultural innovations for sustainable intensification, including yield increases, crop diversification and labour savings. The global or overall effects of an innovation are often positive, but the resulting benefits may be shared in different ways by different social groups, between men and women, and, in extreme cases, even increase gender and social inequalities. The evidence presented suggests that it is critical to address the different needs and constraints of both female and male marginalized farmers in the processes and systems through which agricultural intensification innovations are developed, disseminated and promoted.

Gendered trade-offs need to be considered and assessed in relation to the other expected human and sustainable development impacts of the agricultural innovation in question. There are several pathways to mitigating trade-offs and building on opportunities to enhance gender and social equity. The incorporation of gender-transformative and general empowerment approaches in agricultural research and development interventions can be helpful in this respect. Decisions as to which pathway is chosen should be developed together with male and female stakeholders – of different social groups – and can include engagement with non-traditional partners with the necessary skills and abilities to work at the levels where trade-offs occur. The promoted institutional or technological innovation can also be combined with other technologies which are able to compensate or mitigate trade-offs created by the promoted main technology. In addition, policy interventions can contribute to the stimulation of inclusive development and the reduction of gender constraints related to specific interventions.

Aiming at positive, equity-enhancing development impacts through technology development and innovation diffusion, a holistic farming and food systems approach is recommended which is gender-sensitive and social transformative. Further evidence of the potential but also specific challenges hereof, especially scientifically accompanied case studies, is needed to build broader support for mainstreaming social and equity approaches in agricultural research and development projects.

Acknowledgements The support from the funding initiative "Securing the Global Food Supply-GlobE" of the German Federal Ministry of Education and Research (BMBF), specifically the project "BiomassWeb – Improving food security in Africa through increased system productivity of biomass-based value webs" is gratefully acknowledged. Further thanks go to Lone Badstue and Detlef Virchow for their comments and input.

Open Access This chapter is distributed under the terms of the Creative Commons Attribution-Noncommercial 2.5 License (http://creativecommons.org/licenses/by-nc/2.5/) which permits any noncommercial use, distribution, and reproduction in any medium, provided the original author(s) and source are credited.
The images or other third party material in this chapter are included in the work's Creative Commons license, unless indicated otherwise in the credit line; if such material is not included in the work's Creative Commons license and the respective action is not permitted by statutory regulation, users will need to obtain permission from the license holder to duplicate, adapt or reproduce the material.

References

Ackerman F (2007) The economics of atrazine. Int J Occup Environ Health 13(4):437–445. doi:10.1179/oeh.2007.13.4.437

Alderman H, Chiappori P, Haddad L, Hoddinott J, Kanbur R (1995) Unitary versus collective models of the household: is it time to shift the burden of proof? World Bank Res Obs 10(1):1–19. doi:10.1093/wbro/10.1.1

Amigun B, Musango JK, Stafford W (2011) Biofuels and sustainability in Africa. Renew Sustain Energy Rev 15(2):1360–1372. doi:10.1016/j.rser.2010.10.015

Anand S, Sen A (2000) Human development and economic sustainability. World Dev 28(12):2029–2049

Anderson MD (2008) Rights-based food systems and the goals of food systems reform. Agric Hum Values 25(4):593–608. doi:10.1007/s10460-008-9151-z

Arenas M, Lentisco A (2011) Mainstreaming gender into project cycle management in the fisheries sector. Food and Agriculture Organization of the United Nations, Bangkok

Aw-Hassan AA (2008) Strategies for out-scaling participatory research approaches for sustaining agricultural research impacts. Dev Pract 18(4–5):564–575. doi:10.1080/09614520802181590

Badstue LB (2006) Smallholder seed practices: maize seed management in the Central Valleys of Oaxaca, Mexico. Wageningen University, Wageningen

Bartlett A (2008) No more adoption rates! Looking for empowerment in agricultural development programmes. Dev Pract 18(4/5):524–538. doi:10.2307/27751956

Baudron F, Mwanza HM, Triomphe B, Bwalya M (2007) Conservation agriculture in Zambia: a case study of Southern Province. African Conservation Tillage Network, Centre de Coopération Internationale de Recherche Agronomique pour le Développment, Food and Agriculture Organization of the United Nations, Nairobi

Bayard B, Jolly CM, Shannon DA (2007) The economics of adoption and management of alley cropping in Haiti. J Environ Manage 84(1):62–70. doi:10.1016/j.jenvman.2006.05.001

Bellon MR, Berthaud J, Smale M, Aguirre A, Taba S, Aragon F, Diaz J, Castro H (2003) Participatory landrace selection for on-farm conservation: an example from the Central Valleys of Oaxaca, Mexico. Genet Resour Crop Evol 50(4):401–416

Berti PR, Krasevec J, FitzGerald S (2004) A review of the effectiveness of agriculture interventions in improving nutrition outcomes. PHN 7(05):599–609. doi:10.1079/PHN2003595

Beuchelt TD, Badstue L (2013) Gender, nutrition- and climate-smart food production: opportunities and trade-offs. Food Sec 5(5):709–721. doi:10.1007/s12571-013-0290-8

Beuchelt TD, Camacho Villa CT, Göhring L, Hernández Rodríguez VM, Hellin J, Sonder K and Erenstein O (2015) Social and income trade-offs of conservation agriculture practices on crop residue use in Mexico's central highlands. Agr Syst 134:61–75

Bishop-Sambrook C, Wonani C (2008) The household approach as an effective tool for gender empowerment: a review of the policy, process and impact of gender mainstreaming in the agricultural support programme in Zambia. http://asp.ramboll.se/Docs/stcs/GenderStudy.pdf. Accessed 28 Jan 2013

Carr ER (2008) Men's crops and women's crops: the importance of gender to the understanding of agricultural and development outcomes in Ghana's Central Region. World Dev 36(5):900–915. doi:10.1016/j.worlddev.2007.05.009

CGIAR Consortium. (2012). Redressing gender disparities: a transformativeapproach to our research. http://www.cgiar.org/consortiumnews/redressing-gender-disparities-a-transformative-approach-toour-research/. Accessed 29 January 2013

Cornwall A (2003) Whose voices? Whose choices? Reflections on gender and participatory development. World Dev 31(8):1325–1342. doi:10.1016/S0305-750X(03)00086-X

Cornwall A, Nyamu-Musembi C (2004) Putting the 'rights-based approach' to development into perspective. Third World Q 25(8):1415–1437. doi:10.1080/0143659042000308447

Derpsch R, Friedrich T, Kassam A, Hongwen L (2010) Current status of adoption of no-till farming in the world and some of its main benefits. Int J Agric Biol Eng 3(1):1–25

DFID (1999) Sustainable livelihoods guidance sheets – section 1 + 2. Department for International Development (DFID), London

Doss CR (2001) Designing agricultural technology for African women farmers: lessons from 25 years of experience. World Dev 29(12):2075–2092. doi:10.1016/S0305-750X(01)00088-2

Doss CR (2002) Men's crops? Women's crops? The gender patterns of cropping in Ghana. World Dev 30(11):1987–2000. doi:10.1016/S0305-750X(02)00109-2

Doss CR (2003) Understanding farm-level technology adoption: lessons learned from CIMMYT's micro surveys in Eastern Africa, Economic working paper 03–07. International Maize and Wheat Improvement Center, Mexico

Erenstein O, Sayre K, Wall P, Hellin J, Dixon J (2012) Conservation agriculture in maize- and wheat-based systems in the (sub)tropics: lessons from adaptation initiatives in South Asia, Mexico, and Southern Africa. J Sustain Agric 36(2):180–206. doi:10.1080/10440046.2011.620230

Ewing M, Msangi S (2009) Biofuels production in developing countries: assessing tradeoffs in welfare and food security. Environ Sci Pol 12(4):520–528. doi:10.1016/j.envsci.2008.10.002

FAO (1998) The right to food in theory and practice. Food and Agriculture Organization of the United Nations, Rome

FAO (2008) Making sustainable biofuels work for smallholder farmers and rural households. Food and Agriculture Organization of the United Nations, Rome

FAO (2011) The state of food and agriculture, 2010–2011. Women in agriculture: closing the gender gap for development. Food and Agriculture Organization of the United Nations, Rome

Farioli F, Dafrallah T (2012) Gender issues of biomass production and use in Africa. In: Janssen R, Rutz D (eds) Bioenergy for sustainable development in Africa. Springer Netherlands, Dordrecht/Heidelberg/London/New York, pp 345–361. doi:10.1007/978-94-007-2181-4_28

Giller KE, Witter E, Corbeels M, Tittonell P (2009) Conservation agriculture and smallholder farming in Africa: the heretics' view. Field Crop Res 114(1):23–34. doi:10.1016/j.fcr.2009.06.017

Govaerts B, Sayre KD, Deckers J (2005) Stable high yields with zero tillage and permanent bed planting? Field Crop Res 94:33–42. doi:10.1016/j.fcr.2004.11.003

Haque MM, Little DC, Barman BK, Wahab MA (2010) The adoption process of ricefield-based fish seed production in Northwest Bangladesh: an understanding through quantitative and qualitative investigation. JAEE 16(2):161–177. doi:10.1080/13892241003651415

Hellin J, Keleman A, Bellon M (2010) Maize diversity and gender: research from Mexico. Gend Dev 18(3):427–437. doi:10.1080/13552074.2010.521989

Hellin J, Erenstein O, Beuchelt T, Camacho C, Flores D (2013) Maize stover use and sustainable crop production in mixed crop–livestock systems in Mexico. Field Crop Res 153:12–31. doi:10.1016/j.fcr.2013.05.014

Hobbs PR (2007) Conservation agriculture: what is it and why is it important for future sustainable food production? Paper presented at international workshop on increasing wheat yield potential, CIMMYT, Obregon, 20–24 Mar 2006. J Agric Sci 145(02):127. doi:10.1017/S0021859607006892

Hobbs PR, Sayre K, Gupta R (2008) The role of conservation agriculture in sustainable agriculture. Philos Trans R Soc B: Biol Sci 363(1491):543–555. doi:10.1098/rstb.2007.2169

Hunsberger C, Bolwig S, Corbera E, Creutzig F (2014) Livelihood impacts of biofuel crop production: implications for governance. Geoforum 54:248–260. doi:10.1016/j.geoforum.2013.09.022

IISD (2013) Gender equity. Improving distribution of environmental benefits and burdens. http://www.iisd.org/gender/. Accessed 17 Jun 2013

Karlsson G, Banda K (2009) Biofuels for sustainable rural development and empowerment of women. ENERGIA Secretariat, Leusden

Kassam A, Friedrich T, Shaxson F, Pretty J (2009) The spread of conservation agriculture: justification, sustainability and uptake. Int J Agr Sustain 7(4):292–320. doi:10.3763/ijas.2009.0477

Kettles MK, Browning SR, Prince TS, Horstman SW (1997) Triazine herbicide exposure and breast cancer incidence: an ecologic study of Kentucky counties. Environ Health Perspect 105 (11):1222–1227

Kiptot E, Franzel S (2011) Gender and agroforestry in Africa: are women participating. World Agroforestry Centre (ICRAF), Nairobi. doi:10.5716/OP16988

Klos S (2000) Lessons learned from the Gender Oriented Participatory Extension Approach (GPEA) in Zambia. Newsletter of the emerging platform on services within Division "Rural Development" (45) of GTZ, pp 20–23

Lemke S, Bellows AC (2011) Bridging nutrition and agriculture: local food-livelihood systems and food governance integrating a gender perspective. Technikfolgenabschaetzung – Theorie und Praxis 20(2):52–60

Mangoyana RB, Smith TF, Simpson R (2013) A systems approach to evaluating sustainability of biofuel systems. Renew Sustain Energy Rev 25:371–380. doi:10.1016/j.rser.2013.05.003

Meinzen-Dick R, Quisumbing AR, Behrman J, Biermayr-Jenzano P, Wilde V, Noordeloos M, Ragasa C, Beintema N (2011) Engendering agricultural research, development, and extension. International Food Policy Research Institute, Washington, DC

Momsen JH (2010) Gender and development, 2nd edn. Routledge, London/New York

Moser CON (1993) Gender planning and development: theory, practice & training. Routledge, London/New York

Ndiritu SW, Kassie M, Shiferaw B (2014) Are there systematic gender differences in the adoption of sustainable agricultural intensification practices? Evidence from Kenya. Food Policy 49 (1):117–127. doi:10.1016/j.foodpol.2014.06.010

Nyanga PH (2012) Food security, conservation agriculture and pulses: evidence from smallholder farmers in Zambia. J Food Res 1(2):120–138. doi:10.5539/jfr.v1n2p120

Nyanga PH, Johnsen FH, Kalinda TH (2012) Gendered impacts of conservation agriculture and paradox of herbicide use among smallholder farmers. Int J Technol Dev Stud 3(1):1–24

Palmer-Jones R, Jackson C (1997) Work intensity, gender and sustainable development. Food Policy 22(1):39–62

Paris TR, Chi TTN (2005) The impact of row seeder technology on women labor: a case study in the Mekong delta, Vietnam. Gend Technol Dev 9(2):157–184. doi:10.1177/097185240500900201

Paris TR, Pingali PL (1996) Do agricultural technologies help or hurt poor women? Competition and conflict in Asian agricultural resource management: issues, options, and analytical paradigms: IRRI discussion paper series No 11. In: Pingali PL, Paris TR (eds) International Rice Research Institute, Los Banos, pp 237–245

Peterman A, Behrman J, Quisumbing A (2010) A review of empirical evidence on gender differences in nonland agricultural inputs, technology, and services in developing countries. IFPRI discussion paper 00975. International Food Policy Research Institute, Washington, DC

Ponniah A, Puskur R, Workneh S, Hoekstra D (2008) Concepts and practices in agricultural extension in developing countries: a source book. International Food Policy Research Institute and International Livestock Research Institute, Washington, DC/Nairobi

Quisumbing AR (2003) Household decisions, gender, and development. International Food Policy Research Institute, Washington, DC

Quisumbing AR, Maluccio JA (2000) Intrahousehold allocation and gender relations: new empirical evidence from four developing countries. FCND discussion paper No 84. International Food Policy Research Institute, Washington, DC

Rae I (2008) Women and the right to food. International law and state practice. Right to food studies. Food and Agriculture Organization of the United Nations, Rome

Ragasa C (2012) Gender and institutional dimensions of agricultural technology adoption: a review of literature and synthesis of 35 case studies. Selected poster prepared for presentation at the International Association of Agricultural Economists (IAAE) Triennial conference, Foz do Iguaçu, 18–24 Aug 2012

Ramírez-López A, Beuchelt TD, Velasco-Misael M (2013) Factores de adopción y abandono del sistema de agricultura de conservación en los Valles Altos de México. Agricultura, sociedad y desarrollo 10:195–214

Schroeder RA (1993) Shady practice: gender and the political ecology of resource stabilization in Gambian Garden/Orchards. Econ Geogr 69(4):349–365

Shaw AB (1987) Approaches to agricultural technology adoption and consequences of adoption in the third world: a critical review. Geoforum 18(1):1–19. doi:10.1016/0016-7185(87)90017-0

Skutsch MM (2005) Gender analysis for energy projects and programmes. Energy Sustain Dev 9 (1):37–52. doi:10.1016/S0973-0826(08)60481-0

Socorro Diokno M (2013) Human rights based approach – development planning toolkit. http://www.hrbatoolkit.org/. Accessed 20 Jan 2013

Tipilda A, Alene A, Manyong VM (2008) Engaging with cultural practices in ways that benefit women in northern Nigeria. Dev Pract 18(4–5):551–563. doi:10.1080/09614520802181574

Umali L, Feder G, Umali DL (1993) The adoption of agricultural innovations: a review. Tech Forecasting Soc Chang 43(3–4):215–239

UNDESA (2007) Small-scale production and use of liquid biofuels in Sub-Saharan Africa: perspectives for sustainable development, Background paper No 2 – DESA/DSD/2007/2. United Nations Department of Economic and Social Affairs, New York

UNICEF (2011) Promoting gender equality: an equity-focused approach to programming. United Nations Children's Fund, New York

USAID, IGWG (2011) A summary report of new evidence that gender perspectives improve reproductive health outcomes. USAID and The Interagency Gender Working Group IGWG, Washington, DC

Valbuena D, Erenstein O, Homann-Kee Tui S, Abdoulaye T, Claessens L, Duncan AJ, Gérard B, Rufino MC, Teufel N, van Rooyen A, van Wijk MT (2012) Conservation agriculture in mixed crop–livestock systems: scoping crop residue trade-offs in Sub-Saharan Africa and South Asia. Field Crop Res 132:175–184. doi:10.1016/j.fcr.2012.02.022

Wiskerke WT, Dornburg V, Rubanza C, Malimbwi RE, Faaij A (2010) Cost/benefit analysis of biomass energy supply options for rural smallholders in the semi-arid eastern part of Shinyanga Region in Tanzania. Renew Sustain Energy Rev 14(1):148–165. doi:10.1016/j.rser.2009.06.001

World Bank (2009) Gender in agriculture sourcebook. World Bank, Washington, DC

Chapter 12
Assessing the Sustainability of Agricultural Technology Options for Poor Rural Farmers

Simone Kathrin Kriesemer, Detlef Virchow, and Katinka M. Weinberger

Abstract This chapter presents an analytic framework to identify agricultural inno-vations that are sustainable and suitable for the poorest and most vulnerable parts of the population. The framework contains a set of tools to collect and evaluate information on appropriate innovations based on relevant criteria. It considers the dimensions of environmental resilience, economic viability, and social sustainability, as well as technical sustainability considering important properties of the innovation itself. Information on already available agricultural innovations was collected in ten countries in South and Southeast Asia, as well as from the national and international agricultural research communities. A composite sustainability indicator was constructed to compare the collected innovations and radar charts were computed to visualize their performance in each sustainability criterion.

Keywords Poverty • Vulnerability • Sustainability indicator • Resilience • Innovation assessment

Background

Agriculture is a sector that urgently requires transformative changes to support sustainable development. This is true for several reasons. Firstly, the agricultural sector is important in respect to provision of food. As global population is expected to increase by two billion by 2050, and incomes rise, so will demand for more, more diverse and higher quality food. Secondly, farming also remains a key source of

S.K. Kriesemer (✉)
Horticulture Competence Centre, University of Bonn, Bonn, Germany
e-mail: sk.kriesemer@uni-bonn.de

D. Virchow
Center for Development Research (ZEF), University of Bonn, Bonn, Germany
e-mail: d.virchow@uni-bonn.de

K.M. Weinberger
Environment and Development Policy Section, United Nations Economic and Social
Commission for Asia and the Pacific, Bangkok, Thailand
e-mail: weinbergerk@un.org

© The Author(s) 2016
F.W. Gatzweiler, J. von Braun (eds.), *Technological and Institutional Innovations for Marginalized Smallholders in Agricultural Development*,
DOI 10.1007/978-3-319-25718-1_12

income – 75 % of the world's poor in developing countries live in rural areas, and in developing countries, the sector contributes 29 % of GDP and accounts for 65 % of all employment. And finally, agriculture uses resources that are becoming increasingly scarce – including land, soil, water, nutrients – and contributes to as well as suffers from the consequences of climate change (Godfray et al. 2010).

A transformation of the sector requires the adoption of new and innovative approaches that support sustainable outcomes. Many agricultural research organizations, both from the public as well as the private sector, and at national, regional and international levels, are involved in making solutions available for enhanced agricultural sustainability. Many of these have value beyond the particular local setting for which they were developed. However, decision takers at all levels, including farmers, extension workers and programme managers, require better tools to determine what innovations, i.e., what practices and technologies have relevance in certain settings. Traditional tools based on profit-maximization at the farm level, such as linear programming, do not take into consideration sustainability enhancing aspects and are therefore not sufficient in supporting the sustainability agenda.

The formulation of universal sustainable development goals that were agreed upon in the Rio+20 conference (UN 2012) is based on the principles of economic profitability, social justice and environmental friendliness: "Sustainable development is the management and conservation of the natural resource base and the orientation of technological and institutional change in such a manner as to ensure the attainment and continued satisfaction of human needs for present and future generations. Such sustainable development (in the agriculture, forestry, and fisheries sectors) conserves land, water, plant and animal genetic resources, is environmentally non-degrading, technically appropriate, economically viable and socially acceptable". This definition was adopted in 1989 by FAO (1995).

A decision-making tool that aims for optimizing sustainability outcomes of the use of new technologies and innovations should, thus, take (at least) these three pillars of sustainable development into account. Yet, clearly, this is not an easy task. Sustainability does not have an intrinsic value unto itself, and different stakeholders and interest groups hold different assumptions about values, for instance, the relationship between economic development and human wellbeing, the relationship between present and future needs, the relationship between resource allocation and level of consumption, or views of what should be sustained. We therefore aim to present a tool that supports the decision-making process by making information on different aspects of sustainability available to decision-makers.

Evaluating Technologies for Sustainable Agriculture

The following paragraphs present the most important scientific work and literature reviews with relevance for the task of assessing the sustainability of innovations in agriculture.

Singh et al. (2012) provide an overview of sustainability assessment methodologies. They mention twelve approaches from four different fields that assess sustainability at the level of industries or technologies. The approaches are: (1) composite sustainability performance indices for industries; (2) product-based sustainability indices; (3) environment indices for industries; and (4) energy-based indices. But approaches based on life cycle assessments also play a role (Aistars 1999). Other reviews present international approaches to sustainability assessment (Grenz and Thalmann 2013) and provide an overview of sustainability assessment systems (Doluschitz and Hoffmann 2013).

Although the classical approach to sustainability comprises the three pillars of environment, economy and society, several authors suggest additional aspects of sustainability. Veleva and Ellenbecker (2001) suggest six main aspects of sustainable production: (1) energy and material use (resources); (2) natural environment (sinks); (3) social justice and community development; (4) economic performance; (5) workers; and (6) products. The authors adapt nine principles of sustainable production from the Lowell Center for Sustainable Production, provide recommendations for the development of indicators and suggest using a set of core and supplemental Indicators of Sustainable Production (ISP).

Rigby et al. (2001) present three facets of agricultural sustainability: (1) improved farm-level social and economic sustainability (enhances farmers' quality of life, increases farmers' self-reliance, sustains the viability/profitability of the farm); (2) improved wider social and economic sustainability (improves equity/is 'socially supportive', meets society's needs for food and fiber); and (3) increased yields and reduced losses (while minimizing off-farm inputs, minimizing inputs from non-renewable sources, maximizing use of (knowledge of) natural biological processes, and promoting biological diversity/'environmental quality').

Dunmade (2002) suggests a framework of indices to assess the sustainability of a technology[1] for introduction into a developing country. Adaptability is the primary indicator of sustainability of a technology and is evaluated using four secondary indicators, namely technical, economic, environmental and socio-political sustainability (Fig. 12.1).

---

[1] The terms technology (set) or best practice should be understood in the broadest sense possible as agricultural innovations, as *"an idea, practice, or object that is perceived as new by an individual or other unit of adoption"* Rogers (2003). Although the use of the word 'innovation' would be the most appropriate from the point of view of social science, the term 'technology' is commonly understood and frequently used by colleagues of other disciplines and extension practitioners. Keeping in mind the broad definition of innovation, in this chapter, we use the term technology as a synonym for innovation and best practice.

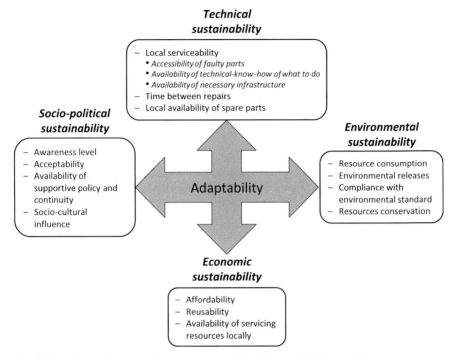

Fig. 12.1 Indices of foreign technology sustainability (Dunmade 2002, p. 464)

Dewulf and Van Langenhove (2005) describe a set of five sustainability indicators for the assessment of technologies based on industrial ecology principles. The indicators are: (1) renewability of resources; (2) toxicity of emissions; (3) input of used materials (reuse of materials); (4) recoverability of products at the end of their use (recoverability of waste materials); and (5) process efficiency. The indicators are based on the second law of thermodynamics. This allows for the quantification of all material and energy flows, exchange rates and conversion rates within a production system in exergy terms.

Dantsis et al. (2010) combine 21 individual indicators that cover the three pillars of sustainability (environment, economy, and society) into a unique indicator using the Multiattribute Value Theory (MAVT).

Many authors agree that sustainability is difficult to define (Kemmler and Spreng 2007). Therefore, Smith et al. (2000) inverse the approach and look at features of a system that are unsustainable, rather than searching for those that are sustainable. In their threat identification model (TIM), they first identify and rate potential hazards to sustainability depending on location-specific conditions, in particular, considering soil conditions and the risks of identified hazards. The final results are location-specific best management practice guidelines that can be developed for the farm level using GIS and that allow users to examine and understand the logic behind recommendations.

Analytical Framework

When analyzing the sustainability of an agricultural technology, the characteristics of the technology itself are of critical importance, because the speed and rate of adoption of an innovation depend on the personal characteristics of the potential adopter, the nature of the social system, the type of adoption decision, the extent of the change agent's promotion efforts and the specific attributes of the innovation itself that determine its usefulness for the potential adopter (Rogers 2003). Therefore, we consider four dimensions of sustainability in the analytical framework, namely the dimensions of environmental resilience, economic viability, social sustainability, and technical sustainability considering important properties of the innovation itself.

Criteria for Sustainable Agriculture

A literature search conducted at the end of 2012 resulted in the identification of 104 sustainability criteria relevant for agricultural technologies. These were reduced by merging similar indicators and deleting criteria with the same meaning but different terms, phrasing or unit of measurement, and eliminating indicators irrelevant for agriculture, or irrelevant in the context of developing countries. Criteria for which data collection would be too costly to collect were eliminated as well. As a result, 27 criteria were identified as highly relevant to the description of various aspects of technologies in the context of sustainability. Due to data limitations encountered during initial rounds of application of the framework, the criteria were further reduced, as shown in Fig. 12.2. The analysis aims to identify technologies that are sustainable, but also appropriate for the poor and vulnerable people, especially women and landless or land poor people. Such technologies are called suitable in this chapter. To address the special needs of poor and vulnerable people, two criteria are included that are not typically sustainability criteria, namely the minimum amount of land area required to adopt the technology and the percentage of female adopters as proxy for the suitability of the technology for vulnerable groups.

To decide on the relative importance of the criteria under consideration, experts were invited to provide weights for individual criteria using the Analytical Hierarchy Process (AHP) developed in the 1980s (Saaty 1990). This approach is a multi-criteria decision-making process that is suitable for involving a group of experts. It was implemented via an online survey that asked experts to compare all criteria in a pairwise manner. For each pair of criteria within the same sub-objective, experts

204   S.K. Kriesemer et al.

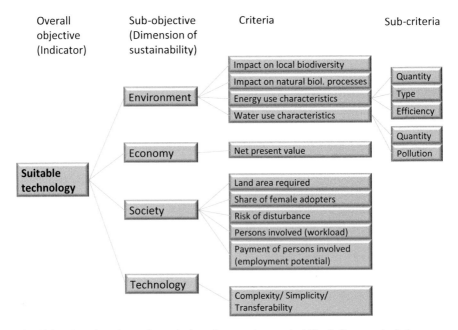

Overall objective (Indicator) — Sub-objective (Dimension of sustainability) — Criteria — Sub-criteria

Fig. 12.2 Hierarchy scheme for analysis and composite sustainability indicator calculation

were first asked which criterion is more important or if they are of equal importance. If one was selected to be more important, experts were then asked how much more important the criterion is. Fifty-one experts were invited to participate in the online survey, out of which 12 took part (23.5 %). The results of this weighting exercise were ambiguous, probably due to the online survey format of the exercise. This format didn't allow for detailed personal oral presentation of the meaning of criteria to experts or questions and answers among survey participants. Although a written introduction was included at the beginning of the survey, the results revealed that respondents had a diverging understanding of some criteria. Further, the AHP method is prone to human error, especially with a large number of pairwise comparisons. For this reason, only five experts had consistency ratios below the recommended threshold. Based on these results, a criteria weight distribution was developed in consultation with a team of interdisciplinary scientists and project partners. Table 12.1 shows all essential criteria and the corresponding weights assigned to them by consensus.

To calculate the composite sustainability indicator, the sum of all weighted and normalized[2] criteria values was built:

---

[2] Where needed.

Table 12.1 Weights assigned to sub-criteria, criteria and sub-objectives of the framework

Sub-objective	Criteria	Weight Sub-objective	Criteria	Sub-criteria
Environment	Water consumption	0.30	0.35	
	Water quantity			0.50
	Water pollution			0.50
	Energy consumption		0.35	
	Energy quantity			0.33
	Type of energy			0.33
	Energy use efficiency			0.33
	Impact on natural biological processes		0.15	
	Impact on local biodiversity		0.15	
Society	Number of people involved (workload)	0.30	0.15	
	Payment of people involved (employment potential)		0.15	
	Risk of disturbance		0.10	
	Share of female adopters		0.30	
	Land area required		0.30	
Economy	Net present value	0.30	1.00	
Technology	Complexity, simplicity, transferability	0.10	1.00	

$$CSI = \sum_i^n w_i \cdot cv_i, \quad (12.1)$$

where CSI is the composite sustainability indicator, w is the weight of criterion i, and cv is the criteria value of criterion i (compare Krajnc and Glavič 2005).

Application of the Tool

Detailed data was collected for 42 technologies, of which 30[3] were included in the data analysis presented here. The following questions were used for selection of technologies for analysis: (1) Can the technology be adopted by an individual or a single household? (2) Is the technology mature and has it been tested successfully in practice many times? (3) Is all information concerning the technology a public good

---

[3] Table 12.2 shows 32 data lines, because two technologies appear twice: the technology leasehold riverbed farming is considered using both original information from the expert and information gathered during an independent validation study on the technology. The sand-based mini hatchery can be used for either chickens or ducks. The tools and items needed for hatching duck or chicken eggs are identical, while handling and economic results differ slightly.

and does the technology have no patent right attached to it? Technologies for which all questions could be answered "yes" were included and are listed in Table 12.2.

Technologies for which no economic data was available were taken into account by using an average normalized net present value.[4] To make economic figures comparable, they were transformed into values per hectare of production. Conversion of monetary values from local currencies to US dollars was done using the OANDA online currency converter,[5] using the conversion rate of 31 March 2013. For criteria for which expert opinions were used, the data was compared among technologies and harmonized, where necessary. For instance, for the amount of water used, three answer options were available for experts to choose from: the technology uses (i) no water (0), (ii) little water (0.5), and (iii) large amounts of water (1). This left little room for distinctions between rainfed crops, intermittently irrigated, and flooded crops. Therefore, all technologies involving wet rice were assigned the score (1), all irrigated crops the score (0.5), and rainfed crops the score (0.2). Where necessary, data was normalized before analysis.

Three technologies will be presented in more detail, namely vermitechnology, broom grass farming on marginal lands, and the mini hatchery model for chickens.

Vermitechnology is a process which uses earthworms to produce good quality compost (vermicompost) through organic waste recycling. The commonly used earthworms include *Eudrillus* sp. *Perionyx* sp., *Eisenia* sp. or any locally-available earthworms living and feeding on the surface of the soil (epigeic worms). A tank of $5 \times 1 \times 1$ m allows about 500 kg of waste to be composted through the activity of worms and microorganisms, producing about 250–300 kg of compost over approximately 1 month. Vermitechnology can either be practiced in tanks or in the ground. However, the major advantage of a tank is the efficiency of composting and keeping the worms captured. The technology requires little investment and technical know-how.

Broomgrass (*Thysanolaena maxima*) is a perennial, high-value, non-perishable Non-Timber Forest Product (NTFP) that can be grown on degraded, steep, or marginal land. Broomgrass is a multipurpose crop: only its panicle is used for brooms. The stems are used by farmers as construction material, fuel, fodder, mulching, or staking crops, or sold to the pulp industry to manufacture paper. The leaves and tender shoots are used as fodder in times of scarcity. Broomgrass farming can generate additional income through cultivation on marginal lands unsuitable for food production. It can also be used as part of an agroforestry system to regenerate degraded land.

A sand-based *mini-hatchery* uses a simple wooden incubator to hatch chicken (and duck) eggs in rural areas to assure a regular supply of chickens (and ducks) for income and food security. The heat that is needed to brood the eggs comes from

---

[4] These technologies are tomato grafting, treadle pump and micro irrigation technology, vermitechnology, windmill, chili and sweet pepper grafting, school gardens, crotalaria, and rainbow trout aquaculture.

[5] http://www.oanda.com/lang/de/currency/converter/.

kerosene lamps: the sand helps to retain and distribute the heat evenly inside the insulated cabinet. Besides the wooden cabinet and wooden or metal trays and racks, it uses cheap local materials such as quilts, sand that can retain the heat, jute sacks, and kerosene. The incubator should be placed in a separate hatchery room. It can assure a regular supply of 1-day-old chickens (or ducks) for income and food security in rural areas.

Results

Technologies are grouped into more suitable and less suitable technologies based on a non-hierarchical cluster analysis (K means) of the Composite Sustainability Index (CSI) (Table 12.2). Groups A to C represent the 13 most suitable technologies based on the data presently available and the analytical assumptions made. Looking at three exemplary technologies, the CSI ranks vermitechnology as the most suitable, followed by broom grass farming, and the mini hatchery for chickens.

The radar chart (Fig. 12.3) reveals more details on the performance of the three examples, vermitechnology (*green line*), broom grass farming (*blue line*), and mini hatchery (*red line*). If the line is close to the outer edge of the diagram, the technology is performing well in terms of the particular criterion. All technologies can be seen to be performing relatively well in terms of water consumption.

The hatchery hardly uses any water (for cleaning only), broom grass farming is a rainfed culture and vermitechnology needs little water to keep the substrate in which the earthworms live and upon which they feed moist. Only the hatchery uses a little energy, which comes, however, from a non-renewable source in the present state of technology design. It has no impact on biological processes, while broom grass prevents soil erosion and vermitechnology positively impacts nutrient cycling. Vermicompost has a better impact on biodiversity than the mini hatchery and broom grass farming. The latter performs not so well in terms of biodiversity, because land areas that were formerly covered with a diversity of wild plants are then cultivated with broom grass alone. All technologies require little input in terms of work; broom grass cultivation even creates local jobs in peak times. None of the compared technologies has a risk of disturbing the neighborhood or creating social conflicts. All technologies are suitable for female adopters, but broom grass farming involves some hard work to prepare the soil during the planting period. The hatchery and vermitechnology can be practiced on a few square meters of land, while only broom grass should be practiced over more extended areas. The composite sustainability indicator is calculated with an area of at least 200 m^2. According to the vermitechnology expert, farmers can sell one kg of compost for 50¢ (US). With initial investment costs of about $50 (US) for a tank that covers 5 m^2, and considering some additional space for charging and discharging the tank, the figures lead to a net present value per hectare of $4.78 million (US) for a 5 year period at 1.5 % interest. This is by far the highest NPV per hectare for the set of technologies included in the analysis. For comparison, the 5 year period NPV of

Table 12.2 Sustainability clusters of selected technologies

Group	Technology name	CSI	Note
A	Vermitechnology	0.494	*example*
B	Vegetable pool	0.15 – 0.23	
	Organic vegetable production in sack		
	Bio intensive school gardens		Mean NPV
	Domestic yam production		
	Stinging nettle for enhancing animal productivity		
	Ecological sanitation		Mean NPV
	Crotalaria against nematode damage of chili		
C	EFSB IPM, Bangladesh (summer + winter crop)	0.10 – 0.13	
	Broom grass farming on marginal lands		*example*
	Backyard poultry farming		
	Sugiharto organic fertilizer (if cows are already available)		
	Leasehold riverbed vegetable production		
D	Treadle pump and micro-irrigation technology for smallholders	0.06 – 0.095	Mean NPV
	Windmill		Mean NPV
	Floating vegetable garden		
	Leasehold riverbed vegetable production (validated)		
	Floating cultivation on organic bed		
	Cricket farming		
	Sandbar vegetable cultivation technique		
	Non chemical IPM technology package for tomato cultivation		
	Hybrid tomato seed and tomato production		
E	Mini hatchery for chickens	0.03 – 0.05	*example*
	Mini hatchery for ducks		
	Open cultivation of off-season tomatos		
	Chili and sweet pepper grafting		Mean NPV
	Integrated rice-duck farming technology		
	Tomato grafting		
F	Cage fish culture	−0.07 – −0.01	
	Improved Kharif paddy production system		
	Improved cultivation of rainfed maize-based cropping systems		
	Himalayan rainbow trout aquaculture technology		Mean NPV

Note: technologies with "*example*" in the right column are presented in more detail in this chapter

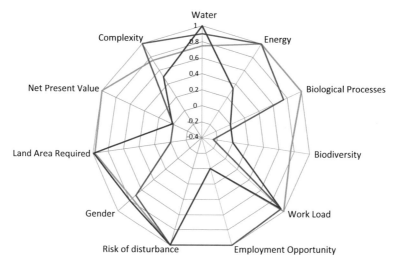

Fig. 12.3 Suitability radar chart for vermitechnology (*green*), broom grass farming (*blue*), and mini hatchery (*red*)

broom grass farming and the mini hatchery is $1,881 and $50,016 (US), respectively. From the amount of knowledge and skills a person has to master for successful operation, the mini hatchery appears to be the most complex and broom grass farming the easiest of the three technologies.

Limitations of the Framework

Although the objective of identifying suitable technologies is soundly justified and well-grounded, there are several inherent issues and limitations that need to be kept in mind when interpreting results and formulating extension recommendations on the suitability of technologies.

Firstly, the analytical framework presented in this brief should not be considered a tool for comparing the sustainability of different technology types against each other. Rather, it provides information on various aspects of sustainability for a given technology and can serve as a decision tool for comparing different but related technologies with each other.

Secondly, combining biophysical information with social and economic information into a single indicator carries the inherent problem of incommensurability between different dimensions of sustainability (Rigby et al. 2001). Another issue in respect to composite indicator calculation relates to compensation between the values of its components. For instance, low or no energy consumption cannot balance a high use of water or a low net present value. However, this issue can be overcome by looking at the underlying data that can be presented visually with radar charts.

Also, our assessment is based on inputs (e.g., energy and water), rather than actual sustainability outcomes. This is due to the limited availability of impact data. *"It is commonly the case that assessments of sustainability operate by prediction rather than direct evaluation of impact. . . . One of the key issues is the extent to which one can map with confidence from inputs to environmental impact."* However, we, as others, believe that the assumptions that we have made on impacts are, while crude, nevertheless robust (Rigby et al. 2001).

Future Research Needs

The research presented here is not final, results present the current state of knowledge, and data and efforts should go on to further increase the data set and refine the methodology. More sophisticated criteria, like the actual amount of water used, waste water produced, actual amounts of inputs and outputs, from recycled farm materials or from offsite, could be included if reliable data could be traced for all technologies. The effort by local experts to collect necessary data for sustainability analysis, and the exchange with the Food Security Center (FSC) to fill data gaps and validate expert data and opinions is ongoing. Furthermore, when additional technologies are included in the comparison, normalized variable values are likely to change, affecting the overall results. For this reason, the presented results are based on the best data available to date. Future findings might change the sustainability ranking and grouping of technologies. Newly emerging tools, like the SAFA small App tool (FAO forthcoming[6]), will allow for cross-checking and validating the generated results when compared with each other.

Open Access This chapter is distributed under the terms of the Creative Commons Attribution-Noncommercial 2.5 License (http://creativecommons.org/licenses/by-nc/2.5/) which permits any noncommercial use, distribution, and reproduction in any medium, provided the original author(s) and source are credited.

The images or other third party material in this chapter are included in the work's Creative Commons license, unless indicated otherwise in the credit line; if such material is not included in the work's Creative Commons license and the respective action is not permitted by statutory regulation, users will need to obtain permission from the license holder to duplicate, adapt or reproduce the material.

---

[6] http://www.fao.org/nr/sustainability/sustainability-assessments-safa/safa-small-app/en/.

References

Aistars GA (ed) (1999) A life cycle approach to sustainable agriculture indicators. In: Proceedings. The center for sustainable systems, University of Michigan, Ann Arbor, p. 118

Dantsis T, Douma C, Giourga C, Loumou A, Polychronaki EA (2010) A methodological approach to assess and compare the sustainability level of agricultural plant production systems. Ecol Indic 10(2):256–263. doi:10.1016/j.ecolind.2009.05.007

Dewulf J, van Langenhove H (2005) Integrating industrial ecology principles into a set of environmental sustainability indicators for technology assessment. Resour Conserv Recycl 43(4):419–432. doi:10.1016/j.resconrec.2004.09.006

Doluschitz R, Hoffmann C (2013) Überblick und Einordnung von Bewertungssystemen zur Nachhaltigkeitsmessung in Landwirtschaft und Agribusiness. In: Hofmann M, Schultheiß U (eds) Steuerungsinstrumente für eine nachhaltige Land- und Ernährungswirtschaft. Stand und Perspektiven, KTBL Schrift 500. Kuratorium für Technik und Bauwesen in der Landwirtschaft e.V, Darmstadt, pp 34–47

Dunmade I (2002) Indicators of sustainability: assessing the suitability of a foreign technology for a developing economy. Technol Soc 24(4):461–471. doi:10.1016/S0160-791X(02)00036-2

FAO (1995) FAO trainer's manual, Vol. 1: sustainability issues in agricultural and rural development policies. Food and Agriculture Organization of the United Nations, Rome

Godfray HCJ, Beddington JR, Crute IR et al (2010) Food security: the challenge of feeding 9 billion people. Science 327(5967):812–818. doi:10.1126/science.1185383

Grenz JAN, Thalmann C (2013) Internationale Ansätze zur Nachhaltigkeitsbeurteilung in Landwirtschaft und Wertschöpfungsketten. In: Hofmann M, Schultheiß U (eds) Steuerungsinstrumente für eine nachhaltige Land- und Ernäh-rungswirtschaft. Stand und Perspektiven, KTBL Schrift 500. Kuratorium für Technik und Bauwesen in der Landwirtschaft e.V, Darmstadt, pp 23–33

Kemmler A, Spreng D (2007) Energy indicators for tracking sustainability in developing countries. Energy Policy 35(4):2466–2480. doi:10.1016/j.enpol.2006.09.006

Krajnc D, Glavič P (2005) A model for integrated assessment of sustainable development. Resour Conserv Recycl 43(2):189–208. doi:10.1016/j.resconrec.2004.06.002

Rigby D, Woodhouse P, Young T, Burton M (2001) Constructing a farm level indicator of sustainable agricultural practice. Ecol Econ 39(3):463–478. doi:10.1016/S0921-8009(01)00245-2

Rogers EM (2003) Diffusion of innovations, 5th edn. Free Press, New York

Saaty TL (1990) How to make a decision: the analytic hierarchy process. Eur J Oper Res 48(1):9–26. doi:10.1016/0377-2217(90)90057-I

Singh RK, Murty HR, Gupta SK, Dikshit AK (2012) An overview of sustainability assessment methodologies. Ecol Indic 15(1):281–299. doi:10.1016/j.ecolind.2011.01.007

Smith C, McDonald G, Thwaites R (2000) TIM: assessing the sustainability of agricultural land management. J Environ Manag 60(4):267–288. doi:10.1006/jema.2000.0384

UN (2012) The future we want. Resolution 66/288. United Nations, Geneva

Veleva V, Ellenbecker M (2001) Indicators of sustainable production: framework and methodology. J Clean Prod 9(6):519–549. doi:10.1016/S0959-6526(01)00010-5

Chapter 13
Land Degradation and Sustainable Land Management Innovations in Central Asia

Alisher Mirzabaev

Abstract Land degradation affects about one-third of global terrestrial area and is having negative impacts on the incomes and food security of agricultural populations. The problem is also acute in the irrigated, rainfed and rangeland areas of Central Asia. There are numerous sustainable land management (SLM) technologies and practices which can help in addressing land degradation. However, many of these technologies have not been adopted at larger scales. The key underlying factors incentivizing SLM adoptions in Central Asia are found to be better access to markets, credit and extension, and secure land tenure. The adoption of SLM technologies can lead to improvements in income among agricultural households, especially the poor. However, SLM technologies alone cannot address land degradation in the region. SLM-friendly policies and institutions are essential.

Keywords Land degradation • Food security • Central Asia • Sustainable land management • Technology adoption

Introduction

Land degradation is a global problem affecting 29 % of the global area across all agro-ecologies and 3.2 bln people around the world (Le et al. 2014), especially the poorest (Nachtergaele et al. 2010). The Central Asian countries of Kazakhstan, Kyrgyzstan, Tajikistan, Turkmenistan and Uzbekistan (Fig. 13.1) are also strongly affected by land degradation, with negative consequences on crop and livestock productivity, agricultural incomes, and rural livelihoods (Pender et al. 2009). The costs of land degradation in the region are substantial (Mirzabaev et al. 2015), with negative implications, especially on the livelihoods of the poorest rural agricultural households (ibid.).

Land degradation in the region is best analyzed along its major agro-ecological zones: secondary salinization is the biggest problem in the irrigated lands, soil erosion in the rainfed and mountainous areas, and loss of vegetation, desertification

A. Mirzabaev (✉)
Center for Development Research (ZEF), University of Bonn, Bonn, Germany
e-mail: almir@uni-bonn.de

© The Author(s) 2016
213
F.W. Gatzweiler, J. von Braun (eds.), *Technological and Institutional Innovations for Marginalized Smallholders in Agricultural Development*,
DOI 10.1007/978-3-319-25718-1_13

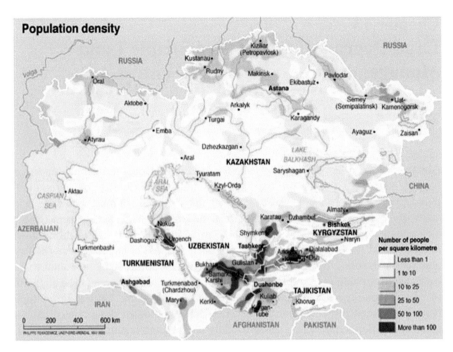

Fig. 13.1 Population density in Central Asia (Source: Philippe Rekacewicz, UNEP/GRID-Arendal, http://www.grida.no/graphicslib/detail/population-density-central-asia_30dd)

or detrimental change in the vegetation composition in the rangelands (Gupta et al. 2009). Secondary salinization is estimated as covering from 40 % to 60 % of the irrigated areas in the region (Qadir et al. 2008), while 11 million ha of rainfed areas in Kazakhstan are affected by wind erosion (Pender et al. 2009). The rangelands cover about 65 % of Central Asia (Mirzabaev 2013), of which 15–38 %, depending on the country, have been found to have degraded between 1982 and 2006 (Le et al. 2014).

Land degradation affects the poorest parts of the region the hardest (Mirzabaev et al. 2015). In spite of this, the adoption of SLM technologies in Central Asia remains inadequate (Gupta et al. 2009), being especially low among poor agricultural households (Mirzabaev et al. 2015). This is despite the availability of many such technologies which have been demonstrated to be economically more profitable than traditional practices (Pender et al. 2009).

In this context, the present study seeks to answer two research questions:

1. What are the key constraints, drivers and impacts of SLM adoption in the region?
2. What are the lessons learnt from previous successful experiences of SLM adoption?

To answer these questions, the existing literature on land degradation and sustainable land management in the region has been analyzed and systematically evaluated.

The Conceptual Framework

The current study is guided by the Economics of Land Degradation (ELD) conceptual framework developed in Nkonya et al. (2015) and von Braun et al. (2013). The conceptual framework (Fig. 13.2) categorizes the causes of land degradation into proximate and underlying, the interactions of which result in different levels of land degradation. Proximate causes of land degradation are those that have a direct effect on the terrestrial ecosystem, such as biophysical natural causes and unsustainable land management practices. The underlying causes of land degradation are those that indirectly affect the proximate causes of land degradation, such as institutional, socio-economic and policy factors (Nkonya et al. 2015).

Inaction against land degradation would lead to continuation, or even acceleration, of land degradation and its associated costs, including the losses in ecosystem services. The lack of appropriate integration of the value of ecosystem services into decision-making – because many of these services are not traded and have no market prices – would mean their value is equalized to zero, leading to more land degradation. However, besides its benefits, action against land degradation also involves costs (von Braun et al. 2013).

The conceptual framework also highlights the role of off-site costs and benefits of land degradation. The actions of individual land users are usually guided by the on-site costs of land degradation and on-site benefits from taking SLM actions. In case on-site costs of land degradation do not exceed the costs of adopting SLM, it may be economically irrational for landusers to adopt SLM practices. However, this lack of SLM adoption may result in significant off-site costs to be borne by third parties or by the society as a whole, necessitating public action for internalizing these externalities. In the case of the poor smallholder farmers often barely eking out their livelihoods from degrading lands, the application of the principle "the polluter pays" may not be feasible. Instead, there may be a need for supportive polices to encourage their adoption of SLM. As long as the social benefits from applying the SLM measures exceed the social costs incurred from incentivizing the land users to adopt them, such public interventions supporting SLM would still be socially more optimal than inaction. Moreover, as the experiences from around the world show, in many instances, poor smallholder landusers do not adopt SLM measures even when the adoption costs are much lower than the on-site benefits from SLM adoption. Thus, public action stimulating SLM is strongly justified, not only in terms of minimizing negative externalities of land degradation, but also for reducing poverty among smallholder landusers.

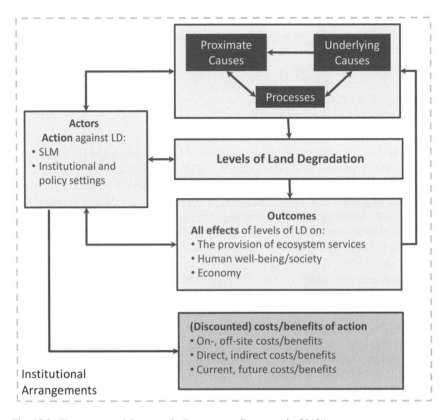

Fig. 13.2 The conceptual framework (Source: von Braun et al. (2013))

Constraints, Drivers and Impacts of SLM Adoption in Central Asia

The constraints to adoption of SLM technologies in the region are numerous and have varying salient features across the major agro-ecological zones (Table 13.1). However, the major constraints across all agro-ecologies seem to be similar. SLM adoption usually does not occur because of one single factor, but is a result of complex interaction of various drivers. For example, in irrigated areas, continued subsidies for irrigation create disincentives to economize on water and adopt water-saving technologies. Across all agro-ecologies, the lack of farmers' and pastoralists' awareness or training in use of appropriate practices, and the lack of adaptation of practices to local conditions, are considered major constraints, especially in combination with poorly functioning extension services (Gupta et al. 2009). The lack of access to credit inhibits the purchase of appropriate equipment, such as, for example, raised bed planters, or conservation agriculture machinery, such that farmers often have to rely on outdated and unproductive equipment from the Soviet

Table 13.1 Factors constraining SLM adoption in Central Asia

Irrigated	Rainfed	Rangeland	Mountainous
Population pressure, low incomes	Shortage of labor, working capital and capital assets, such as new machinery	Population pressure, poverty among small-scale herders	Population pressure, poverty
Subsidized irrigation water	Risk-averseness and slow behavioral change in upgrading to more sustainable agricultural practices (specifically, from excessive tillage to conservation agriculture)	Lack of market access	Lack of market access
Land tenure insecurity	Mono-cropping practices	Breakdown of collective action institutions regulating and facilitating access to remote rangelands	Breakdown of collective action institutions regulating and facilitating access to remote rangelands
Insufficient information on SLM technologies, Poor quality of agricultural extension	Insufficient information on SLM technologies, Poor quality of agricultural extension	Insufficient information on SLM technologies, Poor quality of agricultural extension	Insufficient information on SLM technologies, Poor quality of agricultural extension
Lack of access to credit for adopting SLM		Shortage of herding labor due to rural out-migration	Lack of access to credit for adopting SLM
Production and marketing controls			Shortage of labor due to rural out-migration
Institutional disconnections between various levels of water management			Land tenure insecurity

Sources: CACILM (2006a, b, c, d, e), Pender et al. (2009), Gupta et al. (2009), Kerven (2003)

era (ibid.). Agricultural production and marketing decisions for major crops are controlled by the governments in some countries, and market institutions are underdeveloped or lacking. Coupled with continuing land tenure insecurity, these limit producer incentives and serve as powerful deterrents to SLM adoption (Pender et al. 2009). In rangeland areas, effective pasture management mechanisms are lacking and pasture leasing is not clearly regulated in most countries in the region. Similarly, the interaction of lower disposable farm profits and low access to credit markets prevents farmers from investing in costly, but profitable in the long- term, technologies for sustainable land management. Poverty and low market access, especially in mountainous areas, but also in all other agro-ecologies, increase risk aversion and limit the available resources that could be invested in SLM.

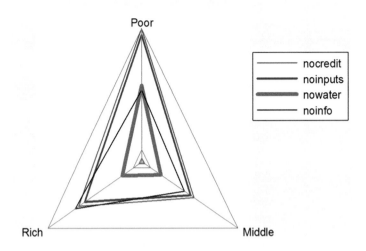

Fig. 13.3 Constraints on SLM adoption in Central Asia (Source: Mirzabaev (2013))

The above constraints and drivers of SLM adoption were based on expert evaluations. In this regard, it would also be important to look at household level evidence of the constraints on the adoption of SLM technologies. Analyzing the household level data from surveys, Mirzabaev (2013) indicates that major constraints for SLM adoption in the region pointed out by farming households themselves are lack of access to credit and affordable inputs, including water, but also lack of information about SLM technologies (Fig. 13.3). These factors seem to be especially constraining for the adoption of SLM technologies by poor farmers.

In this regard, it is also telling that the adoption of SLM practices was found to be lower among the poorest agricultural households (Fig. 13.4), despite the fact that these poorest households, in most cases, seem to be operating more degraded land than the richer households in the sample (Fig. 13.5).

More formally, analyzing these household surveys, Mirzabaev et al. (2015) find that key underlying factors incentivizing SLM adoption in Central Asia are better market access, access to extension, learning about SLM from other farmers, private land tenure among smallholder farmers, livestock ownership among crop producers, lower household sizes and lower dependency ratios. Better market access is likely to provide more incentives for increased production and productivity, making the opportunity cost of foregone benefits due to land degradation much higher, thus incentivizing the households for SLM adoption (ibid.). Similarly, access to extension is found to increase the number of SLM adoptions by increasing farmers' knowledge about SLM practices and their awareness of the benefits of SLM. The greater number of SLM technologies farmers know, the more SLM technologies they adopt (ibid.).

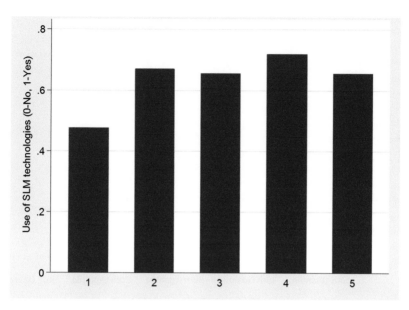

Fig. 13.4 Use of SLM technologies among agricultural households with different incomes. Note: categories: 1-poorest, … 5-richest (Source: Mirzabaev (2014))

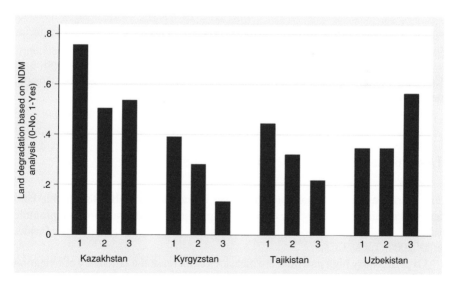

Fig. 13.5 Status of land quality among agricultural households of different incomes. Note: categories: 1-poor, 2-middle, 3-rich (Source: Mirzabaev (2014))

The adoption of SLM technologies could lead to better livelihood outcomes for the agricultural households, specifically the poorest 10 % of them (Mirzabaev 2014). Each adopted SLM technology was found to be likely to increase the monetary value of per capita food consumption by 3 % for the poorest 10 % of

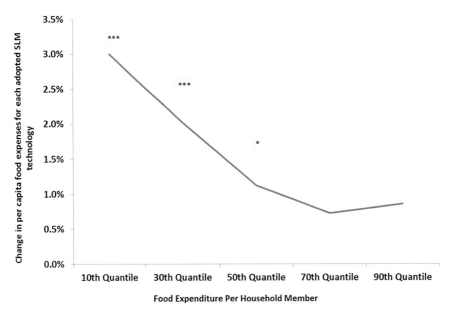

Fig. 13.6 Adoption of SLM technologies and per capita food expenses (Source: Mirzabaev (2014))

agricultural households, while the effect is less pronounced for the richer categories of agricultural households (ibid.: Fig. 13.6).

Summarizing these findings, the key factors incentivizing SLM adoption in Central Asia have been found to be better access to markets, credit and extension, and secure land tenure.

Lessons Learnt from Previous Experiences

The literature points to many available and economically viable sustainable land and water management practices that can help reduce land degradation and promote sustainable crop and livestock production in Central Asia (Gupta et al. 2009; Pender et al. 2009; Table 13.2).

Of particularly high potential are measures to increase the efficiency of irrigation water use: using such technologies as cutback and alternate furrow irrigation, raised bed cultivation, improved leaching methods, conjunctive or alternating use of drainage and fresh water, conservation tillage, and crop rotations and diversification. Other promising measures include use of soil and water conservation measures, organic soil fertility management practices, improved use of fertilizers, use of rock phosphate and phosphogypsum where these are useful and economical and

Table 13.2 Major SLM technologies recommended in the literature for Central Asia

Irrigated	Rainfed	Rangeland	Mountainous
Improved irrigation technologies (cutback, alternate furrow, micro-furrow, drip irrigation)	Zero tillage and direct seeding	Rotational grazing	Strip cropping
Laser land levelling	Mulching	Establishing protected areas	Drip irrigation
Permanent raised beds	Water harvesting and supplementary irrigation	Reseeding	Use of plastic chutes for irrigation
Zero tillage, direct seeding, and mulching	Crop diversification with legumes	Planting halophytic and drought-tolerant plants	Contour furrow irrigation
Crop diversification with legumes	Chemical fallow	Increasing herd mobility	Terracing
Use of phosphogypsum for sodic soils	Continuous cropping without summer fallow	Integrated crop-livestock-rangeland management	
Tree plantations and bio-drainage, Planting halophytic plants (ex, licorice)	Use of fertilizers	Chemical fallow	

Sources: Mirzabaev (2014), who compiled from Gupta et al. (2009), Pender et al. (2009), Toderich et al. (2002), Toderich et al. (2008a, b), Lamers and Khamzina (2008), Lamers et al. (2009), ICARDA (2002, 2003, 2004, 2005, 2006)

improved rangeland and fodder management (ICARDA 2002, 2003, 2004, 2005, 2006).

The impacts of most of these measures are highly context-dependent (Pender et al. 2009). Despite the higher profitability of many of these technologies as compared to traditional practices, the adoption has been limited (ibid). This is due to the large number of factors highlighted in the section above. For example, Pender et al. (2009) indicate that raised bed planters may cost about $4000 (US) to import from India (as there is no local production). Although the annualized cost per hectare of such equipment is low, the high initial cost could be prohibitive for individual smallholder farmers, requiring adequate credit, collective action or development of lease markets to make this equipment accessible (ibid.).

However, despite many constraints, there have been two major successful new technology adoptions in the region over the last two decades, which can provide lessons on the adoption of SLM practices. The first is the planting of winter wheat among standing cotton, instead of the earlier practice of removing the cotton stalks, tilling the land and only then planting the winter wheat. This practice has now been widely adopted in almost all irrigated cotton-winter wheat crop rotations in the region, especially in Uzbekistan. The second is the spread of reduced and zero till technologies in northern Kazakhstan for rainfed production of grain crops (Kienzler et al. 2012). Both of these technologies are now applied on millions of hectares in the region.

In both cases, the wide-scale adoption seems to have been made possible by the confluence of several factors, such as strong government support, strong market incentives, availability of local expertise and the national agricultural research systems actively working to develop these technologies. To illustrate, the Government of Uzbekistan was quick in taking up the technology of direct wheat planting into standing cotton developed by the national agricultural research system and has broadly mandated the use of the technology starting from early 1990s. Until that time, cotton was by far the single most important crop planted in the country. However, with the collapse of trade and mutual exchanges after the break-up of the Soviet Union, the need emerged to develop national wheat production rapidly to maintain food security in the country. Millions of hectares were shifted from cotton to wheat in a matter of a couple of seasons. The crop calendars for cotton and winter wheat left a very narrow window between the harvest of cotton in the fall and the planting of winter wheat. Moreover, additional tillage operations also required massive use of scarce fuel resources. These three factors, lack of time between cotton harvest and winter wheat planting and fuel savings, plus strong Government mobilization, have led to rapid adoption of this technology throughout Uzbekistan. Similarly, conservation agriculture and its elements have been researched in Kazakhstan for many decades. Some elements, such as reduced tillage, were adopted even during the Soviet times. From the 1990s, there have been massive research and demonstration efforts regarding zero tillage by the national and international agricultural centers working in the country, but without much wide-scale adoption until the early-to-mid 2000s, when the Government of Kazakhstan provided subsidies the equivalent of about $7 (US) per ha for the adoption of conservation agriculture practices in the country. Although this amount seems quite small, coupled with significant cost savings in fuel use, especially in the context of super-sized rainfed farms in the north of the country, this incentive has proved to be crucial in rapid spread of conservation tillage in the country, reducing soil erosion and making grain production more resilient to recurrent droughts in northern Kazakhstan (Kienzler et al. 2012).

So, these experiences yield the lesson that the availability of SLM technologies is, of course, vital. However, in the context of Central Asia, at least, but probably also beyond, wide-scale adoption requires cooperation between the Governments, research institutes, and extension services, and all these should be supported by market forces and farmer incentives, and not work against the latter. To give a different example, the Governments in the region have been trying to promote drip irrigation through subsidies and soft loan programs. However, the success of these initiatives has so far been limited. The water is not priced and its supply is highly subsidized in the region. In such a context, drip irrigation loses its major attraction, i.e., saving water resources; because water is free, there are no incentives to save it at the individual farm level, even if there are strong incentives to save water at the national level since, in the context of water scarcity, overuse in upstream areas would mean lack of water in downstream areas.

Conclusions

The key underlying factors incentivizing SLM adoption in Central Asia are found to be better market access, access to extension and credit, access to well-informed peer networks among farmers, private land tenure among smallholder farmers, and livestock ownership among crop producers. Adopting SLM technologies could have positive impacts on rural household food consumption, especially among the poorest. However, SLM technologies alone cannot address land degradation in the region. SLM-friendly policies and institutions are essential. The examples of success stories of sustainable land management reviewed here have occurred as a result of the combination and interaction of technological, social and economic changes, achieved through synergies of bottom-up and top-down approaches in the region.

Open Access This chapter is distributed under the terms of the Creative Commons Attribution-Noncommercial 2.5 License (http://creativecommons.org/licenses/by-nc/2.5/) which permits any noncommercial use, distribution, and reproduction in any medium, provided the original author(s) and source are credited.

The images or other third party material in this chapter are included in the work's Creative Commons license, unless indicated otherwise in the credit line; if such material is not included in the work's Creative Commons license and the respective action is not permitted by statutory regulation, users will need to obtain permission from the license holder to duplicate, adapt or reproduce the material.

References

Central Asian Countries Initiative for Land Management (CACILM) (2006a) Republic of Kazakhstan national programming framework. Prepared by UNCCD National Working Group of the Republic of Kazakhstan. Draft, 01 Feb 2006

CACILM (2006b) Republic of Kyrgyzstan national programming framework. Prepared by UNCCD national working group of the Kyrgyz Republic. Draft, 01 Feb 2006

CACILM (2006c) Republic of Tajikistan national programming framework. Prepared by UNCCD National working group of the Republic of the Republic of Tajikistan. Draft, 14 Mar 2006

CACILM (2006d) Turkmenistan national programming framework. Prepared by Turkmenistan UNCCD national working group. 28 Febr 2006

CACILM (2006e) Republic of Uzbekistan national programming framework. Prepared by Republic of Uzbekistan UNCCD national working group. Draft, 28 Feb 2006

Gupta R, Kienzler K, Mirzabaev A, Martius C, de Pauw E, Shideed K, Oweis T, Thomas R, Qadir M, Sayre K, Carli C, Saparov A, Bekenov M, Sanginov S, Nepesov M, Ikramov R (2009) Research prospectus: a vision for sustainable land management research in Central Asia. ICARDA Central Asia and Caucasus program, Sustainable agriculture in Central Asia and the caucasus series no 1. CGIAR-PFU, Tashkent, p 84

ICARDA (2002) Integrated feed and livestock production in the steppes of Central Asia. Project annual report (2001–2002), IFAD technical assistance grant (TAG): ICARDA-425. International Center for Agricultural Research in the Dry Areas, Beirut, p 207

ICARDA (2003) On-farm soil and water management for sustainable agricultural systems in Central Asia, ADB-RETA 5866. Tashkent

ICARDA (2004) Annual report. Improving rural livelihoods through efficient on-farm water and soil fertility management in Central Asia. Phase II (2004–2006) Project supported by ADB: RETA 6136. Tashkent

ICARDA (2005) Annual report. Improving rural livelihoods through efficient on-farm water and soil fertility management in Central Asia. Phase II (2004–2006) Project supported by ADB: RETA 6136. Tashkent

ICARDA (2006) Annual report. Improving rural livelihoods through efficient on-farm water and soil fertility management in Central Asia. Phase II (2004–2006) Project supported by ADB: RETA 6136. Tashkent

Kerven C (ed) (2003) Prospects for pastoralism in Kazakhstan and Turkmenistan: from state farms to private flocks. Routledge Curzon, London, p 276

Kienzler KM, Lamers JPA, McDonald A, Mirzabaev A, Ibragimov N, Egamberdiev O, Ruzibaev E, Akramkhanov A (2012) Conservation agriculture in Central Asia – what do we know and where do we go from here? Field Crops Res 132:95–105

Lamers JPA, Khamzina A (2008) Woodfuel production in the degraded agricultural areas of the Aral Sea Basin, Uzbekistan. Bois et Forêts des Tropiques 297:47–57

Lamers JPA, Bobojonov I, Khamzina A, Franz J (2009) Financial analysis of small-scale forests in the Amu Darya Lowlands, Uzbekistan. Forests, Trees and Livelihoods (Accepted Vol 19–1)

Le QB, Nkonya E, Mirzabaev A (2014) Biomass productivity-based mapping of global land degradation hotspots. ZEF Discussion Papers No 192. Center for Development Research, Bonn, Feb 2014, p 42

Mirzabaev A (2013) Climate volatility and change in Central Asia: economic impacts and adaptation. Doctoral thesis at Faculty of Agriculture, University of Bonn, Bonn

Mirzabaev (2014) Building the resilience of the poor through sustainable land management. Presentation during the IFPRI 2020 resilience conference side-event on 15 May 2014 in Addis Ababa, Ethiopia

Mirzabaev A, Goedecke J, Dubovyk O, Djanibekov U, Nishanov N, Aw-Hassan A (2015) Economics of land degradation in Central Asia. In: Nkonya E, Mirzabaev A, von Braun J (eds) The economics of land degradation and improvement – a global assessment for sustainable development. Springer, Dordrecht

Nachtergaele F, Petri M, Biancalani R, Van Lynden G, Van Velthuizen H (2010) Global Land Degradation Information System (GLADIS). Beta Version. An information database for land degradation assessment at global level, Land degradation assessment in Drylands technical report No 17. Food and Agriculture Organization of the United Nations, Rome

Nkonya E, Mirzabaev A, von Braun J (2015) The economics of land degradation, and improvement – a global assessment for sustainable development. Springer, Dordrecht

Pender J, Mirzabaev A, Kato E (2009). Economic analysis of sustainable land management options in Central Asia, Final report for the ADB. International Food Policy Research Institute and International Center for Agricultural Research in the Dry Areas, Washington DC/Beirut

Qadir M, Noble AD, Qureshi AS, Raj Gupta, Yuldashev T, Karimov A (2008) Land and water quality degradation in Central Asia: a challenge for sustainable agricultural production and rural livelihoods. (Submitted)

Toderich K, Tsukatani T, Mardonov B, Gintzburger G, Zemtsova O, Tsukervanik E, Shuyskaya E (2002) Water quality, cropping and small ruminants: a challenge for the future agriculture in dry areas of Uzbekistan, Discussion paper No 553. Kyoto Institute of Economic Research, Kyoto University, Kyoto

Toderich K, Tsukatani T, Shoaib I, Massino I, Wilhelm M, Yusupov S, Kuliev T, Ruziev S (2008a) Extent of salt-affected land in Central Asia: biosaline agriculture and utilization of salt-affected resources, Discussion paper No 648. Kyoto Institute of Economic Research, Kyoto

Toderich K, Shoaib I, Juylova E, Rabbimov A, Bekchanov B, Shuyskaya E, Gismatullina L, Osamu K, Radjabov T (2008b) New approaches for biosaline agriculture development, management and conservation of sandy desert ecosystems. In: Abdelly C, Ozturk M, Ashraf M, Grignon K (eds) Biosaline agriculture and high salinity tolerance. Birkhäuser Verlag, Basel

von Braun J, Gerber N, Mirzabaev A, Nkonya E (2013) The economics of land degradation, ZEF working paper 109. Center for Development Research, Bonn

Chapter 14
Biomass-Based Value Webs: A Novel Perspective for Emerging Bioeconomies in Sub-Saharan Africa

Detlef Virchow, Tina D. Beuchelt, Arnim Kuhn, and Manfred Denich

Abstract Growing demand for increasingly diverse biomass-based products will transform African agriculture from a food-supplying to a biomass-supplying sector, including non-food agricultural produce, like feed, energy and industrial raw materials. As a result, agriculture will become the core part of a biomass-based economy, which has the potential not only to produce renewable biological resources but to convert this biomass into products for various uses. The emerging bioeconomy will intensify the interlinkages between biomass production, processing and trading. To depict these increasingly complex systems, adapted analytic approaches are needed. With the perspective of the "biomass-based value web" approach, a multi-dimensional methodology can be used to understand the interrelation between several value chains as a flexible, efficient and sustainable production, processing, trading and consumption system.

Keywords Biomass-based products • Bioeconomy • Biomass-based value webs • Multi-dimensional methodology • Africa

Challenges for the Food and Agricultural System in Sub-Saharan African Countries

Although there are examples in Sub-Saharan Africa (SSA) in which countries have been able to improve their food and nutrition security significantly over the last decade (e.g., Ghana), the general trend in Africa is still worrisome. While both the share and the number of undernourished people worldwide declined in recent decades, SSA largely failed to follow this trend. The share of undernourished

D. Virchow (✉) • T.D. Beuchelt • M. Denich
Center for Development Research (ZEF), University of Bonn, Bonn, Germany
e-mail: d.virchow@uni-bonn.de; beuchelt@uni-bonn.de; m.denich@uni-bonn.de

A. Kuhn
Institute for Food and Resource Economics, University of Bonn, Bonn, Germany
e-mail: arnim.kuhn@ilr.uni-bonn.de

© The Author(s) 2016   225
F.W. Gatzweiler, J. von Braun (eds.), *Technological and Institutional Innovations for Marginalized Smallholders in Agricultural Development*,
DOI 10.1007/978-3-319-25718-1_14

people in SSA is around 25 %, while in the developing world as a whole, the share of the undernourished is around 14 % (FAO 2013). The dramatic situation has long been disguised by sufficient food production at the global scale, but also by the voicelessness of the rural undernourished. Matters would have been worse if world price levels for cereals and other food had not steadily decreased in real terms since the beginning of the 1970s. Low real food prices made staple food affordable for most urban and rural net consumers, while national production shortfalls could be cushioned by imports. Hence, the African food production and consumption situation was seen as tense, but manageable.

The food price crisis of 2007/08, with the underlying trend of the global supply of agricultural produce no longer satisfying the increasing global demand at low price levels, is challenging the agricultural system in many African countries. The annual increase in productivity in food production in Africa is below the 1.7 % required to meet the goal of feeding Africa's rapidly growing population[1] (Global Harvest Initiative 2011). Due to depleted soil fertility, average cereal yields in SSA have stagnated since the 1960s at 1 t ha^{-1}, while in South and East Asia, yields now reach 2.5 and 4.5 t ha^{-1}, respectively (Gilbert 2012).

The longer-term effects of food price hikes since 2007 that originated, among other reasons, from an increased demand for non-food biomass world-wide are still uncertain. This demand increase is caused by the subsidized demand for biofuel use in the USA and the EU, rising oil prices (and predicted future scarcities) and the need for substitutes, and steeply rising feed demand in the livestock sector that responds to the growing demand for meat and dairy products in emerging economies (Keyzer et al. 2005; Headey and Fan 2008). These 'megatrends' are unlikely to abate in the near future and will keep crop prices, but also prices for input factors such as land, machinery and fertilizer, at levels never expected at the beginning of the new millennium. Land and water scarcity in major importing regions, higher and more volatile food prices, and the demand for non-food biomass (especially biofuels) have also induced a global run on land resources that focuses on, but is not limited to SSA (Brüntrup 2011; Deininger 2011).

Energy is another important concern: Africa – with approximately 13 % of the world population – consumes only 5.6 % of the global energy (2001 data, UN-DESA 2004). Its per capita energy use of about 41 % of the global average is expected to rise with growing trade, increasing (urban) affluence and developing infrastructure (UN-DESA 2004). Current energy supply in Africa – dominated by traditional biomass sources – is inefficient and non-sustainable (Hiemstra-van der Horst and Hovorka 2009; Arnold and Persson 2004). Tree-based fuels (wood and charcoal) represent, respectively, 53 %, 78 % and 92 % of total energy consumption in Senegal, Kenya and Tanzania (UN-DESA 2004). Fuelwood is predominantly collected from forests and woodland, which drives deforestation and forest degradation, not to mention the burden on the mainly female work force. Per capita fuelwood production decreased at a similar order of magnitude as food biomass, but

---

[1] From 1.02 in 2010 to 1.56 (2030) and 1.96 billion people in 2050 (UN-DESA 2012).

much less than other non-food crop biomass. Growing populations are currently raising energy demands by 3 % annually. The increasing scarcity of fuelwood and higher prices for fossil fuels, as well as emerging policies for renewable energy, raise demand for alternative energy sources, particularly for modern biofuels (Popp et al. 2014). Accordingly, large-scale commercial biofuel production schemes are under development (Brüntrup and Herrmann 2010; Deininger 2011).

The nexus between biofuel production and food security depends heavily, among other factors, on the type of biomass and use (local, national, for export), size of production, owner of production/production structure (smallholders, large-scale plantations), location of processing, employment possibilities, policies and regulations (weakly regulated or not), and, thus, on the local and national context (see, e.g., Amigun et al. 2011; Baffes 2013; Deininger 2011; German et al. 2013; Giampietro and Ulgiati 2005). There are signs that positive effects from introducing biomass crops, in particular on marginal land, can be expected, with increased direct farming income, more off-farm income opportunities, and positive infrastructure and technology spillovers (Lynd and Woods 2011), especially when directed at the smallholder sector with local processing and consumption (Dufey et al. 2007). However, negative effects of biofuel crops have also been reported, such as increased food insecurity, loss of access to land, no additional employment opportunities, only short-term casual labor contracts with unclear conditions, and unfulfilled promises regarding employment and service provision (Deininger 2011; Dufey et al. 2007; Mohr and Raman 2013). Second generation biofuels, e.g., generated through the enzymatic breakdown of cellulose, pyrolysis or gasification, are likely to be more compatible with food production, since they can also use non-edible parts of food crops, by-products of agriculture and forestry, as well as municipal and construction waste. They, thus, compete less for land and food than first generation biofuels (Amigun et al. 2011; Dufey et al. 2007; Naik et al. 2010; Thompson and Meyer 2013).

In general, availability, use of and access to food in SSA must be considerably improved in the coming decades. In addition to increasing the productivity of agriculture and, hence, food availability, income opportunities need significant expansion to ensure access to food. African countries also have to find solutions to deal with increasing foreign and domestic large-scale investment in their agricultural sectors, especially for export-oriented biofuel production, and the consequences of rising land demand and prices.

Agricultural growth is critical for Africa's economic and social development because of its contribution to food security, employment, income, and wealth creation, thereby enabling more people to live food-secure by making a living through working in the agricultural sector without producing food themselves. This increasing pressure on African agriculture seems to over-stretch its current production potential. A closer look at agricultural potential in Africa, however, leads to the insight that the increasing demand for food and non-food agricultural produce (biomass) is a unique opportunity for many African countries, given their endowment of natural and human resources. The increasing demand could be used as a driving force to stimulate and intensify economic development and to improve domestic food and nutrition security (Amigun et al. 2011; Dufey et al. 2007).

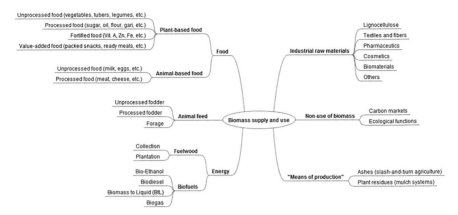

Fig. 14.1 Biomass supply and use

Africa's Biomass Potential

Biomass is biological material from living or recently living organisms. As a renewable resource, it is expected to play an increasing role in the future economies of both low- and high-income countries. It is roughly classified into food, feed, sources of energy and industrial raw materials (Fig. 14.1). Additionally, on-farm or in the cultural landscape, there is also unused biomass that can help generate income through regional or international carbon markets or paid ecosystem services. Another fraction of biomass can be converted to mulch to manage soil fertility or to ashes (slash-and-burn agriculture) to produce crops under resource-poor conditions.

Sub-Saharan Africa, in particular, has a high potential to produce plant biomass (Popp et al. 2014). The total net primary production (NPP)[2] of its land area amounts to 23.8 billion tons per year (Western Europe 3.7 billion tons; in both regions it is ~10 t ha^{-1}yr^{-1}; Krausmann et al. 2008). This quantity sums up the plant biomass of all different types of vegetation cover, such as forests, savannas and cropland. Krausmann et al. (2008) report that, in the year 2000, 3.6 billion tons (15 %) of the continent's NPP was appropriated by humans, as compared to 1.3 billion tons in Western Europe (35 %). In SSA, half of the biomass flow influenced by humans (1.8 billion tons) is destroyed by deliberately set fires (in W. Europe <1 %), predominantly in savanna areas. Biomass imports to and exports from SSA are low and roughly balanced, while Western Europe's biomass imports and exports are 12 times higher than Africa's, and imports exceed exports (Krausmann et al. 2008). Africa has vast land reserves and unused yield potential (Bruinsma 2009). There are limits, however, to the potential of currently unused land from

---

[2] Net primary production (NPP) is the difference between the biomass that plants produce in an ecosystem and the biomass they lose due to metabolic processes (respiration).

environmental perspectives due to important functions of tropical forests and wetlands, biodiversity issues, and high-carbon ecosystems where greenhouse gas emissions larger than the savings could occur upon conversion (Popp et al. 2014).

Focusing on the production and processing of both food and non-food biomass from locally adapted crops within flexible, efficient and sustainable production, processing, trading and consumption systems (hereinafter referred to as "biomass-based value webs") can offer pathways to improve food security and generate jobs and income, particularly in rural areas, to reduce the non-food import bill and even generate urgently needed export earnings, and to maintain environmental assets (Poulton et al. 2006). At first sight, non-food biomass production competes with food crops and feed for land and other resources. However, even within a food security focus, we believe it to be useful to drop the distinction of 'food versus non-food crops', as (1) many crops can be used for both food and non-food purposes and (2) non-food biomass helps generate income, which improves access to food.

Trends in other world regions in comparable geographic latitudes also reveal that neglecting non-food crop biomass is no guarantee for food security. For instance, countries across Asia have managed to increase both the production of food and non-food biomass, resulting in higher agricultural and overall economic growth. This suggests that substantial synergy effects can be harnessed in coordinated, efficient food and non-food biomass production. By contrast, SSA lags behind not only in growth rates of food crop production per capita, but even more so in the production of non-food crop biomass (Fig. 14.2).

The Emerging Bioeconomy

The expected trends in demand for more and more diverse biomass-based products from agricultural land will transform agriculture from a food-supplying to a biomass-supplying sector. This development, which is taking place at different speeds in different countries, has implications for the agricultural sector as a whole. Agriculture will become the core part of a biomass-based economy which comprises farms as producers, as well as the industrial sector and their associated services that produce, process, distribute, consume or in any way use biological resources (Bioeconomy Council 2011).

The application of the bioeconomy concept means transforming "life science knowledge into new, sustainable, eco-efficient and competitive products" (OECD 2008). This is the specific advantage, but also the challenge for African countries: It is the potential not only to produce renewable biological resources but to convert this biomass into products for various uses (such as food, feed, bioenergy and bio-based industrial products), and thereby capture an increasing share of added value. This is relevant for their exports – instead of low-value biomass raw commodities, high value products are exported – but also for the domestic economy: Through value-adding processes, employment and income opportunities can be generated. Inefficient and unsustainable products and consumption patterns can

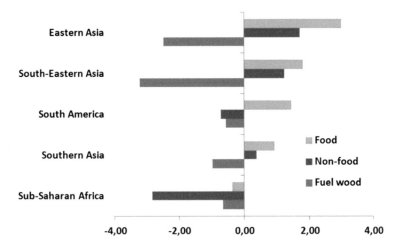

Fig. 14.2 Average annual growth rates of food, non-food, and fuelwood biomass production per capita for selected world regions, 1970–2010 (Source: FAOSTAT Online Database 2012)

be transformed, e.g., a more efficient and less harmful use of fuelwood. Through replacing resource-intensive and environmentally harmful processes through biological processes and new technologies, low-income countries can grow out of the dependency on fossil energy resources, which have to be imported for high value, freeing foreign exchange for other investments. Finally, new and high-value products can be created and utilized on the domestic as well as on the international market.

Together with the transformation from an oil-based to a biomass-based industry, low-income and mainly fossil-oil poor countries will have the opportunity to improve their trade balances by importing less and instead using biomass themselves, including the export of more high-value biomass products. The challenge will be to overcome the traditional division of labor in which low-income countries produce the biomass, while the crucial value addition through processing takes place in high-income biomass-dependent countries (Charles et al. 2007). So many African low-income, but biomass-rich countries have the potential to not only meet their own future demand for biomass-based raw materials (including food) but also to provide high-income and often biomass-dependent countries with biomass (raw and processed) products. For instance, to pursue its bioeconomy strategy, densely populated regions like Europe, being net-consumers, depend on large biomass imports (Erb et al. 2009).

Taking into account that most countries in Africa suffer under a serious unemployment and underemployment problem (AfDB 2013) and that agriculture in Africa will develop to larger management units with a gradual decline in job opportunities at the farm level, alternative and new job opportunities, especially in rural areas, are crucial. With the emerging bioeconomy, rural development opportunities create agricultural employment, and livelihoods are improved through involving small-scale farmers in the production, locating conversion facilities near the feedstock sources in rural areas, and focusing on small-scale processing and, especially, local consumption (Dufey et al. 2007).

Despite these advantages, in Africa, only the Republic of South Africa has devised a bioeconomy strategy so far, whereas outside Africa, this has happened in numerous other developing countries (Goh 2013). However, the shift from an economy mainly using oil-based raw materials to an economy based on renewable resources requires a change of existing approaches to production, consumption, processing, storage, recycling and disposal of goods, and the establishment of some guiding principles. These include, as spelled out in the German bioeconomy strategy, the primacy of food security, the sustainable utilization of natural resources and the environment as a whole, and, finally, the compliance with standards of social responsibility. To ensure that the shift to bioeconomies benefits poor and marginalized farmers and other rural stakeholders, the human and social dimensions also need to be considered (Mohr and Raman 2013). Thus, not only technological approaches but also social aspects, which ensure equitable access to ownership and value along the various biomass value chains, including a fairer income distribution, are required. A biomass-based industry is not necessarily an effective means to link smallholders and the informal economy with the formal economy; adequate policies and support structures are important for fostering this process (Dufey et al. 2007). Furthermore, establishment of a functioning bioeconomy relies strongly on the ecological service function of ecosystems emphasized by the Millennium Ecosystem Assessment (MEA 2005).

The Bioeconomy Will Intensify the Ties Between Biomass Production and Processing

The change of existing technological approaches will imply new and more efficient ways of utilizing biomass, mainly in production and processing, but also in consumption and disposal. The cascading and coupled use of biomass especially will significantly reduce the final "waste", with the ultimate goal of a "zero waste" biomass use.

The increasing demand for diverse biomass products and the intensified utilization of cascading and coupled effects is creating a new, more complex demand structure for biomass products which will determine the biomass production and processing structure. We assume that more and more different biomass resources will be used in the future. To satisfy this increasingly differentiated demand for biomass products, the ties between biomass production and processing will have to be intensified. Furthermore, especially in the processing and trading segment, there will be room for innovations to increase its productivity and efficiency. Consequently, the complexity of value chains of agricultural products is increasing significantly. As described above, the demand side for different biomass types will branch out, with impacts at the handling, processing and trading level leading to an augmented diversity of activities. An example is the current research into second generation biofuels. So-called bio-refineries can simultaneously produce a large range of products, such as fuels, different chemicals and materials, including fertilisers and food ingredients, from many different sources of biomass, including agricultural by-products, residues and waste (Antizar-Ladislao and Turrion-Gomez 2008; Naik et al. 2010). This is an excellent example of how different value chains merge and become linked at one point. However, bio-refineries are complex, highly capital- and knowledge-intensive, and need a vast amount of reliable biomass input. Though it is less directly in conflict with food production, it still may create conflicts for food, feed and fuel requirements of the poor rural population who have used these resources – often for free – and may face difficulties in replacing them (Mohr and Raman 2013). For the poor rural population, bioenergy options, such as small- and medium-scale biogas or gasifiers, and power generators operating with locally available biomass sources, such as biogas from manure and agricultural and forestry by-products, may have a better potential to provide economical and reliable energy services (Antizar-Ladislao and Turrion-Gomez 2008).

The closer ties between biomass production, processing and trading, but also demand and consumption, will lead to an increased merging of the different value chains, forming a partly open system. Especially at the handling, processing and trading level, the feedback loops and cascading effects for utilizing and reutilizing biomass will link or merge existing value chains. The cascades of use and linking-up of value chains can improve resource efficiency, reduce possible areas of competition between uses, and make use of innovation potential (BMEL 2014; BMBF 2011). When, for instance, the bioenergy purchaser (be it for the energy market in Lagos or international markets) wants to buy bioenergy, s/he will be increasingly less interested in whether the biomass source of the bioenergy (in kind of charcoal, for instance) originates from trees or shrubs, or from residues, or other parts from cassava, maize, etc.; the purchaser is mainly interested in buying cheap units of kilojoule. Hence, the charcoal producer, in future, will be interested in optimizing benefits by looking for the biomass resources, which enable him/her to produce charcoal with an optimal cost-benefit ratio.

Change Perspectives Towards an Innovative Approach: Biomass-Based Value Web

Improving food security in Africa is determined through complex and interrelated constraints in availability, access and utilization of food (Hammond and Dubé 2012). To address these constraints, a systems approach is required that matches the complexities (von Braun 2009). Neither a simple agricultural commodity approach nor a non-agricultural industrialization and income generation approach would help overcome food insecurity and its consequences for undernourishment and health. To address the various dimensions of food security while moving towards a sustainable economy based on biological raw materials, completely new approaches to research and development, production, and economy are necessary (Naik et al. 2010) that also consider human and social aspects (Mohr and Raman 2013).

Extending Porter's (1985) classical, firm-based value chain, three main strands of concepts evolved "to explain how global industries are organized and governed, and how, in turn, those relationships affect the development and upgrading opportunities of the various regions and firms involved" (Coe et al. 2008, p. 267). These concepts are the global commodity chain, the global value chain and the global production networks (Coe et al. 2008). However, it is no longer sufficient to analyse the system by following the conventional, more vertical and linear, mainly product-focused commodity or value chain approach. Analytical perspectives are needed which cover the complex pathways of biomass and integrate social, economic and environmental perspectives (Mangoyana et al. 2013). This adds a new perspective to the concepts of value chains and global production networks. Here, the holistic concept of biomass-based value webs becomes instrumental.

A biomass-based value web approach utilizes the 'web perspective' as a multi-dimensional framework to understand the interrelation and linkages between several value chains and how they are governed. Like global production networks, the web approach sees the process of activities that result in a final product as one "in which the flows of materials, semi-finished products, design, production, financial and marketing services are organized vertically, horizontally and diagonally in complex and dynamic configurations" (Henderson et al. 2002, p. 444). Instead of depicting the pathway of one product, the web approach captures the manifold products which are derived from one biomass raw product, but also looks at the whole product mix produced on family farms, the different value chains the households participate in and how they are or could be linked. The web perspective helps to explore synergies between these value chains, identify inefficiencies in biomass use and pinpoint potential for sustainable productivity increases in the entire biomass-based value web of a defined local, national or international system. It includes the analysis of existing and potential recycling processes and cascading uses during the processing phase of biomass, which opens new opportunities to locally capture more of the value-added or create new value (Smith et al. 2000). The cascades of use and interlinking of value chains are instrumental to increasing the efficiency of resources and the sector, reducing possible areas of competition between uses and making use of innovation potential.

An exemplary, very basic biomass-based value web, identified at a stakeholder workshop in Ghana, illustrates basic value web features. Starting from the production/supply side, a crop can be utilized in various ways (e.g., cassava as a multi-purpose crop is used for food, feed, fuel and fiber), leading to a divergent structure and interaction (especially reflecting the fact that first-level outputs may be fed into another processing after utilization). Looking at the system from the demand side, it is obvious that one product in demand (e.g., feed or energy) can be delivered by very different crops from the supply side, i.e., various supplied products converge on one final product. The final product, such as bioethanol, is derived through processing several raw materials or their residues, such as rice, cassava, cotton, sugarcane, and forest biomass. This can be done at different processing locations and enterprises or at the same one.

The structural preconditions, power and social relations embedded in the value chains inevitably shape the value webs and impose a path-dependency; they influence as well as constrain the future trajectory of chain development (Henderson et al. 2002). The web perspective also helps to better identify who participates and benefits in the biomass-based economy (e.g., men or women, small or large producers/processors, national or international actors) and who does not, in which activities and processes, and whether and how the actors co-operate and network with each other. This helps to identify missing links or actors, information gaps, and capacity constraints, as well as governance issues and power relations. Who has the power and the ways in which it is exercised is decisive for how and where more value could be added and captured in low-income countries and, thus, key for a more equitable development (Henderson et al. 2002). The analytical approach contributes to identifying profit and other benefit distributions among the different actors and participants in the whole web. Thus, opportunities can be detected where benefits could be better distributed to be more inclusive, especially towards the poor and marginalized, and where access to food through job and income generation can be increased (Bolwig et al. 2010; Kaplinsky 2000).

The value webs enable a GIS-supported mapping of the spatial distribution of ongoing activities in the web. This allows decision-makers to better identify the locations where new investments in biomass technologies will have a favorable social, economic and environmental impact. For example, decentralized, community-oriented systems contribute to local sustainable development while centralized production systems rather allow for capturing economies of scale (Mangoyana et al. 2013). Using the web approach can also help to analyze and mitigate existing or potential trade-offs of technological innovations in the web. For example, through purposefully introduced cascading uses of cassava peel and maize residues for biochar, jobs could be created which were lost when a processing plant was modernized. Or investments could be directed towards enhanced biomass processing facilities at a women's cooperative to compensate for their current marginalization in the employment market.

Conclusions and Implications

As discussed in this chapter, agricultural growth is critical for Africa's economic and social development because of its contribution to food security, employment, income, and wealth creation. Based on the high potential of biomass in Africa, the increasing demand for biomass as food and non-food agricultural produce is a unique opportunity for many African countries and their populations rather than a risk to their food and nutrition security. However, the system productivity of the food and agricultural sector in SSA has to be increased by maintaining or even improving the economic viability, social acceptability, environmental resilience and technical appropriateness.

The biomass-based value web is a useful scientific and policy perspective for broad, equitable rural development in investigating the potential for agricultural activities and the related economy of a country or region in view of the challenges of coming decades. Three arguments support this position:

- First, given the background of the food versus non-food production problem, the use and further development of multi-purpose crops allows producers to react flexibly to shifting demands. The value chains of multi-purpose crops, and the various forms of demand (food, feed, energy, industrial material) which are jointly satisfied by diverse crops and/or production systems, strongly resemble a web structure.
- Second, we hypothesize that the organizational features of value webs are inherently flexible and, thus, better suited to a volatile price environment, as compared to the classical linear value chains.
- Third, the web perspective better allows for exploring synergies and identifying inefficiencies in an emerging agro-biomass sector and, thus, could be critical to increasing the sector's efficiency.

Increasing the efficiency and throughput in value chains geared towards an "end-product" alone, however, carries the risk of overexploiting natural resources. System management must include waste management and recycling to close material cycles. The value web perspective does not only focus on the creation of narrowly defined 'value added' at different stages. It is also the opportunity to look at non-monetary values, such as material flows, conversion ratios or the use of inputs and production factors like labor and capital. Adopting a web perspective is a systems approach by nature.

Biomass demand is increasing worldwide. Increasing the activities of the domestic processing industry for biomass products in Africa and elsewhere requires the political commitment of governments, as well as international support. Technical and physical infrastructure, a skilled labor force, and financial instruments are part of the solution. However, addressing the human and social aspects should not be detached from bioeconomies and biomass development, but rather combined so that it benefits large shares of the poor population, especially in rural areas. Here, social and gender-responsive approaches need to be incorporated in the planning and design of investments from the beginning. Thus, further research and investment in employment-intensive yet capital-saving processing technologies for biomass commodities in developing countries is important to benefit the poor and unemployed women and men. High-capital and technology-intensive investments, like future biorefineries, will only be an option in a few cases. Small systems focusing on local processing and consumption seem to hold far more promise for the benefit of smallholders and marginalized farmers in particular.

The emerging bioeconomies may help low-income, agriculture-dominated countries to generate jobs and income in the biomass producing, processing and trading sector, in urban and rural areas. The key challenges are to identify pathways for poor countries and poor producers to take advantage of these opportunities, which types of biomass, processing and technologies offer a realistic chance for biomass producers and processors in these countries, and how, at the same time, food security can be enhanced and poverty reduced. Further knowledge gaps exist where the respective value chains and value webs need adjustment and support to ensure that value addition stays in the producing countries and contributes to improving the livelihoods of farmers, fostering small- and medium-sized processors and generating employment opportunities.

Open Access This chapter is distributed under the terms of the Creative Commons Attribution-Noncommercial 2.5 License (http://creativecommons.org/licenses/by-nc/2.5/) which permits any noncommercial use, distribution, and reproduction in any medium, provided the original author(s) and source are credited.

The images or other third party material in this chapter are included in the work's Creative Commons license, unless indicated otherwise in the credit line; if such material is not included in the work's Creative Commons license and the respective action is not permitted by statutory regulation, users will need to obtain permission from the license holder to duplicate, adapt or reproduce the material.

References

AfDB (2013) African statistical yearbook 2013. African Development Bank Group, Abidjan

Amigun B, Musango JK, Stafford W (2011) Biofuels and sustainability in Africa. Renew Sust Energ Rev 15(2):1360–1372. doi:10.1016/j.rser.2010.10.015

Antizar-Ladislao B, Turrion-Gomez JL (2008) Second-generation biofuels and local bioenergy systems. Biofuels Bioprod Biorefin 2(5):455–469. doi:10.1002/bbb.97

Arnold M, Persson R (2004) Reassessing the fuelwood situation in developing countries. Int For Rev 5(4):379–383

Baffes J (2013) A framework for analyzing the interplay among food, fuels, and biofuels. Global Food Secur 2(2):110–116. doi:10.1016/j.gfs.2013.04.003

BMBF (Federal Ministry of Education and Research) (2011) National Research Strategy BioEconomy 2030: our route towards a bio-based economy. Federal Ministry of Education and Research, Berlin

BMEL (Federal Ministry of Food and Agriculture) (2014) National Policy Strategy on Bioeconomy: renewable resources and biotechnological processes as a basis for food, industry and energy. Federal Ministry of Food and Agriculture, Berlin

Bolwig S, Ponte S, Du Toit A, Riisgaard L, Halberg N (2010) Integrating poverty and environmental concerns into value-chain analysis: a conceptual framework. Dev Policy Rev 28 (2):173–194. doi:10.1111/j.1467-7679.2010.00480.x

Bruinsma J (2009) The resource outlook to 2050: by how much do land, water and crop yields need to increase by 2050? Food and Agriculture Organization of the United Nations, Rome

Brüntrup M (2011) Detrimental land grabbing or growth poles? Determinants and potential development effects of foreign direct land investments. Technikfolgenabschätzung – Theorie und Praxis 20(1):3–12

Brüntrup M, Herrmann R (2010) Bioenergy value chains in Namibia: opportunities and challenges for rural development and food security. In: Darnhofer I, Grötzer M (eds) Proceedings to 9th European IFSA symposium: building sustainable rural futures; the added value of systems approaches in times of change and uncertainty, 4–7 July, Vienna, pp 1461–1472

Charles MB, Ryan R, Ryan N, Oloruntoba R (2007) Public policy and biofuels: the way forward? Energy Policy 35(11):5737–5746. doi:10.1016/j.enpol.2007.06.008

Coe NM, Dicken P, Hess M (2008) Introduction: global production networks – debates and challenges. J Econ Geogr 8(3):267–269. doi:10.1093/jeg/lbn006

Deininger K (2011) Challenges posed by the new wave of farmland investment. J Peasant Stud 38 (2):217–247. doi:10.1080/03066150.2011.559007

Dufey A, Vermeulen S, Vorley B (2007) Biofuels: strategic choices for commodity dependent developing countries. Common Fund for Commodities, Amsterdam

Erb K, Krausmann F, Lucht W, Haberl H (2009) Embodied HANPP: mapping the spatial disconnect between global biomass production and consumption. Ecol Econ 69(2):328–334. doi:10.1016/j.ecolecon.2009.06.025

FAO (2013) The state of food insecurity in the world 2013: the multiple dimensions of food security. Food and Agriculture Organization of the United Nations, Rome

FAOSTAT (2012) http://faostat.fao.org, accessed Oct 12, 2014

German Bioeconomy Council (2011) Prioritäten in der Bioökonomie-Forschung. Empfehlungen des BioÖkonomieRats. German Bioeconomy Council, Berlin

German L, Schoneveld G, Mwangi E (2013) Contemporary processes of large-scale land acquisition in Sub-Saharan Africa: legal deficiency or elite capture of the rule of law? World Dev 48:1–18. doi:10.1016/j.worlddev.2013.03.006

Giampietro M, Ulgiati S (2005) Integrated assessment of large-scale biofuel production. Crit Rev Plant Sci 24(5–6):365–384. doi:10.1080/07352680500316300

Gilbert N (2012) African agriculture: dirt poor. Nature 483(7391):525–527. doi:10.1038/483525a

Global Harvest Initiative (2011) 2011 GAP report. Global agricultural productivity report. Global Harvest Initiative, Washington, DC

Goh L (2013) Bioeconomy is the way to go. The Star, Malaysia, 20 Oct 2013

Hammond R, Dubé L (2012) A systems science perspective and transdisciplinary models for food and nutrition security. Proc Natl Acad Sci U S A 109(31):12356–12363. doi:10.1073/pnas.0913003109

Headey D, Fan S (2008) Anatomy of a crisis: the causes and consequences of surging food prices. Agric Econ 39:375–391. doi:10.1111/j.1574-0862.2008.00345.x

Henderson J, Dicken P, Hess M, Coe N, Yeung HW (2002) Global production networks and the analysis of economic development. RIPE 9(3):436–464. doi:10.1080/09692290210150842

Hiemstra-van der Horst G, Hovorka A (2009) Fuelwood: the "other" renewable energy source for Africa? Biomass Bioenergy 33(11):1605–1616

Kaplinsky R (2000) Globalisation and unequalisation: what can be learned from value chain analysis? J Dev Stud 37(2):117–146

Keyzer MA, Merbis MD, Pavel I, van Wesenbeeck C (2005) Diet shifts towards meat and the effects on cereal use: can we feed the animals in 2030? Ecol Econ 55(2):187–202. doi:10.1016/j.ecolecon.2004.12.002

Krausmann F, Erb K, Gingrich S, Lauk C, Haberl H (2008) Global patterns of socioeconomic biomass flows in the year 2000: a comprehensive assessment of supply, consumption and constraints. Ecol Econ 65(3):471–487. doi:10.1016/j.ecolecon.2007.07.012

Lynd LR, Woods J (2011) A new hope for Africa. Nature 474:20–21

Mangoyana RB, Smith TF, Simpson R (2013) A systems approach to evaluating sustainability of biofuel systems. Renew Sust Energ Rev 25:371–380. doi:10.1016/j.rser.2013.05.003

MEA (Millennium Ecosystem Assessment) (2005) Ecosystems and human well-being: synthesis millenium ecosystem assessment. Island Press, Washington, DC

Mohr A, Raman S (2013) Lessons from first generation biofuels and implications for the sustainability appraisal of second generation biofuels. Energy Policy 63:114–122. doi:10.1016/j.enpol.2013.08.033

Naik SN, Goud VV, Rout PK, Dalai AK (2010) Production of first and second generation biofuels: a comprehensive review. Renew Sust Energ Rev 14(2):578–597. doi:10.1016/j.rser.2009.10.003

OECD (2008) Business for development 2008. Promoting commercial agriculture in Africa. OECD Publications, Organisation for Economic Co-operation and Development, Paris

Popp J, Lakner Z, Harangi-Rákos M, Fári M (2014) The effect of bioenergy expansion: food, energy, and environment. Renew Sust Energ Rev 32:559–578. doi:10.1016/j.rser.2014.01.056

Porter M (1985) Competitive advantage: creating and sustaining superior performance. The Free Press, New York

Poulton C, Tyler G, Dorward A, Hazell P, Kydd J, Stockbridge M (2006) All Africa review of experiences with commercial agriculture: summary report. Draft report prepared for the world bank project on competitive commercial agriculture in Africa. Imperial College, London

Smith RW, Broxterman WE, Murad DS (2000) Understanding value webs as a new business modeling tool: Capturing & Creating Value in Adhesives. The ChemQuest Group, Cincinnati

Thompson W, Meyer S (2013) Second generation biofuels and food crops: co-products or competitors? Global Food Sec 2(2):89–96. doi:10.1016/j.gfs.2013.03.001

UN-DESA (2004) Sustainable energy consumption in Africa. United Nations Department of Economic and Social Affairs, New York

Karekezi S, Wangeci J, Manyara E (2004) Sustainable energy consumption in Africa. UN-DESA Final Report. African Energy Policy Research Network, Nairobi, 54 p

UN-DESA (2012) Population division of the Department of Economic and Social Affairs of the United Nations Secretariat: World population prospects: the 2010 revision. UN-DESA, New York

von Braun J (ed) (2009) Food security risks must be comprehensively addressed, IFPRI annual report essay 2008–2009. International Food Policy Research Institute, Washington, DC

Part III
Sustainable Intensification of Agriculture

Chapter 15
Adoption of Stress-Tolerant Rice Varieties in Bangladesh

Akhter U. Ahmed, Ricardo Hernandez, and Firdousi Naher

Abstract This chapter presents results of analyses of survey-based data on the rate of adoption of modern stress-tolerant rice varieties by the beneficiary farmers of the Cereal Systems Initiative for South Asia (CSISA) and compares that with non-CSISA rice farmers who cultivated the CSISA-promoted rice varieties. The study reveals that the adoption of such varieties has been very low. Just 27 % of the farmers in the CSISA beneficiary survey and 9 % of non-CSISA rice farmers grew at least one of the CSISA-promoted rice varieties. Though our survey did not specifically ask the farmers for reasons for non-adoption, education plays a key role in new rice technology adoption and diffusion. Moreover, the role of complementary technologies should not be overlooked when analyzing the adoption of new/modern technologies.

Keywords Technology adoption • Improved seed • Education • Complementary technologies • South Asia

Introduction

Technology is the basis for sustainable agricultural growth. Enhanced agricultural productivity and growth depend, to a large extent, upon the widespread adoption of appropriate technologies by farmers.

Seed, fertilizer and irrigation technologies known as "Green Revolution technologies" have long played major roles in the growth of agriculture production in Bangladesh. The country has made commendable progress in domestic rice production through farmers' adoption of these technologies. In the early 1970s, Bangladesh was a seriously food-deficient country with a population of about

A.U. Ahmed (✉) • R. Hernandez
International Food Policy Research Institute (IFPRI), Dhaka, Bangladesh
e-mail: a.ahmed@cgiar.org; r.a.hernandez@cgiar.org

F. Naher
University of Dhaka, Dhaka, Bangladesh
e-mail: f.naher@cgiar.org

© The Author(s) 2016   241
F.W. Gatzweiler, J. von Braun (eds.), *Technological and Institutional Innovations for Marginalized Smallholders in Agricultural Development*,
DOI 10.1007/978-3-319-25718-1_15

75 million. Today, the population is more than 160 million, and Bangladesh is self-sufficient in rice production, which has tripled over the past three decades.

Upon request from the US Agency for International Development (USAID), the International Food Policy Research Institute's Policy Research and Strategy Support Program in Bangladesh (IFPRI-PRSSP) developed and began a study on the adoption of agricultural technologies in USAID's Feed the Future (FTF) zone of influence in the south and southwest regions of the country. FTF is the US government's global hunger and food security initiative to support country-driven approaches to addressing the root causes of poverty, hunger, and undernutrition. In Bangladesh, FTF's collective efforts aim to improve the livelihood and nutritional status of households through: (1) increased on-farm productivity, (2) increased investment in market systems and value chains, (3) enhanced food security policy and planning capacity, (4) enhanced agriculture innovation capacity, and (5) improved nutritional status of rural poor.

The agricultural technology evaluated in this study is modern stress-tolerant varieties of rice promoted by the Cereal Systems Initiative for South Asia in Bangladesh (CSISA-BD), which is funded by USAID and implemented by the International Rice Research Institute (IRRI) in partnership with CIMMYT and WorldFish, two other CGIAR centers. Bangladesh is one of the countries worst affected by climate change, and cultivating stress-tolerant rice varieties have the potential to become very important in the near future. The CSISA-BD started its operations in the country's 'Feed the Future' zone of influence in 2010. Focusing on rice-based farming systems, the project also promotes the cultivation of cereals such as wheat and maize during the dry season. Moreover, it advocates for rice-fish cultivation, which is a practice of raising fish in conjunction with rice farming.

This study presents results of analyses of survey-based data on the rate of adoption of modern stress-tolerant rice varieties by the CSISA beneficiary farmers and compares that with non-CSISA rice farmers who cultivated the CSISA-promoted rice varieties in the FTF zone.

The report is organized in six sections. Section "Data" describes the surveys that provided the data used in the empirical work. Section "Profile of Survey Farmers" gives a profile of survey households. Section "Usage of Modern Rice Varieties" discusses the findings of analyses of farmers' usage of modern varieties of rice technology. Section "The Determinants of Farmers' Adoption and Duration of Farm Technologies" provides an analysis of the determinants of farmers' adoption of rice technology and the duration of adoption. Section "Conclusions" provides policy conclusions.

Data

The data for the study came from two IFPRI-PRSSP surveys of rice farmers in the FTF zone: (1) a zone-level survey of 2400 rice-farm households, and (2) a CSISA beneficiary farmers' survey of 500 farm households.

Sampling

The FTF zone level survey is statistically representative of all rice farmers in the FTF zone, and its domain included all 120 FTF upazilas within the 20 FTF districts. The sampling process and survey administration included the following steps:

- List all villages in each of the 120 FTF upazilas from the 2011 National Population Census.

  Randomly select two villages in each upazila with probability proportional size (PPS) sampling, using the village-level population data from the 2011 National Population Census.

  Conduct complete census of each of the 240 selected villages.
- List all farm households that cultivated rice in the 12-month period prior to the survey, then randomly select ten farm households from village census list.
- Conduct interviews with selected rice-farm households.

The sampling process and survey administration for the CSISA beneficiary survey included the following steps:

- Randomly select 500 farm households from the USAID-provided list of CSISA beneficiary farm households.
- Conduct interviews with 500 selected farm households (within 344 different villages).

Post-sampling observation demonstrated that the FTF and CSISA samples are independent. That is, there was no overlap between the FTF and the CSISA sample beneficiaries.

Survey Questionnaire

The IFPRI-PRSSP team prepared a draft questionnaire for the rice technology adoption survey, received comments on the draft questionnaire from USAID, and revised the questionnaire by addressing the comments. The questionnaire included six modules: (1) sample household identification; (2) household composition, literacy, and education; (3) roster of land owned or under operation; (4) plot-level information on seeds, irrigation, and fertilizer usage; (5) information on usage of paddy varieties; and (6) information on use rate of paddy seed.

Training and Survey Administration

IFPRI contracted Data Analysis and Technical Assistance (DATA) Limited, a Bangladeshi consulting firm with expertise in conducting complex surveys and data analysis, to implement the survey. DATA worked under the supervision and guidance of senior IFPRI researchers. DATA provided 100 experienced survey enumerators and 20 supervisors to administer the surveys. IFPRI-PRSSP provided the survey questionnaire to DATA for the training of the survey team. DATA translated the questionnaire into Bangla. IFPRI-PRSSP researchers and senior DATA staff pre-tested the questionnaire in the FTF zone and trained survey workers, in both a formal classroom setting and closely monitored practice fieldwork.

The DATA survey team completed the survey of 2900 farm households in 16 working days by engaging 20 teams consisting of 6 members (5 enumerators and a supervisor). The survey was carried out from November 1 to 19, 2013. The enumerators conducted one-on-one, face-to-face interviews with the respondents assigned to them, under the supervision of their field supervisor. To show appreciation for respondents' time, a gift of two plates and one bowl (a 200-taka value) was given to each household. Completed questionnaires were sent to the DATA central office in Dhaka on a regular basis for simultaneous data entry. IFPRI and DATA took extensive care to ensure the quality of the household survey data, and IFPRI researchers made field visits to supervise the fieldwork.

Data Entry, Cleaning, and Analysis

Staff at the DATA office in Dhaka carried out the data entry simultaneously with the data collection, with about 4 days of lag time. DATA used software specialized for data entry (Microsoft Access) that was programmed to identify out-of-range or inconsistent values.

After receiving the cleaned dataset from DATA, IFPRI-PRSSP researchers analyzed the data using Stata software. Senior IFPRI researchers provided guidance for data analysis.

Profile of Survey Farmers

Using household survey data collected through the three surveys, this section presents the profile of rice-farm households living in the FTF zone of influence. Much of the farmer-level analysis in this study disaggregates the sample farmers into four operated farm size groups: (1) marginal (farmers operating less than half an acre of land); (2) small (farmers operating 0.5–1.49 acres); (3) medium (farmers operating 1.5–2.49 acres); and (4) large (farmers operating 2.5 acres or more).

The four farm size groups match the cut-off points of the six operated farm size groups presented in the 2010 Household Income and Expenditure Survey (HIES) report of the Bangladesh Bureau of Statistics (BBS 2011) by aggregating the smallest two HIES farm size groups under the marginal farm category and the largest two groups under the large farm category.

Household Characteristics

Smaller rice-farming households tend to have slightly smaller household sizes than larger farm households; the average household size declines from 5.6 for the large farm group to 4.6 for the marginal farm group in the FTF zone.

The following are additional highlights of household characteristics:

- Primary school-age children (age 6–11) from about 8 % of rice farm households and secondary school-age children (age 11–18) from 25 % of rice farm households in the FTF zone do not go to school.
- Educational attainment in terms of years of schooling of male household head and wife of household head is positively correlated with farm size.
- Educational attainment in terms of years of schooling of adult family members is positively correlated with farm size.
- In the FTF zone, 36 % of adult males and 43 % of adult females in rice farm households never attended school. The rate of no schooling of adults declines as farm size of households increases.
- A person who can read and write a sentence in Bengali is considered to be literate. Overall, the female population has a lower literacy rate than the male population. Literacy rates have strong, positive relationships with farm size.

Inequality in Distribution of Operated Land

Land is the most important factor in agricultural production. However, about one-fifth of the rice farmers in the FTF zone are pure tenants, that is, they do not own any cultivable land. These farmers have either sharecropping or cash-lease arrangements with landlords for their operated land. The sample rice farm households are divided into 20 equal groups and are ranked from lowest to highest according to the size of their total operated land. The survey results indicate that the distribution of operated land is highly unequal. The bottom 25 % of all rice-farm households own only 6.5 % of total operated land in the FTF zone. At the other extreme, the top 5 % of all households own 22.5 %.

Farm-Size Groups and Size of Operated Land

Figure 15.1 shows the distribution of operated land by each of the four farm size groups in the FTF zone and among CSISA beneficiaries as a percentage of all farmers. The distribution is quite similar across the two types of survey samples. Small farmers dominate the FTF zone—about 45 % of all rice farmers are small farmers.

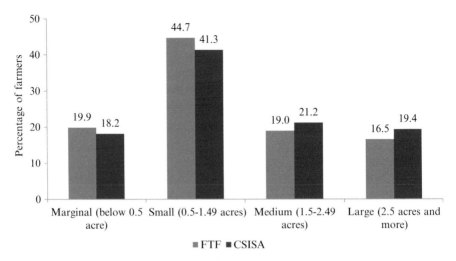

Fig. 15.1 Distribution of operated land, by farm size group. Note: *FTF* Feed the Future zone of influence survey, and *CSISA* Cereal Systems Initiative for South Asia beneficiary survey (Source: IFPRI-PRSSP Agricultural Technology Adoption Survey in the FTF Zone, 2013)

The average size of operated land per rice farm household in the FTF zone is 1.6 acres, ranging from only 0.3 acres per marginal farmer to 4.3 acres per large farmer.

Patterns of Land Tenure

About one-fifth of all rice farm households do not own any cultivable land. About 49 % of rice farmers cultivate only their own land. The proportion of mixed-tenant farmers—those who cultivate their own land and also take land in as sharecroppers and/or leaseholders—is 31 %.

The dominant tenurial arrangement in the FTF zone is sharecropping, where the produce is shared between the cultivator and the landowner in different proportions that have been agreed upon prior to cultivation. About 31 % of the rice farmers are sharecroppers. This group of sharecroppers includes those who do not own any cultivable land (that is, "pure tenant"), as well as those who own land and sharecrop other people's land. About 15 % of the rice farmers have cash-lease arrangements, either as pure tenants or as those with their own land plus cash-leased land. The proportion of rice farmers operating both sharecropped and cash-leased land (either as tenants or landowners) is about 5 %.

Approximately 56 % of all marginal farmers in the FTF zone do not have any land lease arrangements; they cultivate only their own land. This is perhaps a manifestation of their risk aversion. For the marginal farmers who are pure tenants (33.2 %), the sharecropping arrangements represent an overwhelming majority—about 70 % of all pure-tenant farmers are sharecroppers. Only about 13 % of the large farmers are pure tenants, and 48 % of them opt for sharecropping as the mode of renting land. It is interesting to note that about 47 % of the large farmers supplement their own land with some form of sharecropping and/or cash leasing.

Irrigation

Irrigation is one of the most critical factors for agricultural production in Bangladesh. Tripling rice production in the country since the early 1970s would not have been possible without irrigation. It plays three crucial roles in increasing foodgrain production in Bangladesh: (1) irrigation enables farmers to grow an additional boro rice or wheat crop during the dry winter season, and thus increases cropping intensity and eases the land constraint; (2) irrigation complemented with fertilizers and modern high-yielding rice varieties significantly raises rice yields in comparison to rain-fed rice cultivation; and (3) supplemental irrigation can take much of the risk out of the two predominantly rain-fed rice seasons—aus and aman (Ahmed and Sampath 1992).

Bangladeshi farmers use both traditional and modern methods of irrigation. Traditional methods include *done* (a water-lifting devise), swing basket, and dug-well. Modern techniques include shallow tubewell, deep tubewell, low-lift pump, hand pump, and sophisticated canal gravity-flow irrigation schemes. Among these, *done*, swing basket, and low-lift pump use surface water while dug-well, shallow tubewell, deep tubewell, and hand pump use groundwater as irrigation sources.

The shallow tubewell is the predominant method of irrigation used by about 59 % of rice farmers in the FTF zone for boro rice cultivation. The second most important method is the low-lift pump, used by more than one-fifth of the rice farmers. About 5 % of the rice farmers cultivate boro rice without irrigation (Fig. 15.2).

Share of Rice Crops on Total Rice Land

Figure 15.3 shows the share of aman, aus, and boro rice crops on the total rice land of farmers in the FTF zone. Aman and boro rice show opposite patterns across the four farm size groups: while the share of aman rice area increases as farm size increases, the relationship is negative in the case of boro rice crops.

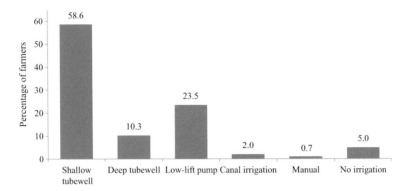

Fig. 15.2 Methods of irrigation used by farmers for boro rice cultivation in the FTF zone. Note: *FTF* Feed the Future zone of influence (Source: IFPRI-PRSSP Agricultural Technology Adoption Survey in the FTF Zone, 2013)

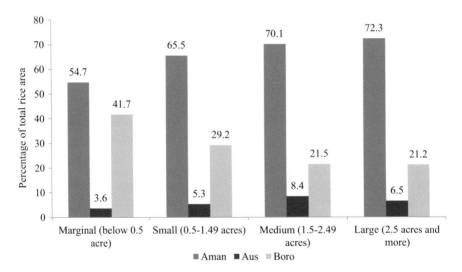

Fig. 15.3 Share of aman, aus, and boro rice crops on total rice land, by farm size groups among rice farmers in the FTF zone (Source: IFPRI-PRSSP Agricultural Technology Adoption Survey in the FTF Zone, 2013)

Usage of Modern Rice Varieties

In close collaboration with the Bangladesh Institute of Nuclear Agriculture (BINA), the Bangladesh Rice Research Institute (BRRI), and the Bangladesh Agriculture University (BAU), CSISA-BD produced large quantities of breeder seeds for different stress-tolerant varieties during the boro 2012, aman 2012, and boro 2013 seasons. The CSISA-promoted rice varieties include varieties that are salt-tolerant

(BRRI dhan 47, 53, 54, 55, and 61; BINA dhan 8 and 10), submergence-tolerant (BRRI dhan 51 and 52; BINA dhan 11 and 12), and drought-tolerant (BRRI dhan 56 and 57). Varieties with both high grain quality and high market value were also promoted, including BRRI dhan 50, BRRI dhan 61, BINA dhan 11, and BINA dhan 12. Two major approaches were adopted for varietal promotion: (1) seed production from demo trials and seed mini-kits (2.5 kg of seeds) to promote awareness among farmers; and (2) improving local availability of quality seed of new varieties for area expansion and farmer-to-farmer dissemination.

This section presents the results of the FTF zone level survey and the CSISA beneficiary farmers' survey on the usage by rice farmers in the FTF zone of the modern rice varieties promoted by CSISA.

More than one-fourth (27.1 %) of the farmers in the CSISA beneficiary survey grew at least one of the CSISA-promoted rice varieties in 2013. In the FTF-zone survey, however, less than one-tenth (8.8 %) of the farmers grew at least one of these varieties. Across the four farm size groups, adoption of CSISA varieties is higher among the medium and large farmers in both the FTF zone and the CSISA beneficiary surveys. Among the marginal farmers in the FTF zone, barely 5 % grew any CSISA-promoted varieties. The figure is much higher for the marginal farmers in the CSISA survey sample, with 26 % adopting them.

A plausible explanation for the relatively low adoption of the stress-tolerant varieties by marginal farmers could be that these farmers are risk-averse given their lack of a diversified portfolio, which severely constrains their ability to offset any crop loss. It is understandable that these farmers prefer to stick to time-tested varieties with assured yields rather than experiment with new ones.

In each of the three rice-cultivation seasons, a greater percentage of farmers in the CSISA survey sample cultivated one or more rice varieties promoted by CSISA than those farmers in the FTF survey sample. The most popular CSISA-promoted variety in the FTF zone survey sample was BINA 7, which is a short-duration transplant aman variety, grown by about one-third (34 %) of the farmers. The ones next in popularity included BR 49 and BR 41. Among CSISA beneficiaries, the most popular aman variety was BR 49, which is grown by one-quarter of the farmers. The second most popular aman variety is BR 52, followed by BINA 7.

During the 2013 boro season, the most popular variety in the FTF zone survey was a salt-tolerant variety, BR 47, with approximately 16 % of the CSISA-variety growing farmers using this variety. An aromatic rice variety, BR 50 (known locally as "Banglar Moti") came next, followed by another salt-tolerant variety, BINA 8. In the CSISA beneficiary survey sample, the three most popular boro varieties were BINA 8, BR 47, and BR 50, in decreasing order of popularity among farmers.

Among the CSISA-promoted rice varieties, Jessore, Magura, and Gopalganj districts have the highest rates of adoption of the CSISA-promoted stress-tolerant varieties among the CSISA beneficiaries, with 49–56 % of the farmers in these districts growing these varieties. Interestingly, in the Barguna, Madaripur and Meherpur districts, there were no farmers in the CSISA beneficiary survey sample growing any of the CSISA-promoted varieties. In the FTF zone survey, the uptake of the stress-tolerant varieties promoted by CSISA is much lower, ranging from less than 1 % in the Shariatpur district to 19 % in the Magura district.

In both the FTF zone and among the CSISA beneficiaries, a wide variety of paddy is grown with no overwhelming majority of any one variety over another. We identified the ten most popular rice varieties. When looking at the percentage of farmers cultivating the ten most popular rice varieties in 2013, we find that the highest adoption for aman is Guti Shorna, which is of Indian origin and may have been introduced to Bangladesh through informal cross-border exchange. This is followed by Mukta BR 11. In the boro season, 39 % of the farmers grow BR 28, followed by BR 29 in the FTF zone survey. Among the CSISA beneficiaries, 52 % of farmers grow BR 28 during boro season and 17 % grow BR 29. Overall, in both the FTF and CSISA samples, BR 28 and BR 29 dominate. BR 28 and BR 29 are rice varieties that were first grown in Bangladesh 20 years ago, and, despite the introduction of many varieties since then, farmers continue to grow them, often preferring to do so. A reason for the continued popularity of these two varieties is their higher yield and shorter growth cycle (for BR 28), which allows farmers to increase their cropping intensity.

Since 2000, in both the FTF zone and the CSISA beneficiary surveys, most farmers have taken to growing BR 28 at different points of time. It is interesting to note that the uptake of BR 28 has been higher among the CSISA beneficiaries compared to the FTF zone, despite the promotional activities of CSISA in favor of the stress-tolerant varieties.

However, it's worth noting that about 2 % of the FTF zone survey farmers and 3.4 % of the CSISA beneficiaries gave up growing BR 28 from 2009 to 2013. The variety that the maximum number of farmers have given up since 2000, however, is BR 11 (Mukta). BR 11 is a first-generation high-yield variety (HYV) rice that was introduced into Bangladesh more than three decades ago. These HYVs, though effective in areas of high potential, do not perform well under stressful agro-ecologies as soil salinity or submergence conditions. Given that a substantial part of the FTF zone falls into these categories, it is understandable why farmers are giving up these varieties.

Usage of CSISA-Promoted Rice Varieties by Number of Rice Farmers in the FTF Zone

Since the FTF zone level survey is statistically representative of all rice farmers in the FTF zone, it is possible to estimate the number of rice farmers who cultivated CSISA-promoted stress tolerant rice crops harvested in 2013. According to the 2011 Population and Housing Census conducted by the Bangladesh Bureau of Statistics, there are 6,272,196 rural households in 120 FTF zone upazilas in 20 FTF districts. IFPRI-PRSSP conducted the 2011–2012 Bangladesh Integrated Household Survey (BIHS), which is a nationally representative survey of 6500 rural households. The survey is also statistically representative of the FTF zone of influence. From the BIHS dataset, we have calculated that 51.9 % of all households in the FTF zone cultivated rice in 2011. Therefore, the total number of rice farm households in the FTF zone is 3,255,270 (that is, 6,272,196 × 0.519], of which 8.8 % or 286,464 households have grown at least one rice variety promoted by CSISA in 2013.

Area Covered by CSISA-Promoted Rice Varieties

In terms of acreage, the surveys revealed that the coverage of area under stress-tolerant varieties promoted by CSISA is very low. Only 3.29 % of the FTF-zone's area under rice is cropped with these varieties. According to the Bangladesh Bureau of Statistics, 2,794,571 ha were under rice cultivation in the 20 FTF districts in 2012/13. Therefore, the area coverage of CSISA-promoted rice varieties is esti-mated at 91,941 ha (that is, 2,794,571 × 0.0329) in the FTF zone of influence. Among the CSISA beneficiaries, the area coverage is about 11 %. In the case of the majority of varieties, the area coverage is less than even 1 %, except for BR 49, BINA 8, BR 47 and BR 52, for which the shares are 2.5, 2.0, 1.5 and 1.1 %, respectively.

Paddy Yields

Yields of CSISA-promoted varieties are 6–19 % higher than their non-CSISA counterparts in both the FTF and CSISA surveys. The yields for boro paddy are, in general, higher than those for aman, since most of the former is planted with HYVs. If we look at the yields of CSISA-promoted varieties, disaggregated by variety type, HYVs in the CSISA survey sample have higher yields than HYVs in the FTF survey sample. This is also true of hybrids. It appears that the CSISA promotional activities are being manifested in higher yields.

Patterns of Seed Use

The surveys showed that 91 % of the farmers in both the FTF and CSISA surveys make their own seedbed prior to transplantation of the seedlings. In doing so, the use of saved seeds from the previous harvest predominates for the summer and monsoon crops of aus and aman. For the dry season boro crop, a lower percentage of farmers use saved seeds and a higher percentage of farmers use purchased seed compared to farmers growing aus and aman crops.

households. The survey is also statistically representative of the FTF zone of influence. From the BIHS dataset, we have calculated that 51.9 % of all households in the FTF zone cultivated rice in 2011. Therefore, the total number of rice farm households in the FTF zone is 3,255,270 (that is, 6,272,196 × 0.519], of which 8.8 % or 286,464 households have grown at least one rice variety promoted by CSISA in 2013.

Area Covered by CSISA-Promoted Rice Varieties

In terms of acreage, the surveys revealed that the coverage of area under stress-tolerant varieties promoted by CSISA is very low. Only 3.29 % of the FTF-zone's area under rice is cropped with these varieties. According to the Bangladesh Bureau of Statistics, 2,794,571 ha were under rice cultivation in the 20 FTF districts in 2012/13. Therefore, the area coverage of CSISA-promoted rice varieties is estimated at 91,941 ha (that is, 2,794,571 × 0.0329) in the FTF zone of influence. Among the CSISA beneficiaries, the area coverage is about 11 %. In the case of the majority of varieties, the area coverage is less than even 1 %, except for BR 49, BINA 8, BR 47 and BR 52, for which the shares are 2.5, 2.0, 1.5 and 1.1 %, respectively.

Paddy Yields

Yields of CSISA-promoted varieties are 6–19 % higher than their non-CSISA counterparts in both the FTF and CSISA surveys. The yields for boro paddy are, in general, higher than those for aman, since most of the former is planted with HYVs. If we look at the yields of CSISA-promoted varieties, disaggregated by variety type, HYVs in the CSISA survey sample have higher yields than HYVs in the FTF survey sample. This is also true of hybrids. It appears that the CSISA promotional activities are being manifested in higher yields.

Patterns of Seed Use

The surveys showed that 91 % of the farmers in both the FTF and CSISA surveys make their own seedbed prior to transplantation of the seedlings. In doing so, the use of saved seeds from the previous harvest predominates for the summer and monsoon crops of aus and aman. For the dry season boro crop, a lower percentage of farmers use saved seeds and a higher percentage of farmers use purchased seed compared to farmers growing aus and aman crops.

Our survey results on seed rates for preparing seed beds for the top 10 popular rice varieties, and prices of seeds show that hybrid rice requires a substantially lower quantity of seed per-hectare compared to its inbred (HYV) rice, but the price of hybrid seeds is much higher than HYV seeds. The seed rate for hybrid variety Hira is about 41.7 kg/ha in the FTF survey and 43.4 kg/ha in the CSISA beneficiary survey. Hira is grown mostly in the boro season. The seed rate ranges from 59 to 72 kg/ha for inbreds.

The Determinants of Farmers' Adoption and Duration of Farm Technologies

This section presents an analysis of adoption, retention, and diffusion of a set of modern paddy varieties promoted by the CSISA project. It presents the cohort analysis, followed by the econometric analysis of rice technology adoption.

It is important to mention at the outset that, while CSISA has promoted many varieties within the FTF zone, as reported in the previous section, the analysis of adoption, retention, and diffusion that follows focuses on the top five varieties—BR 41, BR 47, BR 49, BR 50, and BINA 7— used by sample farmers.

Cohort Analysis and Survivor Functions of CSISA Varieties

The main results from a cohort analysis of CSISA varieties for both the FTF and CSISA samples are as follows:

First, farmers who used the CSISA varieties once continued on as consistent users of such varieties. In the FTF zone survey, the average annual retention rate of the new varieties is 98 %, which implies that once farmers adopt the new varieties, they continue to cultivate those varieties for a long period of time. This is a very encouraging result for CSISA, because it suggests a low dropout rate among farmers who adopt the new varieties. The CSISA sample similarly showed high retention and low dropout rates.

Second, in the FTF sample, 7.3 % of farmers included in the FTF sample and 12.8 % of farmers in the CSISA sample used the CSISA varieties at some point during the observation period (2009–2013). The diffusion was very slow, however. Farmers began using CSISA varieties soon after being exposed to the "possibility" of adoption, but this process has been quite slow. That result is to be expected when analyzing the adoption of new/improved crop varieties in developing countries.

Interestingly, once farmers use CSISA varieties, they continue using the new technology and do not withdraw rapidly. There is minimal evidence of dropout of CSISA varieties, which supports the results of the cohort analysis. By the fourth year of use, more than 95 % of the users have remained adopters of CSISA varieties uninterruptedly. CSISA beneficiary farmers adopted CSISA varieties slightly faster than farmers in the FTF zone survey sample.

Determinants of Time-to-Adoption of CSISA Varieties

From the results of regressions explaining time-to-adoption and duration (time-to-withdrawal) models we discussed the statistically significant results and highlighted the variables we had expected to be significant that, in reality, were not. The likelihood ratio test of significance of the regressions (chi-squared statistics) and the *p* values associated with these statistics show the overall significance of both the adoption and withdrawal spell models to be significant at 1 % for the FTF survey results and 5 % for the CSISA survey sample.

Several results are noticeable for the determinants of time-to-adoption.

- Farm size (determined by operated land) plays a significant role in determining adoption of CSISA-promoted paddy varieties; larger operated land significantly shortens the time to adoption in the both the FTF and CSISA samples. Again, this result is to be expected, because farmers with more operated land tend to have more intensive productive systems and are, therefore, less risk-averse and more willing to try new farm technologies.
- Education of the head of the household reduces the time-to-adoption of CSISA varieties in the FTF zone survey and CSISA survey sample.
- Interestingly, female-headed households are *more likely* to adopt CSISA varieties than other sample households.

Determinants of Duration of CSISA Varieties Technology

The main findings for determinants of duration of CSISA-varieties use are presented below.

- The size of total operated land has a positive effect on duration of CSISA varieties use; it increases duration.
- Having access to irrigation increases duration of time using CSISA varieties. Similar to total operated land, this result is related to production system intensification.
- Being married also has a positive effect on CSISA-varieties duration of use.

Conclusions

Using data collected through two distinct household surveys, the IFPRI-PRSSP study analyzed modern rice technology promoted in USAID's Feed the Future (FTF) zone of influence in the south and southwest regions of Bangladesh. The technology assessed the uptake of a set of modern stress-tolerant paddy varieties promoted by the Cereal Systems Initiative for South Asia (CSISA) project.

The fragile agro-ecology of Bangladesh warrants that farmers grow more stress-tolerant rice varieties. Indeed, for Bangladeshi farmers who are largely risk averse, adapting to these new varieties alongside the more popular ones should be relatively easy. However, our study reveals that the adoption of such varieties has been very low.

Less than one-third (27.1 %) of the farmers in the CSISA beneficiary survey grew at least one of the CSISA-promoted rice varieties. In the FTF zone survey, less than one-tenth (8.8 %) of the farmers grew at least one of these varieties.

Across farm size groups, adoption of CSISA varieties is higher among the medium and large farms than among marginal and small farms in both the FTF zone and the CSISA beneficiary surveys. A plausible explanation for the relatively low adoption of the stress-tolerant rice varieties could be that small and marginal farmers are risk-averse, given their lack of a diversified portfolio, which severely constrains their ability to offset any crop loss. It is understandable that these farmers prefer to stick to time-tested varieties with assured yields rather than experiment with new ones. In terms of acreage, the surveys revealed that the coverage of area under stress-tolerant varieties promoted by CSISA is very low, and only about 3 % of the FTF-zone's area under rice is cropped with these varieties. Among the CSISA beneficiaries, the area coverage is about 11 %.

Though our survey did not specifically ask the farmers for reasons for non-adoption, education plays a key role in new rice technology adoption and diffusion. This is evidenced by the rate at which educated heads-of-household tended to adopt new rice technologies in comparison to less-educated heads-of-household. Those with more education also continued using the new technologies for longer periods of time.

The role of complementary technologies should not be overlooked when analyzing the adoption of new/modern technologies. Our survey shows that farmers who have adopted irrigation systems tend to adopt faster and continue, for longer periods of time, as adopters of the new rice varieties. Having an irrigation system might mitigate the risks resulting from aversion to the uncertainty related to the adoption of new rice varieties.

Projects and programs that promote new farm technologies should be strongly encouraged to monitor retention of those technologies.

Acknowledgments We gratefully acknowledge the United States Agency for International Development (USAID) for funding the Policy Research and Strategy Support Program (PRSSP) in Bangladesh under USAID Grant Number EEM-G-00-04-00013-00. This report is an output of the PRSSP.

At IFPRI, we received excellent analytical support from Arifeen Akter and Nusrat Zaitun Hossain, and we thank them for their efforts. We thank Samita Kaiser for her help with the production of this report.

The study would not have been possible without the dedication and hard work of the survey enumerators and other staff of the Data Analysis and Technical Assistance Limited (DATA), a Bangladeshi consulting firm that carried out the Agricultural Technology Adoption Survey. We are particularly grateful to Zahidul Hassan of DATA for his overall support.

Open Access   This chapter is distributed under the terms of the Creative Commons Attribution-Noncommercial 2.5 License (http://creativecommons.org/licenses/by-nc/2.5/) which permits any noncommercial use, distribution, and reproduction in any medium, provided the original author(s) and source are credited.

The images or other third party material in this chapter are included in the work's Creative Commons license, unless indicated otherwise in the credit line; if such material is not included in the work's Creative Commons license and the respective action is not permitted by statutory regulation, users will need to obtain permission from the license holder to duplicate, adapt or reproduce the material.

References

Ahmed AU, Sampath RK (1992) Effects of irrigation-induced technological change in Bangladesh rice production. Am J Agric Econ 74(1):144–157. doi:10.2307/1242998

BBS (2011) Report of the household income and expenditure survey 2010. Bangladesh Bureau of statistics, statistics division, ministry of planning, government of the people's Republic of Bangladesh, Dhaka

Chapter 16
More Than Cereal-Based Cropping Innovations for Improving Food and Livelihood Security of Poor Smallholders in Marginal Areas of Bangladesh

Mohammad Abdul Malek, Mohammad Syful Hoque, Josefa Yesmin, and Md. Latiful Haque

Abstract Following the marginality approach developed at ZEF, we identified five underperforming sub-districts in Bangladesh, where poverty and other socio-economic dimensions of marginality are widespread, but agricultural potential is also high. Results from extensive quantitative and qualitative surveys suggest that development strategies in these areas should focus on three pathways: agricultural intensification, income diversification and agricultural diversification based on options available for the smallholders in the localities. Cereal-based technology under agricultural innovations could be part of the solution, but should be integrated into other income diversification and agricultural diversification strategies. Intensive crop systems, hybrid seeds, water management technologies, non-crop farming, non-farming enterprises are suggested as potential technology innovations for the study areas.

Keywords Marginality • Agricultural potential • Development strategies • Agricultural intensification • Income diversification

M.A. Malek (✉)
BRAC Research and Evaluation Division (RED), Dhaka, Bangladesh

University of Bonn-Center for Development Research (ZEF), Bonn, Germany
e-mail: malekr25@gmail.com

M.S. Hoque • J. Yesmin • Md. L. Haque
BRAC Research and Evaluation Division (RED), Dhaka, Bangladesh
e-mail: israhat@gmail.com; yesmin.josefa@gmail.com; latiful214@gmail.com

© The Author(s) 2016   257
F.W. Gatzweiler, J. von Braun (eds.), *Technological and Institutional Innovations for Marginalized Smallholders in Agricultural Development*,
DOI 10.1007/978-3-319-25718-1_16

Introduction

Although Bangladesh has made some remarkable achievements in reducing poverty and improving social and economic outcomes in recent decades, about one-third of the rural population still lives below the upper poverty line, most of whom depend on agriculture as their primary source of income. Compared to favorable areas, a quite dismal picture prevails among the marginal areas in Bangladesh. One of the reasons for their poverty is the low productivity that results from sub-optimal use of inputs and other technologies in agriculture. To foster agricultural productivity and rural growth in those lagging regions, technological innovations have to reach all strata of the poor among small farming communities, who we will refer to hereinafter as smallholders (SHs), in rural Bangladesh. For that purpose, opportunities in technology need to be brought together with systematic and location-specific actions related to technology needs, agricultural systems, ecological resources and poverty characteristics to overcome the barriers that economic, social, ecological and cultural conditions can create. As the first step of an ex-ante assessment of technology innovations for inclusive growth in agriculture (TIGA), a project at the Center for Development Research (ZEF), Bonn, in collaboration with BRAC and partners in India, Ethiopia and Ghana, we followed the mapping approach and identified underperforming areas, hereinafter referred to as marginality hotspots with agricultural potential. Those areas are underperforming areas, i.e., rural areas in which the prevalence of poverty and other dimensions of marginality are high, but agricultural potential is also high, since in such areas, yield gaps (potential minus actual yields) are high and productivity gains (of main staple crops) are likely to be achieved (Malek et al. 2013). The marginality mapping presented in the analyses attempted to identify areas with a high prevalence of societal and spatial marginality – based on proxies for marginality dimensions representing different spheres of life – and high (un(der)utilized) agricultural (cereal) potential. The overlap between the marginality hotspots and the high (un(der)utilized) agricultural potential shows that Rajibpur (Kurigram), Dowarabazar (Sunamgonj), Porsha (Naogaon), Damurhuda (Chuadanga), Hizla (Barisal), Mehendigonj (Barisal), Bauphal (Patuakhali) and Bhandaria (Pirojpur) are the marginal areas where the greatest productivity gains could be achieved.

As the next step of TIGA, those identified marginality hotspots with agricultural potential could be used in combination with other instruments in order to improve targeting and priority setting for an agricultural growth productivity program. Thus, this paper aims to address the following research questions: (1) Why has the agricultural potential in those areas not yet been made use of? (2) Who are the poor SHs? Which income strata and segments of the rural poor (by agri-ecological and socio-economic clusters) can be found in those areas? (3) What are the strategic options already available for each segment? (4) Which segments of poor SHs could be eligible for agricultural (crop) productivity programs? (5) What are the technology innovations for each segment of the poor?

To address these research questions, we followed the conceptual framework and theory of change for the TIGA project, as elaborated in section "Conceptual Framework and Methods for Analysis". Then, a selection of study areas, a sample for assessment, and survey methodology are discussed in section "Selection of Study Areas, Sample for Assessment and Survey Methodology". Results with analytical techniques are elaborated in section "Results and Discussion". And, finally, we draw some conclusions for institutional and technological innovation to take place.

Conceptual Framework and Methods for Analysis

With reference to the conceptual framework and theory of change as developed for the Agricultural Technology Innovations for Inclusive Growth in Agriculture (TIGA) project, once the marginality hotspots with agricultural potential are identified, then the poor SHs (the eligible population for any agricultural growth productivity program) are identified in those areas, and stratification according to income criteria is carried out, e.g., subjacent poor are those with incomes between $1 and $1.25/day, medial poor: between 75¢ and $1/day, and ultra-poor: below 75¢/day.[1] Those stratifications of the poor SHs are validated by participatory wealth-ranking and/or self-reported perceptions. At this stage, the poor SHs from each stratum are allocated to five broad strategic options (Fig. 16.1):

(A) agricultural intensification through improving current farming system performance by means of innovations (yet to be identified),
(B) agricultural diversification through changing current farming system and/or shifting to another,
(C) income diversification through progressing along the value chain, for example, by shifting from being a farmer to working as an agro-dealer, or diversifying income from the non-agricultural sector (e.g., by non-farm wage employment or migrating to other areas/abroad)
(D) leaving the agricultural sector completely

This allocation of poor SHs from different strata is carried out parallel to the livelihood assets and need assessment. As it is widely recognized that development strategies for sustainable intensification in marginality hotspots with agricultural potential need a careful adjustment of resource use at the field farm, household and village levels, we need to look for a portfolio of activities and technologies that guarantee input efficiency and labor productivity (Ruben et al. 2007). The sustainable livelihoods framework (SLF) developed by DFID (2000) is used to improve our understanding of the livelihoods of the poor SHs. The livelihoods approach places households and their members at the center of analysis and decision-making, with the implication that the household-centered methods of analysis must play a

---

[1] This stratification needs to be adjusted to national poverty lines in each study country.

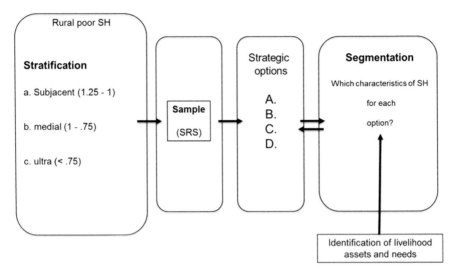

Fig. 16.1 From stratification to segmentation (Source: Personal communication with Franz Gatzweiler)

central role in developing an understanding of livelihood strategies. Applying SLF highlights the multilayered interactions between technologies and the vulnerability context of households – their asset base, access to social capital, and livelihood strategies. However, additional aspects of culture, power, and history are also integrated to understand the role of agricultural research in the lives of the poor (DFID 1999; OECD 2001; Carney 1998).

The sustainable livelihood framework

- provides a checklist of important livelihood issues, with particular focus on current farming practices and agricultural technology use, and sketches out the way these link to each other;
- draws attention to core influences and processes; and
- emphasizes the multiple interactions between the various factors which affect the livelihoods.

The framework is centered on people. It does not work in a linear manner and does not try to present a model of reality. Its aim is to help stakeholders with different perspectives to engage in structured and coherent debate about the many factors that affect livelihoods, their relative importance and the way in which they interact. This, in turn, should help in the identification of appropriate entry points for support of livelihoods (DFID 1999). People and their access to assets are at the heart of livelihood approaches. In the original DFID framework, five categories of assets or capitals were identified, these original categories being: Human capital, natural capital, financial capital, physical capital, social capital- these livelihood assets are the locked potentials of the SHs.

Within the framework, assets are both destroyed and created as a result of the trends, shocks and seasonality of the vulnerability context. Farmers' livelihood assets are affected by the vulnerability context: critical trends, shocks and season-ality – over which they have limited or no control and which are parts of the barriers identified in the next step:

- Critical trends may (or may not) be more benign, though they are more predict-able. They have a particularly important influence on rates of return (economic or otherwise) to chosen livelihood strategies.
- Shocks can destroy assets directly (e.g., in the case of floods, storms, civil conflict). They can also force people to abandon their home areas and dispose of assets (such as land) prematurely as part of coping strategies.
- Seasonal shifts in prices, employment opportunities and food availability are some of the greatest and most enduring sources of hardship for poor people in developing countries.

The livelihood analysis tries to develop a full understanding of all dimensions of the vulnerability context, the aim being to identify those capital assets, trends, shocks and aspects of seasonality that are of particular importance to livelihoods of the poor SHs. Efforts can then be concentrated on understanding the impact of these factors and how negative aspects can be minimized. A need assessment can, in addition, identify demands, wants and requirements for improving the quality of current livelihoods. Such needs can be discrepancies between current and needed or desired conditions of SHs, and they are assessed to ensure that technological innovations which are economically possible also match the wants and aspirations of the poor – an important aspect which is also captured by allocating the strategic options to the surveyed SHs.

Then, allocation of the different strategic options to the poor SHs is done in a participatory manner and supported by agronomic calculations based on household data from the livelihood assets and needs assessment to ensure that the options are realistic (no wish lists) and economically viable for each of the actors from different strata. Trade-offs may need to be made between subjective and rational choices. The SHs being allocated different strategic options come from different strata. By means of their characteristics, the segments are defined for each strategic option. Segmentation is necessary to identify suitable technology innovations – innovations which match the characteristics of each segment and thereby contribute to achiev-ing the overall goal of increasing productivity. For example, all SHs allocated option A own land, or lease land, or are sharecroppers, and each belong to a different income category. Land and income, for example, define different seg-ments which can be further defined by additional characteristics, such as family members, level of education and social status. After this step in the assessment, we know which strategic options are available for which strata of the poor and which characteristics the poor have in each option category (segment). Finally, poor SHs from different strata are segmented to the strategic options stemming from all-inclusive assessment of household attributes, using cluster analysis for this purpose. Some systematic tabulation of perception study and qualitative assessments has been used for identifying technological innovations.

Selection of Study Areas, Sample for Assessment and Survey Methodology

The marginal areas identified for the assessment are usually bypassed by policy-makers due to a generalized convention about the Agro-Ecological Zones (AEZs) as a whole, causing them to receive less attention (Malek et al. 2013). Therefore, marginal (or less-favored or laggard) regions, especially in poor developing countries and emerging economies in Sub-Saharan Africa and South Asia, have recently gained much attention in the development literature (Conway 1999; Fan and Hazell 2000; Pinstrup-Anderson and Pandya-Lorch 1994; Ruben et al. 2007; Pender 2007; Reardon et al. 2012). As mentioned earlier, the first step towards designing systematic interventions is to identify underperforming areas. Identification has been based on a high prevalence of societal and spatial marginality, using proxies for marginality dimensions representing different spheres of life and an overlapping high (un(der)utilized) agricultural (cereal) potential. The available secondary data and household survey data from various sources have been used for the exercise. Figure 16.2 shows that Rajibpur (Kurigram), Dowarabazar (Sunamgonj), Porsha (Naogaon), Damurhuda (Chuadanga), Bhandaria (Pirojpur), Hizla (Barisal), Mehendigonj (Barisal) and Bauphal (Patuakhali) are the marginal sub-districts where the highest productivity gains can be achieved through suitable agricultural technology intervention. These areas are in different AEZs – most of which are agro-ecologically fragile/unfavorable. Among them, Patuakhali, Pirojpur and Barisal are in the Coastal region, Kurigram is in the Northern Char region, Sunamgong is in the Haor region and Naogaon is in the Drought prone areas. Only Chuadanga, among these seven districts, is not in an agro-ecologically vulnerable region, but it is in a food insecure region (HKI and JPGSPH 2011). Another point to note is that 4 out of these 8 sub-districts are adjacent to India's borders, whereas the other 4 sub-districts are located in the coastal region.

Thus, among those eight sub-districts, the first four represent different regions while the latter four represent similar regions (the coastal belt), and among these four, Bhandaria (Pirojpur) would be comparatively less difficult to reach with agricultural technology interventions. Thus, we selected the following five sub-districts to be the study sites for our ex-ante assessment: Rajibpur (Kurigram), Dowarabazar (Sunamgonj), Porsha (Naogaon), Damurhuda (Chuadanga) and Bhandaria (Pirojpur).

Then, we, the research team, visited the localities, assessed the situation, and prepared a list of all marginal villages. Finally, we randomly selected 16 marginal villages for the detail quantitative sample survey. Prior to conducting the in-depth quantitative sample survey, we conducted qualitative surveys in five villages (one

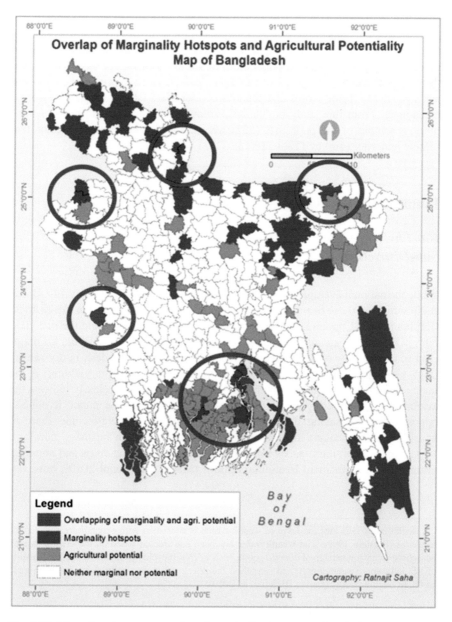

Fig. 16.2 Map of study areas – overlap of marginality hotspots and agricultural potential in Bangladesh

village per sub-district) that included several PRA methods (social and resource mapping, participatory wealth ranking, in-depth interview, focus group discussion) for livelihood assets and needs assessment. Those qualitative data were analyzed through contents analysis, which helps to identify the issues for detailed quantitative investigation. At the beginning of the quantitative sample survey, we first conducted a household census (5,855 households) in all 16 villages, collecting some basic information mainly related to household assets for the primary purpose of identifying poor SHs (study population) for the assessment. For this, we analyzed the census data and developed a wealth index[2] calculated from principal component analysis (PCA) factor scores and found 862 poor SHs[3] (study population) for the assessment. From this study population, following a proportionate random sampling, a sample of the poor SHs (357) were drawn for an in-depth quantitative sample household survey (Table 16.1).

Results and Discussion

Bio-Physical Conditions for the Poor SHs in the Marginal Sub-Districts in Bangladesh: Unused Potentials

While the national average for cropping intensity is about 180, it is only 144 for the study sample in those five sub-districts- it is extremely low for certain sub-districts (Rajibpur under Charland, Dowarabazar under the Haor basin) – and the rice yield rates in those areas are also very low (Tables 16.2, 16.3, and 16.4). While the major crop season in the so-called typical favorable areas in Bangladesh is dry season (high yielding) irrigated rice, *Aman* (wet-season) rice (moderate yielding) is the major crop season for three of the five sub-districts. Our results clearly indicate the availability of unused potential for cereal crops. If we see major livelihood opportunities (by seeing the household members engagement/income share to household total income) in a favorable rural area, non-farm business, non-farm wage employment, remittances from abroad and high yielding crops and non-crop farming are the dominant livelihood options (Malek and Usami 2010); however,

---

[2] A wealth index indicates the level of wealth which is consistent with expenditure and income measures (Rutstein 1999). The wealth index has been constructed based on the census data on household assets (ownership of durable goods, such as TVs, bicycles and landholdings) and quality of life indicators (water supply and sanitation facilities). A single wealth index has been done based on the following equation (Balen et al. 2010):

$$Ai = \hat{\gamma}1 \ \alpha i1 + \ldots + \hat{\gamma} \ n\alpha in,$$

where Ai is the standardized wealth index score for i^{th} households; $\alpha in = (xin - x^-n)/SDn$; γ^n = Weight (factor score); xin = nth asset for household I; x^-n = Mean of nth asset for all households; SDn = Standard deviation for nth asset for all households.

[3] Poor smallholders: Though we considered a farm size of 2.47 acres to be the ceiling, the average farm size in our sample was 0.53 acres, of which 60.78 % were functionally landless (<0.50 acre) farm households, 28.85 % were marginal farm households (0.51–1.00 acre) and 10.36 % were small farm households (1.01–2.50 acres).

Table 16.1 Selection of sample survey households (poor SHs) from the selected villages

| Upazila | Total pop | Under poverty line | Cut off point wealth Index | Poor SHs-study population | Sample poor SHs | Poorest 10 % | | Non-farm households |
		Total pop calculated as per HIES[a] 2010 poverty line				Total	Cut-off index	
Damurhuda	1428	39 % (557)	0.5682703	302 (54 %)	125	102	−1.490896	64 (63 %)
Rajibpur	1299	59 % (766)	0.1626211	163 (21 %)	67	130	−2.74765	79 (61 %)
Dowarabazar	899	29 % (261)	−1.469494	89 (34 %)	37	90	−2.236264	55 (61 %)
Porsha	1021	49 % (500)	−0.4671243	188 (37 %)	78	102	−1.868688	99 (97 %)
Bhandaria	1208	34 % (411)	−1.270219	120 (29 %)	50	121	−2.115256	88 (73 %)
Total	5855	(2945)		862 (29 %)	357[b]	545		385 (71 %)

Source: Authors estimation from TIGA Bangladesh Household Census 2013

[a]Household income and expenditure survey

[b]At 4 % error and 95 % CI

Table 16.2 Farm size, cropped area and cropping intensity of poor SHs in the marginal sub-districts of Bangladesh: 2013

Sub-districts	hh_farm_size (acre)	Cropped area (acre)	Cropping intensity
Damurhuda	0.58	0.94	159.40
Rajibpur	0.56	0.57	100.61
Dowarabazar	0.79	0.95	121.99
Porsha	0.63	0.97	156.10
Vandaria	0.93	1.49	163.57
Total	0.66	0.96	144.03

cereal (predominantly rice) farming and low productive agricultural day laboring are the major livelihood options in these sub-districts. The poor SHs in these areas are unable to develop the opportunities of high yielding cereal- and non-cereal-based farming, non-crop farming, non-farm business activities, non-farm wage employment and international migration, realities which came from both qualitative investigation and sample survey. This is not only a result of their adverse geographical location but also their poor capital bases and the unavailability of innovative development interventions in the locality, as will be further explained in a later section. The qualitative investigations suggest that the poor SHs in the marginality hotspots are vulnerable due to their agro-ecological vulnerability-almost all five areas face, to some extent, natural calamities (flood, drought, salinity by tidal flow) that discourage farmers from thinking that innovative process and technology might be useful for agricultural production for their livelihoods (Box 16.1). The poor SHs in all areas (except Damurhuda) are usually less motivated for agricultural intensification and also lack agricultural knowledge. Almost all areas face water management and irrigation problems with varying degrees of severity. They are also constrained by their limited connectivity with the main growth centers, poor physical irrigation and extension/communication infrastructure, and power shortages.

Box 16.1: Farmers Are Physically Weak and Naturally Vulnerable

As can be seen from our qualitative field data collected during March, 2013, most of the farmers who belong to poor or ultra-poor strata groups are physically vulnerable in regards to farming. The majority of them suffer from severe backbone/waist pain and physical weakness at some point during working hours. Abul Hashem is a farmer from Poromesshoripur, Sunamganj who has been living with waist pain for over 12 years. Though it's overburdening for a 50-year-old farmer to do hard work in the agro-field, there is no other way for him to fulfill his function as a household head. To describe his physical condition, Hashem opines, *"I am sick and suffering*

(continued)

Table 16.3 Distribution of cropped land (acre/household) for different crops of poor SHs in marginal sub-districts of Bangladesh: 2012–2013 (N = 313)

Sub-districts	All_cereal_cropped areas	Rice cropped area	Maize croppedarea	Wheat cropped area	Other cropped area	Total cropped land
Damurhuda	0.81	0.64	0.14	0.02	0.14	0.94
Rajibpur	0.30	0.30	0.00	0.00	0.27	0.57
Dowarabazar	0.87	0.87	0.00	0.00	0.08	0.95
Porsha	0.95	0.89	0.01	0.05	0.02	0.97
Vandaria	1.43	1.43	0.00	0.00	0.06	1.50
Total	0.83	0.76	0.05	0.02	0.12	0.96

Table 16.4 Yield rate for cereals for poor SHs in marginal sub-districts of Bangladesh (N = 313)

Sub-districts	Rice_yield (t/ha)	Maize_yield (t/ha)	Wheat_yield (t/h)
Damurhuda	4.50	8.87	2.93
Rajibpur	2.79	–	–
Dowarabazar	3.42	–	–
Porsha	5.15	–	3.16
Vandaria	2.67	–	–
Total	4.01	–	–

Box 16.1 (continued)

from bone decay on the left side of my waist. This doesn't make me feel good in any way. Also I am so weak from working hard in the field because of my age." He seems sicker compared with the other villagers of his age. Tobacco use could be one of the foremost factors affecting his state. Besides this sort of physical sickness and the inability to do the sort of hard work demanded by farming, sometimes farmers have to face a natural barrier to cultivation. According to Hashem, he got a lower amount of production than he had in the previous year. As he states, *"Disaster and flood have damaged a large amount of [the] crop this year, which has driven [the] economic and household conditions into a vulnerable state."*

Number and Characteristics of the Poor at Each Poverty Strata

National sources (BBS 2011 show that the population under the upper poverty line, regardless of their farming involvement in those five sub-districts, varies from 34 % to 59 %, except in Dowarabazar (haor area) where the figure is nearly equal to national averages (31 %). Results from the TIGA Bangladesh household census 2013 conducted in 16 villages of 5 marginal sub-districts show that about 3,135 households (54 % of 5,855 total) are SHs, of which about 862 households (27 % of SHs and 15 % of total) are poor SHs who could be eligible for an agricultural productivity improvement program in the marginal sub-districts. From this study population, a sample of 357 SHs has been drawn for the detailed investigation. Then, the sample households have been stratified by quantitative income criteria and validated by participatory wealth ranking and self-reported perceptions. For income criteria, we use both US dollar classification and PPP dollar classification, finding that US dollar classification (e.g., subjacent poor being those with incomes between $1 and $1.25/day, medial poor: between 75¢ and $1/day, and ultra-poor: below 75¢/day) is more consistent with self-reported perception (Table 16.5). Table 16.5 suggests that about 12.32 % of the sample belongs to the non-poor category of US dollar income criteria (equivalent to 8.4 % of self-reported

Table 16.5 Surveyed poor SHs' stratifications in marginal sub-districts with agricultural potential in Bangladesh (N = 357)

Household status	Self-reported perceptions (%)	As of US $ (@ 80.00 BDT)	As of PPP $ (@33.53)
non_poor	8.4	12.32	63.02
subjacent_poor	20.17	13.73	11.2
medial_poor	55.18	17.93	8.4
ultra_poor	16.25	57.7	18.77
Total	100	100	100

Table 16.6 Distribution of poor SHs among selected sub-districts (as of US $ classification) (N = 313)

Sub-districts	Ultra poor	Medial poor	Subjacent poor	All samples
Damurhuda	63 %	23 %	15 %	36 %
Rajibpur	71 %	17 %	15 %	19 %
Dowarabazar	63 %	17 %	20 %	11 %
Porsha	72 %	20 %	8 %	20 %
Vandaria	58 %	21 %	23 %	14 %
Total	66 %	20 %	15 %	100 %

perception), and thus, the latter analyses are centered on this sample (313 poor SHs). It is also found that the number of subjacent poor is almost the same in both USD income criteria and self-reported perception, but varies significantly for medial and ultra-poor households. Our qualitative participatory wealth ranking exercise also shows that the majority of the households in the sample should be in the ultra-poor category. Thus, we followed the latter analyses based on the USD income classification. Sub-district-wise distribution (Table 16.6) shows that the number of subjacent and medial poor SHs does not differ significantly, but the number of ultra-poor SHs is comparatively higher in Porsha and Rajibpur than it is in the other three sub-districts. Though the overall economic condition in Damurhuda is much better compared to that in Dowarabazar, the similar number of ultra-poor SHs in those two sub-districts may be a result of the fact that, in Damurhuda, poor SHs are more marginalized compared to the better off households. A later section will furnish us with a greater explanation of these facts.

Poor SHs Livelihood Capitals as Per Stratification

Table 16.7 shows that the poor SHs' capital bases are very poor, but these capitals don't significantly differ quantitatively from different strata (subjacent, medial and ultra-poor). However, qualitative investigations suggest that the majority of the community defined by ultra-poor categories are differentiated from medial to subjacent poor in terms of landholdings/access to farmland, livelihood engagement,

Table 16.7 Descriptive statistics for poor SHs five capitals in marginal sub-districts in Bangladesh (N = 313)

Variables	Ultra poor (N = 206)		Medial Poor (60)		Subjacent poor (47)		1 vs 2		1 vs 3		2 vs 3	
	Mean	SD	Mean	SD	Mean	SD	Diff	P-value	Diff	P-value	Diff	P-value
Human capital												
Members schooling years	2.42	1.78	2.74	1.88	3.00	1.90	−0.31	0.72	−0.58	0.15	−0.26	1.00
Household head schooling years	1.47	2.54	1.80	2.82	2.02	2.85	−0.33	1.00	−0.55	0.60	−0.22	1.00
Financial C												
Total Income (BDT)	40,700.60	26,023.22	77,931.79	25,863.62	10,2152.53	42,244.83	−37,231.19	0.00	−61,451.93	0.00	−24,220.74	0.00
Loan (BDT)	6253.41	14,673.95	5599.17	13,191.64	6166.32	8711.8451	654.25	1.00	87.09	1.00	−567.15	1.00
Savings (BDT)	2294.41	5180.86	2581.67	7355.42	4871.28	15,861.92	−287.26	1.00	−2576.87	0.15	−2289.61	0.44
Natural capital												
Farm_size	62.14	43.69	68.85	59.64	77.23	48.94	−6.71	1.00	−15.10	1.00	−8.38	1.00
Physical C												
Total Physical assets (BDT)	60,059.35	45,817.72	64,810.35	52,450.87	61,914.64	57,008.44	−4751.00	1.00	−1855.28	1.00	2895.71	1.00
Farm productive assets	20,977.82	23,065.22	20,905.83	23,893.92	22,288.30	45,070.22	71.98	1.00	−1310.48	1.00	−1382.47	1.00
Non-farm productive assets	5017.18	11,923.47	2433.33	3545.13	3483.19	5825.46	2583.85	0.24	1533.99	1.00	−1049.86	1.00
Durables	34,064.35	33,240.93	41,471.18	48,273.61	36,143.15	32,451.77	−7406.83	0.50	−2078.80	1.00	5328.03	1.00
Social Capital												
NGO Involvement	1.91	0.29	1.88	0.32	1.98	0.15	0.02	1.00	−0.07	1.00	−0.10	0.25
Social safety net involvement	1.93	0.26	1.95	0.22	1.98	0.15	−0.02	1.00	−0.05	0.55	−0.03	1.00

technology adoption, credit accessibility, cell phone use, motivation and communication/networking skills, and physical fitness (Annex 1). Poor SHs are also insecure and vulnerable (Box 16.1).

Poor SHs' Livelihood Opportunities and Income Pattern Across Poverty Strata

The livelihoods of poor SHs and their households' working members include farming, non-agricultural enterprises, wage employment in the locality, and migration (Annex 2). Rice during the Boro and Aman seasons is a common cereal crop for all strata of SHs in marginal areas. Additionally, the subjacent poor SHs in the *Charland* produce a limited scale of maize and wheat, while the poor SHs produce maize in food insecure zones at a larger scale and wheat in drought prone areas of *barind tract* areas at a limited scale. Other crops that the SHs produce are jute, sweet potato, pulses, spices, sugarcane, mung bean, and several types of vegetables. Most of the poor SHs are engaged in non-crop farming, include poultry and cattle rearing, beef fattening, goat rearing, fruit gardening, commercial fishing, and plantations. Raising poultry is a common non-crop practice among SHs, for the purpose of both consumption and commerce. Fishing is mostly done by poor SHs who live in the coastal belt areas. Poor SHs are engaged in non-agricultural enterprises/businesses, like renting tractors and spray machines, working in groceries and sweet shops, or serving as local transport drivers (*korimon*). The wage employment opportunities available in certain areas for poor SHs are day labor (e.g., agricultural day labor or work in a break field), masonry, rickshaw pulling, or wood cutting. In-country migration is familiar among the poor SHs. In a particular time of the year, they migrate from their own areas to different areas so as to be able to earn additional income for their livelihoods and purchae agricultural inputs. While Rajibpur and Porsha's SH household members don't migrate to other countries, members from the other three sub-districts do migrate, especially in the Middle East and southeast Asia (Malaysia) in limited scale.

The sample for this study was drawn from the population of poor SHs, and thus, their income is naturally very low compared to the national rural average and also the national rural average of poor households. As shown in qualitative investigation, their income comes mainly from that of farm and non-farm day labor and cereal crop farming (Table 16.8). The income differences are observed along the different strata of poor SHs. While ultra-poor SHs' income is differentiated from that of the medial and subjacent poor mainly by the income from the cereal crop and day-labor, and also partly from non-cereal crop income, the subjacent poor SHs' income is also differentiated from business income. That means that the medial poor and subjacent poor SHs, when compared to ultra-poor SHs, are taking some advantage of livelihood opportunities other than cereal-based farming. However, compared to the livelihood opportunities available in a typical advanced rural

Table 16.8 Poor SHs' pattern of income from different sources (N = 313) as of US$ classification (BDT)

Variables	Ultra poor (N = 206)		Medial Poor (60)		Subjacent poor (47)		1 vs 2		1vs 3		2 vs 3	
	Mean	SD	Mean	SD	Mean	SD	Diff	P-value	Diff	P-value	Diff	P-value
Farm income	13,635.13	16,130.46	26,916.88	19,266.31	27,196.36	21,152.30	−13,281.75	0.00	−13,561.23	0.00	−279.49	1.00
Cereals	9439.37	10,698.95	18,286.13	15,976.01	16,758.47	14,341.72	−8846.75	0.00	−7319.09	0.00	1527.66	1.00
Non-cereal crops	1885.98	6413.65	3447.17	7146.92	5255.98	12,477.53	−1561.19	0.51	−3370.00	0.02	−1808.81	0.70
Non-crop farming	2309.78	10,605.04	5183.58	14,551.04	5181.91	7707.40	−2873.80	0.24	−2872.13	0.33	1.67	1.00
Non-farm income	27,065.47	23,237.25	51,014.92	30,709.21	74,956.17	39,256.93	−23,949.45	0.00	−47,890.70	0.00	−23,941.25	0.00
Business	2094.66	7142.93	1833.33	9521.01	11,085.11	27,575.06	261.33	1.00	−8990.45	0.00	−9251.77	0.00
Home-based non-farm activities	7.14	62.49	19.17	148.46	42.55	291.73	−12.03	1.00	−35.42	1.00	−23.39	1.00
Wage employment	22,787.62	23,397.01	47,275.33	31,074.28	57,843.40	43,239.86	−24,487.71	0.00	−35,055.78	0.00	−10,568.07	0.18
Day laborers income	22,365.29	23,382.76	43,715.33	30,771.58	52,992.34	45,030.45	−21,350.04	0.00	−30,627.05	0.00	−9277.01	0.31
Salaries	364.08	3897.28	2960.00	13,884.80	4851.06	19,164.03	−2595.92	0.24	−4486.99	0.02	−1891.06	1.00
Rentals	1458.09	4563.71	1205.00	4284.12	293.62	1764.31	253.09	1.00	1164.48	0.26	911.38	0.80
Migration	72.82	751.93	0.00	0.00	3617.02	17,622.31	72.82	1.00	−3544.21	0.00	−3617.02	0.02
Social safety nets	703.40	2048.47	1282.08	5564.17	2074.47	9267.65	−578.69	1.00	−1371.07	0.20	−792.38	1.00
Total income	40,700.60	26,023.22	77,931.79	25,863.62	10,2152.50	42,244.83	−37,231.19	0.00	59,907.67	0.00	−24,220.71	0.00

location, the income sources for poor SHs are limited only by the low productive nature of their activities. Thus, it is evident that the poor SHs in those areas are marginalized not only in the national context but also within the community.

Segmentation of Poor SHs: Findings from Cluster Analysis

To suggest which types of agricultural growth productivity program seem most promising for the improvement of agriculture and livelihoods of poor SHs in the marginality hotspots with agricultural potential in Bangladesh, we used cluster analysis to group the poor SHs according to appropriate dimensions leading to different strategic options. For this purpose, cluster analysis (a major technique for classifying data) is used. Cluster analysis assigns observations to groups (clusters) so that observations within each group are similar to one another with respect to variables or attributes of interest and each group stands apart from one another. In other words, it divides the observations into homogeneous and distinct groups. This is achieved by assigning all similar observations according to the degree of proximity (closeness) among the cluster elements by calculating the shortest possible distance between observations, referred to as the Euclidean distance. The Euclidean distance between observations $\{X_{1i}, X_{2i}, \ldots, X_{ki}\}$ and $\{X_{1j}, X_{2j}, \ldots, X_{kj}\}$ is estimated as:

$$D(i, j) = \sqrt{\left(X_{1i} - X_{1j}\right)^2 + \left(X_{2i} - X_{2j}\right)^2 + \ldots \left(X_{ki} - X_{kj}\right)^2}. \qquad (16.1)$$

Observations with the closest distance are then grouped into one cluster. Allocation of the different strategic options to the farmers is done using both hierarchal and k-means cluster analysis. At first, cluster analyses are performed using a sequence of a common hierarchal and exchange algorithm using variables and attributes containing both dichotomous and categorical values. A cluster dendogram (cluster tree) reveals the appropriate number of clusters (in our case, 5). Then, we used K-means clustering, which aims to partition 313 observations into 5 clusters in which each observation belongs to the cluster with the nearest mean. K-means cluster analysis is a well-accepted exploratory statistical technique in social science research that creates natural, internally similar groups from rating scale questionnaire data. The statistical program identifies the centroid for each cluster by running the algorithm until a stable solution with minimum variability within each cluster and maximum variability between each cluster results. Through the focus group interviews and key informant discussions, the respondents are characterized into five strategic groups (Table 16.9). Based on the findings, the clusters are homogeneous in the sense that most are male-headed, have a relatively small family size, represent a very low number of schooling years, have similar non-land agricultural productive assets, have a low per capita income, and have insignificant salaried and remittance income, as well as all clusters benefitting from some form of social

Table 16.9 Segmentation of poor SHs in marginal sub-districts in Bangladesh 2012–2013 (N = 313): Results of cluster analysis

Clusters	Freq. (%)	Characteristics	Strategic options
1	36 (11.5)	Farm size medium, CI low, moderate ownership of the land, everybody sells their produce, non-land physical assets and household durables high, cereal income medium, other crop income high, business and day labor income medium, savings low, cereal technology adoption low, access to the cereal inputs/markets low	Non-cereal crops and day labor
2	107 (34.19)	Zero ownership of the land but farm size high (good access to the tenancy market), CI low, about 75 % sell their produce, non-land physical assets low and household durables medium, cereal income high, other crop income moderate, no business income but day labor income high, savings medium, cereal technology adoption medium, access to cereal inputs/markets medium	Both cereal and non-cereal crops and day labor
3	98 (31.31)	Farm size high, CI high, high ownership of the land, almost everybody sells their produce, non-land physical assets and household durables high, cereal income high, other crop, business and day labor income medium, savings low, cereal technology adoption high, access to cereal inputs/markets high	Cereal crops
4	33 (10.54)	Farm size low, CI low, low ownership of the land, about 23 % sell their produce, non-land physical assets and household durables low, crop income low, business income moderate but day labor income high, savings low, cereal technology adoption medium, access to cereal inputs medium but output market low	Day labor, business
5	39 (12.46)	Farm size medium, CI medium, low ownership of the land, about 62 % sell their produce, non-land physical assets and household durables medium, cereal income medium but other crop income low, business income high but day labor income low, savings high, cereal technology adoption medium, access to cereal inputs medium but output market medium	Business and cereal crops
Total	313		

safety net and having taken out some form of loan. On the other hand, ownership of the land, farm size, cropping intensity, agricultural crop sales, household durables, cereal income, other crop income, business income, day labor income, household savings, cereal technology adoption, access to the agricultural market, play a decisive role in making the clusters distinct from one another. Thus, among the five groups of poor SHs, *non-cereal and non crop farming with day labor* and *day labor with business* could be appropriate strategic options for two groups, while the other three appropriate strategic options could be farming (crop and non-crop) *with day labor*, *cereal crops*, and *business with cereal crops* (Table 16.10). The meanings of these results are: (1) For a productivity growth program geared towards individual poor SHs, day – labor cannot be a strategic option, although poor SHs naturally utilize it as a survival strategy; (2) Among poor SHs, though about 97.78 % of households cultivate cereals as a way of accruing the majority of their household's income share, they are still living under the poverty line, and subsequently need alternative options that could increase their income and livelihood security. Thus, only cereal-based productivity programs will be insufficient for improving the food and livelihood security of poor SHs, and the growth productivity program should be designed in a way that the SHs could have the opportunity to explore their human capability in farming (cereal and non-cereal crops and non-crop farming) and business that creates both backward and forward linkages with those farming in the locality. Therefore, we should extend our focus on crop technology innovations to include non-crop farming and non-farm businesses that could better link SHs with the market.

Technology Innovations for Poor Small Holders and the Barriers: Beyond Crop Technology Innovations

Initially, we focused on cereal crop technology innovations; later, it was expanded from cereal crops to all crops, non-crop framing and non-farm innovations required for growth productivity programs for poor SHs in the selected areas. For identifying technological innovations, we did not follow the traditional pipe-line approach, that is, scientists develop technology and then it is given to the extension agents for adoption among the farmers. Rather, we took a bottom-up approach that matched available technologies with the needs, aspirations and potentials of poor SHs and the projected costs (barriers), i.e, the matching available technological innovations usually require to enable conditions to work for poor SHs. In our approach, the focus of the innovation packages should be related to current farming practices and cropping technology use by SHs covering all stages of production (pre-production, production, harvesting, processing and marketing) – it could be newly introduced goods and services for most of the farmers but should be readily available in the locality (despite having potential, some farmers are adopting certain technology innovations, others are not; in a similar context, some farmers are getting very good returns, others are getting far less).

Following literature/document review, and consultation with scientists, both at national and regional levels, and local level extension workers/officials from both

Table 16.10 Distribution of strategic options for poor SHs in marginal sub-districts in Bangladesh (N = 313)

Strategic options	Damurhuda		Rajibpur		Dowarabazar		Porsha		Vhandaria		All sample	
	Freq.	%	Freq.	%	Freq.	%	Freq.	%	Freq.	%	Freq.	%
SO1: Non-cereal crops and day labor	7	6.25	22	37.29	1	2.86	5	7.81	1	2.33	36	11.5
SO2: Both cereal and non-cereal crops and day labor	28	25	22	37.29	12	34.29	19	29.69	26	60.47	107	34.19
SO3: Cereal crops	57	50.89	6	10.17	7	20	26	40.63	2	4.65	98	31.31
SO4: Day labor, business	5	4.46	5	8.47	10	28.57	6	9.38	7	16.28	33	10.54
SO5: Business and cereal crops	15	13.39	4	6.78	5	14.29	8	12.5	7	16.28	39	12.46
Total	112	99.99	59	100	35	100.01	64	100.01	43	100.01	313	100.00

GOs and NGOs, we prepared a list of more than 50 technology innovations (Annex 3) and conducted a perception study. The perception study addressed several key questions: (1) Are the SHs aware of this technological innovation? (2) How many SHs of those who are aware are currently using it? (3) Which technologies (for the farmers who are aware) are most important?

Poor perception by SHs about those technologies (following the frequencies and percentages of their responses) can be grouped in several ways: (1) all three indicators, awareness, adoption and further importance of certain technologies (for example, power tiller/tractor, machine for pesticide use, seed plantation in line with definite spacing), are very high, which means that even though these technologies have already been intensively adopted, awareness of their necessity prevails; (2) for some technologies (rice mill (diesel driven), shallow tube well (STW), rice mill (electricity driven), etc.), awareness and importance are high but adoption is not high, which means that adoption of the second group of technologies needs to increase significantly; (3) for some technologies, awareness, adoption and importance are all low – most of these technologies have only recently been developed at the research station, and the farmers in those areas are not quite aware of their importance. At the second stage, mainly in regard to the third group of technologies, we consulted with BRAC in-house technology experts/practitioners who are knowledgeable about those technologies and those study areas, and found some technologies that could be useful, for example, short-duration *aman* rice varieties, hybrid maize and stress-tolerant wheat varieties, handy kits for using guti urea, etc. At the final stage, we again validated our study results with the local level stakeholders, for example, extension workers (both public and NGOs), input dealers, processors, model farmers, poor SHs and made the lists of technological innovations for the future growth productivity program (Table 16.11).

Conclusions

Under a collaborative project entitled "Technology assessment and farm household segmentation for inclusive poverty reduction and sustainable productivity growth in agriculture (TIGA)" conducted by the Center for Development Research (ZEF), Bonn, in four partner countries from South Asia and Sub-Saharan Africa, this paper discusses the results generated from the Bangladesh country study. Following a marginality approach developed at ZEF, we identified five marginal sub-districts in Bangladesh, i.e., underperforming areas in which the prevalence of poverty and other socio-economic dimensions of marginality are high and agricultural potential is also high, since, in such areas, yield gaps (potential minus actual yields) are high and productivity gains (of main staple crops) are likely to be achieved. Then, we conducted a household census of 5,855 households in 16 marginal villages from those five sub-districts and drew a sample of 357 poor SHs for an in-depth quantitative sample survey. Some qualitative surveys (focus group discussions, in-depth interviews) were also conducted. Then, we developed the analytical

Table 16.11 Suggested technological innovations for marginal areas in Bangladesh

Theme	All locations	Rajibpur	Dowarabazar	Porsha	Damurhuda	Bhandaria
Intensive crop system technologies		Maize + chilis, Chilis + vegetables,		Wheat/rice + orchard	Maize + chilis + vegetables, maize + sugarcane + chilis	Group based fish + poultry + vegetable farming
Seed Technology	Hybrid and short duration rice varieties, quality seeds	Hybrid maize and stress-tolerant wheat varieties	Quality seed, especially for flash flood-tolerant rice varieties	Maize and stress-tolerant wheat varieties through shifting from *Boro* rice; drought-tolerant, short duration aman rice varieties	Maize and hybrid vegetables	Hybrid rice varieties, saline resistant rice varieties, sun flower, hybrid vegetable seeds
Technology related to water management and irrigation	Water management/saving practices	Improved *fita* pipe	STW, LLP, rubber dam	Pond digging or re-excavation	STW, AWD	Low lift pump, STW, rubber dam
Mechanical innovations	Power tiller, power tiller operated seeder, thresher (both paddle and mechanical), rice miller	Power tiller, thresher (both paddle and mechanical), rice miller	Power tiller, thresher (both paddle and mechanical), rice miller	Power tiller, thresher (paddle), rice miller	Handy USG (Guti urea) applicator, power tiller, power tiller-operated seeder, thresher	Power tiller (rental cost is high for cultivation), handy USG (Guti urea) applicator, power tiller, power tiller-operated seeder, thresher

| Non-crop innovations (non-crop farming, non-farm enterprises/businesses, migration) | Businesses/enterprises: seasonal crop (to sell surplus at reasonable price thereby instigating others), livestock and poultry rearing, seed business (distribution channel and awareness building), extension service | Businesses/enterprises: seasonal crops, livestock and poultry rearing, fishing, boat, rice milling; seed business | Extension service Seed distribution channel with awareness building Seasonal crop business (creating forward linkage for duck/fish/crops to sell surplus) | Businesses/enterprises: Mango cultivation/orchard, water harvesting/mini-pond digging/re-excavation; | Businesses/enterprises: seasonal crop/vegetable business, beef fattening, poultry, small scale fruit gardening, goat farming, power tiller and threshing service; agro-machinery | Extension service (to make people aware of potential in the area) Commercial enterprises for sunflower production with backward and forward linkage Business for seasonal vegetables, Fishing + poultry, livestock and poultry, agro-machinery |

methodology to create a thorough understanding of the interactions between technology needs, farming systems, ecological resources and poverty characteristics in the different strata of the poor SHs, and to link these insights with technology assessments in order to guide action for overcoming current barriers to technology access and adoption under the common approach for technological innovations for inclusive growth in agriculture developed at ZEF jointly with its partners. Results suggest that five marginal sub-districts with agricultural potential are very different from each other. Sufficient potential exists in those sub-districts, and enough scope to develop that potential, to ensure farm intensification and livelihood diversification. Regarding adverse agro-ecological vulnerability, almost all five areas are facing to some extent, natural calamities (flood, drought, salinity by tidal flow). This discourages poor SHs from thinking that innovative processes and technology might be useful for their agricultural intensification and livelihoods. Poor SHs' income mainly accrues from cereal crop income and low productive non-farm sources (say, agricultural day labor) and their capital bases being very poor do not differ significantly from different strata quantitatively, though qualitatively, some differences among the capital bases have been observed. Cluster analysis gives meaningful segmentation of poor SHs. Development strategies should focus on three pathways: agricultural intensification, income diversification and agricultural diversification based on options available for the SHs in the localities. Cereal-based technology under agricultural innovations could be part of the solution, but that could also be integrated with other income diversification and agricultural diversification strategies. Intensive crop system, hybrid seeds, water management technologies, non-crop farming, non-farm enterprises/businesses are the suggested potential technological innovations for the study areas. The technological innovations could be promoted through introducing strategic development programs that include promotion of crop and non-crop farming production and related (backward and forward) non-farm businesses in the localities.

Open Access This chapter is distributed under the terms of the Creative Commons Attribution-Noncommercial 2.5 License (http://creativecommons.org/licenses/by-nc/2.5/) which permits any noncommercial use, distribution, and reproduction in any medium, provided the original author(s) and source are credited.

The images or other third party material in this chapter are included in the work's Creative Commons license, unless indicated otherwise in the credit line; if such material is not included in the work's Creative Commons license and the respective action is not permitted by statutory regulation, users will need to obtain permission from the license holder to duplicate, adapt or reproduce the material.

Annex 1: Characteristics of Poor SHs with Regard to Crop Technology Innovations in Marginality Hotspots with Agricultural Potential in Bangladesh

Areas/sub-districts/villages	Poverty strata		
	Subjacent poor	Medial poor	Ultra poor
Charland/ Rajibpur/ Borober	Own some agricultural land and do farming mainly through tenancy *Cropping pattern*: Boro rice/Chilis/Vege-tables- Maize/Maize+ vegetables + Chilis/veg-etables/Chilis+ Vegeta-bles- Vegetables/ Sugarcane+ Maize/Nuts/ Chilis+ Vegetables *Non crop farming*: Poultry and cattle rearing, beef fattening, fruit gardening, etc. Affected by riverbank erosion Physically able to do hard work	Agricultural day labor and farming, mainly through ten-ancy, are the main occupation *Cropping pattern*: Bororice-Maize-Vegetables, Mung Beans/Maize *Non-crop farming*: poultry and cattle farming Have no homestead land, live on lease land Physically able to do hard work	Main occupation is agricultural day labo but very few do tenant farming *Cropping pattern*: Sugar Cane/Boro rice-vegetables/jute/ Maize-vegetables/ maize/maize+ sugarcane+ chilis *Non-crop farming*: Nil Have no homestead Mostly old people (separated from their children)- physically not very able to do hard work
Haor basin/ Dowarabazar/ Poromessorpur	Own agricultural land <3 acres *Cropping pattern*: Boro rice/ Vegeta-bles- Fallow-Aman rice Non-crop farming: nil Level of education: above primary level Influential and respect-able in the village	Own agricultural land <2 acres *Cropping pattern*: Boro rice/vegeta-bles- Fallow-Aman rice Non-crop farming: poultry (consump-tion) Level of education: below primary level Physically not well fit to do hard work	Own agricultural land<1 acre and some people do not own even homestead land *Cropping pattern*: Boro rice/vegetables-Fallow-Aman/Fallow Non-crop farming: poultry (consumption) Level of education: illiterate Aged and physically not fit to do hard work
Barind tract/ Porsha/ Bhobanipur	Own agricultural land – 1–1.5 acres *Cropping pattern*: Boro rice/Wheat-Aus/ vegetables/Jute- Aman/ Potatoes Non-crop farming: gardening	No agricultural land except homestead land Farm size through tenancy<1 acre *Cropping pattern*: Boro rice/wheat-vegetables-Aman/ vegetables Non-crop farming: gardening, commer-cial fishing	No agricultural land, not even homestead Main occupation is agricultural day labor but very few do tenant farming *Cropping pattern*: Boro/Fallow/ Wheat-Aus/Fallow-Aman/Vegetables Non-crop farming: gardening, commercial fishing
Barind tract and food insecure	Amount of owned land <0.70 acre and able to	Amount of owned land <0.30 acre	Own land<0.05 acre Main occupation is

(continued)

Areas/sub-districts/villages	Poverty strata		
	Subjacent poor	Medial poor	Ultra poor
zone/ Damurhuda/ ChotoDudpatila	take some land on rented in. *Cropping pattern*: Boro rice/chilis/vege-tables- maize/maize+ vegetables + chilis/vege-tables/chilis+ vegeta-bles- vegetables/ sugarcane+ maize/nuts/ chilis+ vegetables Non-crop farming: poul-try and cow fattening Unable to manage household expenditure by farming and so day labor is needed for livelihood	Difficult to take land as tenant *Cropping pattern:* Boro – Hybrid rice/ Maize-Vegetables, Mung Beans/Maize Non-crop farming: Fruit gardening (small scale), poultry, cow fattening, goat farming	agricultural day labor but very few do tenant farming Unable to take loan to do agriculture *Cropping pattern:* Sugar Cane/Boro rice-vegetables/jute/ maize- vegetables/ maize/ maize+ sugarcane+ chilis Non-crop farming: poultry, goat farming
Coastal belt/ Bhandaria/ OttorJunia	Own farm land: 30–40 decimals *Cropping pattern*: Boro rice/vegetables (hybrid)-Aus (fallow)- Aman rice/fallow/ Spices/lentils *Non-crop farming:* Fishing + poultry, fruit gardening Level of education: above primary level Aware and always trying to change economic con-dition High communication skills and maintain good relationship with agricul-tural extension officers/ workers Capable of giving fertil-izer, irrigation, and pes-ticides in time Receive credit from dif-ferent sources and repay the loan on time -Everyone has mobile phone -Send children to school	Own farm land (excluding home-stead): 10–20 decs. *Cropping pattern* Boro rice-Aus (fal-low)- aman rice/ spices/lentils/robi Crop Non-crop farming: Fishing, poultry, fattening beef cattle Level of education: primary level Rate of return on investment is rela-tively better than ultra-poor Have enough agricul-tural knowledge and experiences Use traditional tech-nology in farming Able to deposit a small amount for renting a piece of land -Preserve seed for next crop season	Have no farm land except homestead *Cropping pattern:* Bororice-Ausrice-Aman+vegetables+s-pices+lentils Non-crop farming: Fishing, poultry, fattening beef cattle Illiterate and unable to adapt to new technol-ogy Limited opportunity to improve economic conditions Partly involved in agriculture and poor social capital Family size: usually large

Source: Extracts from qualitative survey conducted for TIGA Bangladesh: April 2013

Annex 2: Strata of Specific Livelihood Options for Poor SHs in Marginality Hotspots in Bangladesh

Areas/sub-district/Villages	Strata	Livelihood options				In-country migration	Out-country migration
		Crop farming	Non-crop farming	Non-agricultural enterprises/businesses	Wage employment (farm and non-farm)		
Charland/Rajibpur/Borober	Subjacent poor	Rice, maize, wheat, jute, sweet potatoes, mustard, vegetables (chilis, brinjal)	Poultry and cattle rearing, beef fattening, fruit gardening	Grocery shop, jewelry business, cattle business, boat man, fishing, handy craftsshop, rice milling	Poultry farm	Mainly Dhaka	Nil
	Medial poor			Cattle and poultry business, fishing	Agri-business, poultry farm, cloth shop	Dhaka, Barisal, Dinajpur, Rajshahi, Comilla, Sylhet, Tangail	Nil
	Ultra poor			Nil	Agriculture day labor	Dhaka, Tangail, Gazipur, Bhoirob	Nil
Haor basin/Dowarabazar/Poromessorpur	Subjacent poor	Boro rice, vegetables, Aman rice			Day labor (both agricultural and non-agricultural)	Comilla, Dhaka	Dubai, Oman
	Medial poor	Boro rice, Aman	Poultry (Consumption)		Day labor (both agricultural and non-agricultural)	Sylhet, Comilla	Saudi Arabia, Dubai, Oman
	Ultra poor	IRRI rice Boro rice, Aman	Poultry (Not commercially)		Day labor, agricultura; day labor	Comilla, Dhaka, Sylhet, Shunamganj	Dubai, Oman

(continued)

Areas/sub-district/Villages	Strata	Livelihood options				In-country migration	Out-country migration
		Crop farming	Non-crop farming	Non-agricultural enterprises/businesses	Wage employment (farm and non-farm)		
Barind tract/ Porsha/ Bhobanipur	Subjacent poor	Boro rice, wheat, Aus rice, vegeta-bles, jute, Aman rice, potatoes	Gardening	Grocery	Grocery shop, sweet shop, seasonal van driving	Nil	Nil
	Medial poor	Boro rice, wheat, vegetables, Aman rice, vegetables	Gardening, com-mercial fishing	Grocery, sweet shop, seasonal van driver	Agricultural day labor, working in grocery shop, Rick-shaw pulling	Nil	Nil
	Ultra poor	Boro rice, wheat, Aus rice, Aman rice, vegetables	Gardening, com-mercial fishing	Grocery, sweet shop	Non-agricultural day labor, Rickshaw pulling, seasonal van driving, shop assistant	Nil	Nil
Barind tract and food in secured zone/ Damurhuda/ ChotoDudpatila	Subjacent poor	Boro, chilis, vege-tables, maize, chilis, sugarcane, nuts	Cow fattening, poultry, no planta-tion fisheries-unavailability of pond and pool	Power tiller, thresher, korimon (local name) owner and provides this as rent	Korimon driver, masonry	Dhaka	Malaysia, Saudi Arab, Dubai, (2 %)
	Medial poor	Boro rice, hybrid rice, maize, vege-tables, mung beans	Fruit gardening (small scale), poul-try, cow fattening, goat farming		Day labor, masonry	Dhaka, Faridpur, Mymensing, Rajbari	Malaysia, Saudi Arab, Dubai,
	Ultra poor	Sugarcane Boro rice, vegetables,	Poultry, goat farming		Day labor (agricul-tural day laborer,	Faridpur, Mymensing, Rajbari	Malaysia, Saudi

					work in break field), masonry		Arab, Dubai,
Coastal belt/ Bhandaria/ OttorJunia	Subjacent poor	Boro rice, Aman rice, spices, lentils	Fishing + poultry, vegetables (hybrid), fruit gardening	Tractor, spray machine, fertilizer and seed sales	Day labor, Rickshaw pulling, business, masonry,	Dhaka, Faridpur, Khulna, Bagerhat, Chittagong.	Malaysia, Middle East
	Medial poor	Boro rice, Aman rice, spices, lentils, Robi crop	Fishing, vegetables, poultry, fattening beef/cattle	Tractor	Day labor, rickshaw pulling, business, masonry, wood cutting, work in break field,	Dhaka, Faridpur, Khulna, Bagerhat, Munsiganj	Malaysia, Middle East
	Ultra poor	Boro rice, Aus rice Aman rice, vegetables, spices, lentils		Spray machine, business	Day labor, rickshaw pulling, masonry, wood cutting, work in break field	Dhaka, Faridpur, Feni, shylet, Chittagong	Malaysia, Middle East

Source: Extracts from qualitative survey conducted for TIGA Bangladesh: April 2013

Annex 3: Technology Innovations for Marginal Areas: Results from the TIGA Perception Study 2013

Technologies	Are the SHs aware of this? (%)	What % of the aware SHs are using it?	Which technologies (for the aware farmers) are most imporant?
1. All three indicators- awareness, adoption and further importance of some technologies are very high			
Power tiller/tractor for preperation of land	98.88	95.52	95.52
Machine for pesticide use	79.27	75.35	75.63
Seed plantation in line with definite spacing	83.19	74.79	75.07
Rice mill (diesel-driven)	88.24	69.19	70.59
Shallow tubewell (STW)	85.99	53.22	68.35
Rice mill (electricity-driven)	84.03	52.38	63.03
Thresher/Bomaor Auto Machine	63.31	49.30	56.58
Deep tubewell (DTW)	67.51	38.10	52.38
Irrigation in dry season	42.3	35.29	39.22
Herbicides	34.45	27.45	28.57
Pedal thresher	35.01	25.77	26.05
Irrigation by Fita Pipe	25.21	17.65	18.49
2. Awareness and importance are high but adoption is not high			
Hybrid paddy	84.03	27.45	61.9
Irrigation from pond/river using power driven pump (LLP)	64.99	41.18	50.7
Using money instead ofcrops in share cropping	46.22	33.05	35.29
Hybrid maize	36.69	16.25	28.57
HYV wheat	40.62	15.41	26.89
Plastic container/drum/polythene for seed collection	26.61	22.41	25.21
Compost	30.53	17.65	22.41
SRI in rice cultivation (young seed-ling, half plant in one bundle, space between bundles, irregular irrigation)	27.17	18.49	22.13
3. Awareness, adoption and importance are all low			
Drugs for seed preservation	20.45	10.36	16.25
Guti urea (urea tablet/USP/UDP)	27.73	5.60	13.17
Rainwater reserved by pond digging and irrigation in dry season	21.01	6.44	10.36
Aromatic Boro variety	12.61	5.60	9.52
Short-duration Aman variety	12.04	3.92	7
Mobile phone for exchanging agri-cultural information (information	6.44	4.48	5.32

(continued)

Technologies	Are the SHs aware of this? (%)	What % of the aware SHs are using it?	Which technologies (for the aware farmers) are most imporant?
dissemination, price of fertilizer, price of crop)			
Inter cropping of maize	6.72	1.12	4.76
Rice-fish mixed cropping	10.92	0.84	3.92
Using large water reservoir to hoard rainwater for irrigation i n the dry season	5.04	1.96	2.52
Irrigation by Barid Pipe/alternative to Fita Pipe	3.08	1.40	2.24
Mechine for using Guti urea	4.2	1.68	2.24
Combined thresher	6.44	3.36	3.92
Water tolerant Aman variety	8.12	2.52	3.08
Inter cropping of rice	4.76	0.28	2.52
Seeder machine for land preparation, seeding and weeding	3.92	1.40	1.96
Water hoarding using Ruber Drum Reservoir	3.08	1.68	1.96
Inter cropping of wheat	1.68	0.28	1.4
Bed pl anter mechine for plant, fertilizer and seeding	2.52	0.84	1.12
Drought-tolerant wheat variety	1.4	1.12	1.12
IPM (Integrated Pest Management)	1.96	0.56	1.12
If you become aware of any technology which is not mentioned above (specify. . ..)	1.12	0.84	1.12
Drought-tolerant and short-duration A man variety	1.96	0.56	0.84
Drought-tolerant maize variety (instead of paddy)	0.56	0.56	0.56
Short-duration maize variety	0.84	0.56	0.56
Early maturing maize variety	0.84	0.56	0.56
Water-tolerant maize variety	0.84	0.00	0.56
Introducing more short-duration crop variety	0.56	0.28	0.28
Leaf color chart (LCC)	0.56	0.56	0.28
Drought-tolerant wheat variety	1.4	0.00	0
Short-duration wheat variety		0.00	0
Early ma turing wheat variety	0.28	0.00	0
Magic Pipe – (AWD)	0.56	0.56	0

References

Balen J, McManus DP, Li Y, Zhao Z, Yuan L, Utzinger J, Williams GM, Li Y, Ren M, Liu Z, Zhou J, Raso G (2010) Comparison of two approaches for measuring household wealth via an asset-based index in rural and peri-urban settings of Hunan province, People's Republic of China. Emerg Themes Epidemiol 7(1):7. doi:10.1186/1742-7622-7-7

BBS (2011) Report of the Household Income and Expenditure Survey 2010. Bangladesh Bureau of Statistics, Statistics Division, Ministry of Planning, Government of the People's Republic of Bangladesh, Dhaka

Carney D (ed) (1998) Sustainable livelihoods: what contribution can we make? Department for International Development, London

Conway G (1999) The doubly green revolution: food for all in the twenty-first century. Cornell University Press, Ithaca

DFID (1999) Sustainable livelihoods guidance sheets. Department for International Development, London

DFID (2000) Sustainable livelihoods guidance sheets. Department for International Development, London

Fan S, Hazell P (2000) Should developing countries invest more in less-favoured areas? An empirical analysis of rural India. Econ Polit Wkly 35(17):1455–1464

HKI, JPGSPH (2011) State of food security and nutrition in Bangladesh, 2010. Helen Keller International and James P Grant School of Public Health, Dhaka

Malek MA, Usami K (2010) Do the non-farm incomes really matter for poverty among small households in Rural Bangladesh? A case of advanced villages. J Dev Agric Econ 2(7):250–267

Malek MA, Hossain MA, Saha R, Gatzweiler FW (2013) Mapping marginality hotspots and agricultural potentials in Bangladesh. Working paper No. 114. Center for Development Research, Bonn

OECD (2001) Poverty reduction – the DAC guidelines. Organisation for Economic Co-operation and Development, Paris

Pender J (2007) Agricultural technology choices for poor farmers in less-favored areas of South and East Asia. IFPRI discussion paper 00709. International Food Policy Research Institute, Washington, DC

Pinstrup-Anderson P, Pandya-Lorch R (1994) Alleviating poverty, intensifying agriculture and effectively managing natural resources. Food agriculture and the environment discussion paper no 1. International Food Policy Research Institute, Washington, DC

Reardon T, Chen K, Minten B, Adriano L (2012) The quiet revolution in staple food value chains-Enter the Dragon, the Elephant, and the Tiger. Asian Development Bank and International Food Policy Research Institute, Manila/Washington, DC

Ruben R, Pender J, Kuyvenhoven A (2007) Sustainable poverty reduction in less-favoured areas: problem, options and strategies. In: Ruben R, Pender J, Kuyvenhoven A (eds) Sustainable poverty reduction in less favored areas. CABI, Wallingford

Rutstein S (1999) Health nutrition and population country fact sheets. DHS (Demographic and Health Surveys) Programm, Macro International, Calverton

Chapter 17
Integrated Rice-Fish Farming System in Bangladesh: An Ex-ante Value Chain Evaluation Framework

Abu Hayat Md. Saiful Islam

Abstract Rice and fish are an important source of food and nutrition security, income, and livelihood options for many people in Bangladesh. Integrated rice-fish farming systems are a potential option which respond to scarce land and water resources but their potential has not been fully explored in the country. Thus, this study assesses the ex-ante socio-economic competitive potential of this technology, as well as the crucial factors for its widespread adoption and diffusion. To assess the true performance of an activity, we take into account its upstream and downstream actors who are directly or indirectly related to that activity. The overall quantitative results from gross margin, partial budgeting and gendered employment analyses show positive benefits due to the introduction of rice-fish technology instead of rice monoculture in Bangladesh.

Keywords Integrated rice-fish farming system • Ex-ante assessment • Value chain evaluation framework • Partial budgeting • Bangladesh

Introduction

With more than 150 million inhabitants in an area of 147,570 km^2, Bangladesh is characterized as one of the most densely populated countries in the world (about 964 persons/km^2, or only 0.06 ha available per head), with rapid population growth (1.37 % per annum) and low per capita income ($848 (US) per year) (BER 2012; World Bank 2012). Although the poverty level declined in the last decade at an impressive rate, the absolute number of people below the poverty line remains significant. Around 53 million people still live below the poverty line and most of them (about 75 %) live in rural areas (World Bank 2012; BER 2012). Agriculture (including fisheries) is still one of the major contributors to the economy,

A.H. Md. Saiful Islam (✉)
Center for Development Research, University of Bonn, Bonn, Germany

Department of Agricultural Economics, Bangladesh Agricultural University, Bonn, Germany
e-mail: saiful_bau_econ@yahoo.com

© The Author(s) 2016
F.W. Gatzweiler, J. von Braun (eds.), *Technological and Institutional Innovations for Marginalized Smallholders in Agricultural Development*,
DOI 10.1007/978-3-319-25718-1_17

accounting for 20 % of GDP and growing at 5 % over the years. Most of the rural people directly or indirectly engage with agriculture for their daily livelihood and about 48 % of labour is employed in this sector (BER 2012). From ancient times, agriculture, including fisheries, has been an integral part of the life of the Bangladeshi people, and plays a major role in food security, employment, nutrition, foreign exchange earnings and other aspects of the economy. Fish with rice is the national diet, giving rise to the proverb *Maache-Bhate Bangali* ("A Bengali is made of fish and rice"): fish alone supplies about 60 % of animal protein intake and rice alone supplies 70 % of direct human calorie intake (Alam and Thomson 2001; DOF 2010; Sarder 2007). Bangladesh is one of the top nations in terms of producing and consuming rice and fish, and both are associated with the daily food culture of the Bangladeshi people, especially for poor rural people.

Due to high population growth, economic development and urbanization demand for rice and fish is increasing day by day. On the other hand, the supply is threatened due to conversion of agricultural land, climate change and the environmental impact of overuse of fertilizer and pesticides during the green revolution period. Thus, there is an urgent need for a sustainable option which can produce rice and fish in a sustainable manner. Integrated rice-fish farming systems (IRFFS) seem to be such an option, producing more rice and fish with less use of land and water in a sustainable way. Since its inception, different researches have shown that it is ecologically/environmentally friendly, works as IPM, increases soil fertility, optimally uses scarce land and water resources complementarily, increases productivity, environmental sustainability, system biodiversity, intensification, farm diversification and household nutrition, and is a sustainable option for producing rice and fish through less use of land and water (Frei and Becker 2005; Fernando 1993; Nhan et al. 2007; Ahmed and Garnett 2011; Ahmed et al. 2011; Berg 2001, 2002; Halwart and Gupta 2004; Halwart et al. 1996; Little et al. 1996; Lightfoot et al. 1992; Giap et al. 2005; Rothuis et al. 1998a, b, 1999; Haque et al. 2010; Dugan et al. 2006; Coche 1967; Gurung and Wagle 2005). Although the potentiality of this technology has been widely documented, rice-fish farming systems are still not widespread in Bangladesh, remaining a marginal farming system (Ahmed and Garnett 2011; Ahmed et al. 2011; Nabi 2008). This issue gives major impetus to properly assessing the potential socio-economic benefit of this system compared to rice-monoculture, as well to identifying the factors which facilitate and hinder rice-fish technology adoption and diffusion.

For widespread diffusion of and proper policy-making in regard to integrated rice-fish farming systems, it is necessary to know the adoption pattern as well as its impacts (Noltze et al. 2012; Becerril and Abdulai 2010). Doss (2006) mentioned that, after 20 years, technology adoption studies have made substantial progress in examining the intensity of adoption (not just dichotomous choices) and addressing the simultaneity of adoption of different components of a technology package. However, the issues of how institutional and policy environments affect the adoption of new technologies and how the dynamic patterns of adoption affect the distribution of wealth and income remain unanswered. To the best of our knowledge, no studies have been done considering those aspects in the case of integrated rice-fish farming systems in Bangladesh, although value-chain analysis, along with such useful tools as partial budgeting and SWOT analysis, can work as an ex-ante framework for assessing the details of performance of this farming system by considering the upstream and downstream actors (like Macfadyen et al. 2012; Veliu et al. 2009; Christensen et al. 2011) compared to the rice-monoculture value chain. Value chain analysis is a strong qualitative as well as quantitative approach widely applied to pro-poor economic development. It can assess economic viability and sustainability and identify the critical issues and impasses for different actors, and then generate robust and effective policies and development strategies (Coles and Mitchell 2011). Thus, this study is an attempt to fill this research gap by using the powerful value chain analysis as a framework for assessing the comparative performance of integrated rice-fish farming systems for indigenous peoples in Bangladesh. The article's most important contribution is its use of the three above-mentioned forms of analysis to assess the performance of rice-fish technology in marginalized, extreme poverty settings, which, in turn, will help to design and execute pro-poor agricultural interventions to reduce extreme poverty and marginality in the developing world. To do so, the paper continues as follows: the next section presents the research methodology, including data and analytical research, employed in this paper. A result and discussions section comes next, including subsections regarding value chain mapping, gross margin analysis, partial budgeting and SWOT of the integrated rice-fish farming system value chain. The paper finishes with a discussion of the conclusions and policy implications of our findings.

Methodology of the Study

Data and Study Area

Data were collected between August 2012 and January 2013 at 12 Upazilas in the Dinajpur, Rangpur and Joypurhat Districts in the northwestern region and at 4 Upazilas in the Netrokona and Sherpur Districts in the northern region (Fig. 17.1). The study sites were chosen because there was an EU-funded *adivashi* fisheries project conducted by the WorldFish Center with its partner organizations from 2007 to 2009. These farmers received training and other facilities in rice-fish farming, and other actors also received training and initial financial support from the Adivashi Fisheries Project,[1] funded by the European Union. These districts were therefore selected for the study.

A field survey for collecting primary data was done by using two types of detail-structured interview, scheduled for a period of 6 months by the trained enumerators with the supervision of one of the primary authors of this paper. The interview schedule was prepared and finalized based on relevant literature reviews, pretesting and expert consultation. These two finalized interview schedules were used by a

---

[1] See Pant et al. (2014) for a detailed discussion about the *Adivashi* fisheries project.

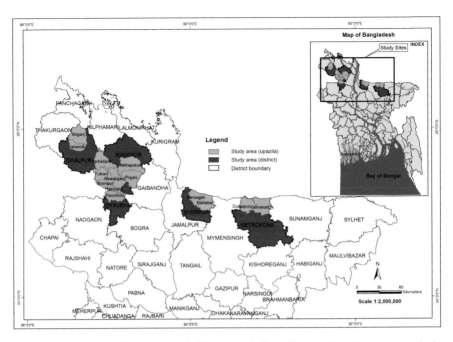

Fig. 17.1 **Study areas**: districts and sub-districts are indicated by *purple* and green, respectively (Islam et al. 2015)

trained enumerator in the field to conduct the surveys. Data were collected from integrated rice-fish farming system value chain participants. The details of the sample size used in this study are shown in Table 17.1. The author got the participant list from the WorldFish Center, and a sample was chosen randomly from that list. Farmers were interviewed at their houses and/or farm sites. After data was collected by the enumerator, the supervisor checked the data in the field and, if there was any indication of error or confusion, cross-validation was done by the supervisor with the same farmers. During the entire period of fieldwork, observation (direct observation, passive deception) by participants was used to triangulate the information gathered through interviews (Bernard 2006).

In addition to primary data, secondary data were collected whenever necessary from various government sources like the Department of Fisheries (DOF), the WorldFish Center (WFC) and other relevant ministries in Bangladesh, as well from review of the extensive published literature.

Table 17.1 Sample size by category

Category	Sample size	Percent
Fingerling trader	17	4.02
Rice-fish	48	11.35
Rice-monoculture[a]	311	73.52
Fisherman	19	4.49
Fish trader	28	6.62
Total	423	100.00

[a]Within the rice-monoculture category, there are 132 samples from non-indigenous farmers, collected randomly from the neighboring indigenous households, with the rest of the rice-monoculture farmers and other categories coming solely from indigenous households

Analytical Methods

Value Chain Analysis

Since value chain analysis became widely used in the early 1990s as a novel methodological tool for understanding the dynamics of a system, there has been no hard and fast definition of the concept of the value chain. Definitions vary widely depending on the authors and their fields and scopes of study. Thus, it is necessary to clarify briefly how we define the term 'value chain' in this study. According to Kaplinsky and Morris (2001), a value chain "describes the full range of activities which are required to bring a product or service from conception, through the different phases of production (involving a combination of physical transformation and the input of various producer services), delivery to final consumers, and final disposal after use" (Kaplinsky and Morris 2001, p. 4; Kaplinsky 2000). Value chain analysis focuses on 'vertical', as well as 'horizontal', linkages among different actors and the movement of goods or services from producer to consumer along the chain. Value chain analysis is widely used throughout the whole industry, and more recently, in agricultural sector research and policy fields, as an analytical tool, even in environments of more complex production networks (Kaplinsky 2000; Dolan and Humphrey 2004; Gereffi 1994; Sturgeon 2001). Value chain analysis can analyse values and value addition within the chain, the nature of power relations and power distributions based on governance of the supply chain, and potential points of entry or exclusion (especially in the case of small farmers), as well as the distribution of revenues and benefits among the actors (Walters and Lancaster 2000; Doland and Humphery 2004; Wood 2001). In addition, value chain analysis allows us to integrate the gendered (e.g., Barrientos et al. 2003), nutrition (e.g., Fan and Pandya-Lorch 2012), welfare, poverty, inequality and environmental concerns (e.g., Bolwig et al. 2010; Kaplinsky 2000; Riisgaard et al. 2010; Gereffi et al. 2001; Trifković 2014).

Many methods of value chain approach have evolved over the years as it has been used in various disciplines, such as economics, environmentalism, political science, etc. (Fasse et al. 2009). Broadly, it can be categorised into two groups: one that is more descriptive and qualitative emphasized (Kaplinsky and Morris 2001), and another that refers to specialized tools with an analytical focus with is more quantitatively oriented, such as modelling and simulation, especially in business administration, e.g., optimizing chain logistics (Ondersteijn et al. 2006). The blending of qualitative and quantitative methods in value chain analysis can include a combination of surveys, focus group interviews, participatory rapid appraisals, informal interviews, and secondary data sourcing. It is also important to look at the institutions, their arrangement and how they are embedded in the chain to get to know the economic, social and political implications. This sort of analysis is especially affected by certain norms, working rules, and property relations, which have a big influence on the choice of the individual, meaning the particular internal or external stakeholder of the chain. Different actors decide whether they are willing to agree and act on the next step or not. Here, the term 'governance' comes into consideration. Because under these circumstances, 'governance' means the 'transformation' of institutions driven by the actors. With this regard, governance (systems) shows whether institutions become effective or not (Hagedorn 2008, p. 360). As goods and services move along the chain from actors to actors, every time a good or service is transferred between the actors (transaction), costs emerge, which might be fixed or variable. So, coordination problems arise. According to Williamson (1985), three determining factors of transaction exist: Asset specificity, uncertainty and frequency. Asset specificity is related to the specific investment for the transaction and how costly the investment is in comparison to an alternative use of the good/service. The more difficult it is to reallocate the resource to another use, the more specific the transaction. Uncertainty means the uncertain action or behaviour of the contract partner. Frequency indicates the repetition and number of transactions. The more frequent it is, the more trust exists between the actors, and the less probability there is of opportunistic behaviour.

Various types of analysis can be undertaken through the use of the value chain approach, such as functional analysis (Bahr et al. 2004; Guptill and Wilkins 2002), institutional analysis (FAO 2005a), social network analysis (Kim and Shin 2002), financial analysis (FAO 2005b), input–output analysis (Hecht 2007), social accounting matrix (Courtney et al. 2007; Adelman et al. 1988), life cycle analysis (Rebitzer et al. 2004), input–output-life cycle analysis (Lenzen 2001), material flow analysis (Finnveden and Moberg 2005), energy analysis (Finnveden and Moberg 2005) and an integrated ecological-economic modelling approach (Pacini et al. 2004; Baecke et al. 2002; Kledal 2006). One method could not hope to cover all relevant aspects, so in this study, we used a combination of methods: functional analysis, which depicts the interaction between actors of the value chain, describing their full activities from node to node along the chain; institutional and social network analysis, which presents an overview of the various chain actors the and relationships between people, groups and organizations in value creation; financial and input–output analysis, which determines the financial costs and benefits of the individual agents along the chain and traces the flow of goods and services between actors; and material flow analysis, which assesses the physical units of input and output involved in the production, processing, consumption and distribution.

Gross Margin Analysis

Gross return (GR) is calculated by multiplying the total volume of output by the average price in the harvesting period (Dillon and Hardaker 1989). The following equation was used to estimate GR:

$$GR_i = \sum_{i=1}^{n} Q_i P_i.$$

Where,

GR_i = Gross return from the ith product (Tk/ha)
Q_i = Quantity of the ith product (kg)
P_i = Average price of the ith product (Tk/kg)
$i = 1, 2, 3. . . . ., n$

Gross Margin

In farming, the financial performance of an activity is usually expressed in terms of a gross margin, defined as the difference between gross return and total variable costs. Fixed costs are not included (Nix 2000).
   That is,

$$GM = GR - TVC,$$

Where,

GM = Gross margin
GR = Gross return
TVC = Total variable cost.

Benefit Cost Ratio (BCR)

The BCR is a relative measure, which is used to compare benefit per unit of cost. The BCR was estimated as a ratio of gross returns and total variable costs. The formula for calculating BCR (undiscounted) is shown below:

$$\text{Benefit cost ratio} = \frac{\text{Gross return (GR)}}{\text{Total variable cost (TVC)}}.$$

Partial Budget Analysis

Partial budgeting is normally used to re-evaluate the economic viability when there is a minor change in a production technique resulting in a partial change in the cost-return structure (Shang 1986; Barnard and Nix 1979). Partial budget analysis assesses the incremental technological change at the field level (Holland 2007; Roth and Hyde 2002). It includes only those resources that will be changed, leaving out those that are unchanged (e.g., fixed assets), and supports the assessment of alternatives. Partial budget is a balance which measures the positive and negative effects of a change in the existing activities (Kay et al. 2008). It shows how adopting a new technology affects profitability by comparing the existing one with the new or alternative methods. It is based on the concept that technological change will have positive and/or negative economic effects. On the positive side, it is assumed that the adoption of technological innovation will eliminate or reduce some costs and/or will increase returns. On the negative side, it is assumed that technological change will cause some additional costs and/or reduce some returns. The net effect of the introduction of technological innovation is measured by the net change between positive and negative economic effects. A positive and negative net change indicates a potential increase and decrease in income/profit, respectively, due to the introduction of the new technology (William et al. 2012). In this article, we used partial budgeting technique to re-evaluate the economic viability of an integrated rice-fish farming system instead of a rice monoculture system.

SWOT Analysis

SWOT analysis as a framework is uncritically widely used due to its simplicity and practicality. SWOT (the acronym stands for Strengths, Weaknesses, Opportunities and Threats) analysis is used for analyzing internal and external factors in order to attain a systematic approach and support for decision-making. It is a valuable tool for addressing some of the weaknesses of quantitative analyses. The aim of this type of analysis is to try and maximize the future position of an organization/business/ enterprise/activity, in our case, a rice-fish farming system in Bangladesh (Kurttila et al. 2000).

SWOT analysis is a strategic planning tool consisting of two parts (FAO 2006):

1. An analysis of the internal situation (strength and weakness). This only discusses actual strengths and weaknesses rather than speculative, future strengths and weaknesses.
2. An analysis of the external situation (opportunities and threats). This includes the actual situation, i.e., existing threats, as well as unexploited opportunities and probable trends.

New technologies like the rice-fish system are promoted as having the potential to improve the economic, environmental, and health conditions in developing countries. However, the adoption rates of these new technologies are often disappointing and are not uniform (Feder et al. 1985). In this article, we use SWOT analysis to identify and analyze the constraints and facilitating factors for adoption and diffusion of integrated rice-fish farming system technology in Bangladesh, in general, and among the indigenous communities in particular.

Results and Discussion

Value Chain Mapping

Value chain analysis determines how the linkages between the production, distribution and consumption of products are interconnected along the value chains that represent a network of activities and actors (Kaplinsky 2000; Sturgeon 2001). The value chain approach identifies the input–output structure or the value-added sequence in the production and consumption of a product; dispersion of production and marketing; a governance structure or the power relations that determine how financial, material and human resources are distributed within the chain; and an institutional framework that identifies how local, national, and international contexts influence activities within chains (Gereffi 1994, 1995). Governance structures determine how the benefits of participation are distributed along the chain (Gereffi et al. 2001; Gibbon 2000; Humphrey and Schmitz 2001). Governance structures can be producer-driven and/or buyer-driven (Gereffi 1995). These structures are helpful for identifying how the power is exercised within chains (Barrientos et al. 2003).

The rice-fish value chain maps in Fig. 17.2 provide a schematic snapshot of the key value chain actors and the product and information flows at a given point in time. The horizontal product flows indicate the alternative supply channels, while each vertical level in the value chain describes the productive function. Value chains encompass a network of competing supply/marketing channels. The chain of actors through which the transaction of goods takes place between producer and consumer constitutes a marketing/supply channel. In other words, a marketing channel refers to a pathway composed of various marketing intermediaries who perform such functions as needed to ensure smooth and sequential flow of goods and services from the producers to consumers. Marketing/supply channels are alternative routes of product flows from producers to consumers (Kohls and Uhl 2002). In Bangladesh, fish produced in a rice-fish system moved from the producer-intermediaries to consumers through the channel, i.e., through some market intermediary, such as fish wholesalers and retailers. It was observed that fish produced in a rice-fish system needed to move a short distance from the point of production to the consumers due to its perishable nature and small-scale production, as well as the high demand in the local market. Within the value chain, marketing channels through which the fish produced in a rice-fish system moved in Bangladesh are observed in the study areas, which are depicted in Fig. 17.2. Here, we only discuss

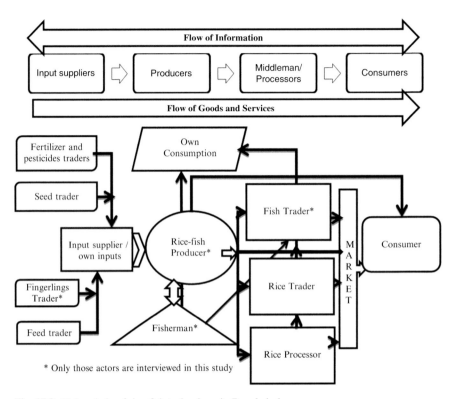

Fig. 17.2 Value chain of rice-fish technology in Bangladesh

the fish value chain under the rice-fish system value chain rather than both rice and fish, although both are shown in Fig. 17.2, because we assume that the fish value chain brought extra benefit to the rice value chain, the actors of which are working within a rice-fish system rather than rice monoculture. For a detailed value chain analysis of rice, please see Minten et al. 2011, 2013; Reardon et al. 2012.

In Fig. 17.2, we observe only a few channels that are used for rice-fish production and distribution, channels that are very short. Rice-fish farmers produced rice and fish by using their own inputs or buying them from input suppliers like fingerling traders; they also use the services of fisherman. Fish produced under a rice-fish system are consumed by the rice-fish producing households fully or partly, with the rest being sold to neighbouring households or in the nearby market; alternatively, they may sell it to a fish trader, who then sells it to the consumers through a market and possibly consumes some portion of it themselves. In these channels, major actors of fish production and distribution under a rice-fish system are fingerling traders, rice-fish small-scale farmers, fisherman, fish traders and consumers, the latter group of which may itself contain fingerling traders, rice-fish small-scale farmers, fisherman, and fish traders, all of whom were interviewed in this study.

All these actors play an important role along the chain. The input suppliers supply input, mainly fish fingerlings, to the small-scale rice-fish farmers through credit or cash. In our case, the farmers said they only bought the fingerling from fingerling traders, and mostly used feed they made themselves at home or sometimes that which they would buy from the nearby feed traders/market. The fish traders collect the fish from farms using their own transportation to take them to the nearby market, or sometimes buy and sell the fish in the same market. From the value chain map in Fig. 17.2, it is evident that rice-fish producers have several options for making good use of their fish, whether it's through their own consumption, or directly selling it to the market, neighbouring consumers or fish traders, and similarly, consumers also have the opportunity to buy from different sources.

Actors, Value Addition, Governance, Institutional Framework and Gendered Employment in the Value Chain

Major actors in the rice-fish value chain are input suppliers (fertilizer and pesticide traders, seed traders, feed traders and fingerling traders), rice-fish producers, fisherman, rice and fish traders, rice processor/millers and consumers. Normally, many functions/services are offered by different actors along the chain, such as exchange functions (buying and selling), physical functions (transport, storage, processing) and facilitating functions (standardization, financing, risk-bearing and market intelligence) (Kohls and Uhl 2002). As we have observed, the fish value chain under a rice-fish system is neither that long nor that simple. Almost all farmed and wild fish are sold either live, fresh on ice, or fresh without ice; there is no primary or secondary processing at all. For producing fish in the rice field, farmers who do not use much feed primarily rely on natural food (phytoplankton, zooplankton, periphyton and benthos). Some farmers use additional supplementary feed that they often make themselves at home (like cow-dung, waste rice, rice and wheat bran, etc.). In addition, some farmers, especially those on relatively large farms, sometimes buy feed (like mastered oilcake, poultry manure, fishmeal, industrially manufactured pelleted feeds etc.) from the feed traders in the nearby market. Major species cultivated under a rice-fish system tend to be major Indian carp species, both prevalent and exotic, like common carp (*Cyprinus carpio*), catla (*Katla katla*), rohu (*Labeo rohita*), mrigal (*Cirrhinus mrigala*), bata (*Labeo bata*), Nile tilapia (*Oreochromis niloticus*), silver carp (*Barbodes gonionotus)* and naturally occurring small indigenous species (SIS). Normally, all rice-fish farmers use a slightly larger size of fingerling in a rice-fish system, believing that fingerlings who start out a bit bigger will grow faster than those that start out smaller. In general, under a rice-fish farming system in Bangladesh, farmers will not stock any specific ratio of different fish species (Ahmed and Garnett 2011). Almost all the actors participating in the production and distribution channel consumed fish throughout the production and distribution period. Fish are harvested by fisherman, and sometimes by the farmers themselves, then sold to neighbors, or the wholesalers and retailers in the nearby market. Farmers sometimes grade by species or by size of the fish to get higher and differentiated prices. Most of the time, naturally occurring SIS of fish by-pass the established market and are both sold to neighbors and consumed by the farmers themselves, due to good testing and greater demand for smaller and cheaper fish, as the purchasing power of the local population is weaker. Different studies show that SIS of fish are very nutritious and have the potential to ensure food and nutrition security in the developing world (Roos et al. 2003, 2007a, b; Thilsted et al. 1997). Fish traders (wholesalers or retailers) either collect fish from the farmers or have the farmers deliver the fish to them, and then sell it to retailers, consumers, and restaurant owners.

Once fish have been harvested from a rice-fish farm, there are no distinct value-chains for different species, i.e., individual traders/wholesalers and retailers deal in all fish species, rather than in particular ones. All rice-fish farmers reported that they produce and sell a mix of fish species, dominated by sales of carp, but also including tilapia and SIS. Almost all fish is sold live, with some being sold fresh on ice (in the summer months or if sales are conducted relatively far from the harvest area). There is a growing trend in the country's consumers for live and fresh fish and a preference for wild fish over farmed fish. Thus, a rice-fish system is a potential technology for fulfilling those demands in Bangladesh and other developing countries. In the rice-fish value chain, we see the fish distributed through a few short channels and very few actors involved in the performance of different marketing functions, such as buying, selling, transportation, processing/grading, cooling/ icing, pricing, etc., adding value to the product by working those functions and taking a portion of the marketing margin (discussed in the next section) as a result; the reason there are so few value addition functions is that the fresh fish are typically being harvested and sold on the same day. Fish produced under a rice-fish system value chain are governed by the spot market transactions involving a large number of small traders, which is the case in many traditional agricultural commodity value chains in developing countries, while the modern value chains' governance is based on the use of high standards and safety throughout the chains, high levels of vertical coordination (such as contract-farming), a high degree of coordination of the supply base, and agro-industrial processing (Maertens and Swinnen 2010). Power relations among the actors are almost balanced/equal because all the actors' operation units are small in nature. So, the price of the fish determined through the bargaining power that exists between the actors depends on the supply and demand for fish, as well as the number of buyers and sellers. Fish traders have more bargaining power when they sell to consumers than when they buy from producers, because the bargaining power between producer and trader is almost equal, owing to the number of fish traders and producers being limited and small-scale in nature. Information flow within the rice-fish value chain is quite transparent. All the actors get information from each other as almost all have the access to mobile phones (whether their own or nearby neighbours/shop/market). Although there are several laws and law-enforcing regulatory agencies to check the quality and standard of daily food (like fish) in Bangladesh, the fish value chain, especially in the urban areas, is still adulterated (especially by the traders) through the use of different poisonous chemicals (like formalin) to keep the fish fresh (Rahman 2013; Uddin et al. 2011). Commercial aquaculture, which passes through a long value chain, faces this problem in particularly severe terms. Fish produced under a rice-fish system are generally sold live/fresh and within a short time to the nearby market, nearby consumers or is consumed by the producing farmers without any major processing. Thus, large scale adoption and diffusion of the rice-fish system could be an instrument for tackling fish adulteration and the related health risks in Bangladesh. Farmers and fish traders report that they are not familiar with any standard issues. There are several (mainly government) institutions in Bangladesh that are related to the rice-fish value chain actors, like the Ministry of Agriculture (MOA), which supports farmers by providing extension services and technical knowhow other than aquaculture and fisheries, and the Ministry of Fisheries and Livestock, which provides more or less the same services to fisheries and livestock sectors. The Department of Fisheries (DOF) is specifically responsible for the fishery sector's overall activities (like extension, quality and standard inspection, etc.). There are several national organizations, like the Bangladesh Rice Research Institute (BRRI) and the Bangladesh Fisheries Research Institute (BFRI), international research organizations, like the WorldFish Center, and non-governmental organizations, like CARE, who also develop and disseminate rice and fishery-related technologies in Bangladesh. Most of the actors report that they perform their activities individually rather than in any association or group (like farmers or traders association, cooperatives). There are some community-based organizations among the indigenous people, but most of them deal with socio-cultural problems which are sometimes indirectly linked with rice-fish value chain activities. Rice-fish value chain actors are not satisfied with the services of most of governmental organizations. So there is plenty of scope to strengthen the governmental organizations that could facilitate the adoption and diffusion of rice-fish systems in Bangladesh.

Table 17.2 demonstrates the differences in labor use between rice cultivation and rice-monoculture. In a traditional double or triple subsistence rice monoculture system, total labor requirements are estimated at 209.46 person-days/ha, which is significantly lower than the labor requirements for a rice-fish system at 287.57 person-days/ha. Interestingly, the labour requirement for different operational activities in producing a rice-monoculture and a rice-fish system also shows significant differences between the systems. As with land preparation, cop establishment, feeding, harvesting, threshing, cleaning and processing and buying and selling of input–output all required significantly more labour under a rice-fish system compared to a rice monoculture. In contrast, fertilizer application, pesticide application and weeding required more labour under a rice-monoculture system compared to a rice-fish system. It is important to note that gendered employment opportunities in different operational activities under both systems are largely dominated by male and household labour supply. But a rice-fish system creates a significantly higher amount of gendered employment opportunities than a rice-monoculture system. Some of the operational activities in an extensive rice-fish

302 A.H. Md. Saiful Islam

Table 17.2 Labor allocation pattern (person day/ha/year) for rice and rice-fish farming household

Activities	Gender pattern	Rice-fish (RF)		Rice-monoculture (RM)		Total labour		Diff
		Hired labour	Family labour	Hired labour	Family labour	Rice-fish	Rice-monoculture	
Land preparation	Male	6.71	36.97	2.65	8.07	47.37	11.05	−36.32***
	Female	0.00	3.69	0.02	0.32			
Crop establishment	Male	18.45	47.86	19.83	26.52	75.00	53.21	−21.79***
	Female	4.65	4.04	1.80	5.07			
Fertilizer application	Male	1.13	7.52	1.72	8.03	8.99	10.22	1.24
	Female	0.00	0.33	0.03	0.44			
Weeding	Male	6.66	17.46	13.72	20.85	28.89	40.86	11.97**
	Female	0.73	4.04	1.09	5.20			
Pesticide application	Male	0.24	1.91	0.74	4.80	2.15	5.70	3.56***
	Female	0.00	0.00	0.00	0.16			
Feeding	Male	0.00	4.57	0.00	0.00	14.91	0.00	−14.91***
	Female	0.00	10.34	0.00	0.00			
Harvesting	Male	12.62	37.57	22.75	22.78	63.13	52.67	−10.46**
	Female	3.32	9.62	1.50	5.64			
Threshing, cleaning and processing	Male	4.72	18.47	7.78	15.29	32.04	29.08	−2.95
	Female	1.87	6.98	0.94	5.07			
Selling and buying of input and output	Male	0.91	14.14	0.64	5.72	15.09	6.66	−8.44***
	Female	0.00	0.05	0.00	0.30			
Total	Male	51.45	186.47	69.84	112.06	237.92	181.90	−56.02***
	Female	10.57	39.09	5.37	22.20	49.65	27.57	−22.09***
N		48		311		48	311	

farming system (such as homemade feed preparation, feeding, and supervision) are associated with somewhat less drudgery compared to a rice monoculture system and are a source of employment for women labourers, especially for household women labourers.

Interestingly, value chain analysis findings indicate that fish produced under a rice-fish system and distributed within a very short time-period from harvest to final consumption due to the live/fresh nature of all sales are generally sold to consumers on the same day as the harvest, with almost zero post-harvest losses (which stands in contrast to many agricultural product value-chains, in which significant post-harvest losses often occur in developing countries). Thus, it is evident that the rice-fish value chain is an efficient and opportunistically gendered distribution system that produces close to the ultimate consumers.

Gross Margin Analysis of Value Chain Actors

Following a system-level approach for an entire agricultural year, detail cost and return from rice-fish and rice monoculture farming system are presented in Table 17.3. In the table, the economic feasibility of different farming systems is displayed on a yearly and per hectare basis. It is evident from the table that rice-monoculture farmers use human labour, seed/seedlings, ploughing, manure and chemical fertilizers, irrigation, pesticides and others as variable cost items which vary with the level of production. In addition to these variable cost items, rice-fish farmers use feed and fish fingerlings as additional variable inputs.

In the table, there is a significant difference in labor input between the farming systems. An integrated rice-fish farming system requires higher labor input due to additional works necessary to make the land suitable for a rice-fish system, like strengthening dikes and excavating refuges,[2] as well as feeding and other operational activities down the line which are detailed in Table 17.2. The costs of ploughing and seedlings does not differ much between the farming systems. The stocking density of fish fingerlings for integrated rice-fish farmers is 92.48 kg/ha. Farmers reported that they prefer the comparatively larger size of fish fingerling because of their high survival rate, as well as their high growth rate. Although most of the indigenous farm household rice-fish systems contain abundant natural foods like phytoplankton, zooplankton, periphyton, and benthos, most farmers, nevertheless, use supplementary homemade/on-farm feed, like rice bran, wheat bran, mustard, oilcake and cow dung. Some farmers also use fishmeal and industrial concentrate feed along with their supplementary home-supplied feed. Interestingly, there is a significant difference in fertilization rate among culture systems. Integrated rice-fish farmers use less chemical fertilizers but more manure/inorganic

---

[2] A form of ditch, sump or small-size pond in a low-lying part of the rice field where fish can go when water is unavailable in the rest of the field, and therefore a good shelter for the fish.

Table 17.3 Average quantity and cost/return of inputs and outputs of different farming systems

Items	Quantity (mean in kg or ML or Human man-days/ha/year)		Total value (mean in BD Taka^a/ha/year)		Diff.	Factor share of total revenue (%)	
	RF	RM	RF	RM		RF	RM
Variable costs							
(i) Family labour	225.56	134.26	47,066.50	26,772.12	−91.30***	18.20	19.69
(ii) Hired labour	62.02	75.21	12,940.69	14,997.13	13.19	5.00	11.03
Total labour (i + ii)	287.57	209.46	60,007.18	41,769.24	−78.11***	23.21	30.72
Ploughing	–	–	9913.52	11,421.12	1507.60	3.83	8.40
Seeds/seedlings	–	–	6718.35	6869.81	151.46	2.60	5.05
Fish fingerlings	92.48	0.00	17,382.20	0.00	−92.48***	6.72	0.00
Manure (e.g., cow dung)	5604.60	4810.44	5170.21	3358.98	−794.15	2.00	2.47
Chemical fertilizer	359.37	495.37	8410.50	11,547.48	136.00**	3.25	8.49
Irrigation	–	–	11,639.09	10,228.21	−1410.88	4.50	7.52
Feed	2604.60	0.00	11,316.63	0.00		4.38	0.00
Concentrate Pesticides/insecticides	0.70	3.70	1041.72	2101.77	3.01***	0.40	1.55
Liquid pesticides/insecticides	300.76	672.66			371.89**	0.00	0.00
Miscellaneous (e.g., machine cost for dike preparation, lime)	–	–	1815.05	180.85	−1634.20***	0.70	0.13
(A) Total variable cost (TVC)	–	–	133,414.40	87,477.47	−45,936.98***	51.59	64.33
Output							
Rice (in Mound = 40Kg)	159.82	233.19	92,589.10	128,371.10	73.38***		
Rice byproduct	–	–	7354.79	7613.93	259.13		
Cultured fish	1149.92	0.00	128,628.70	0.00			
Indigenous fish	110.76	–	19,868.18	0.00			
Others (e.g., vegetables)	–	0.00	10,155.10	0.00			
(B) Total return from output			258,595.90	135,985.00	−122,610.9***		
C. Gross margin (B-A)			125,181.50	48,507.53	−76,673.94***		
G. Benefit cost ratio (B/A) (Undiscounted)			2.01	1.84			

a$1 (US) = 79 Bangladeshi Taka (BDT) in 2012

***, ** and * stand for significance at the 1%, 5% and 10% levels, respectively fertilizers, whereas the rice-monoculture farmers do the opposite. The integrated rice-fish farmers use of less chemical fertilizer may be due to the presence of fish in the rice field, which increases soil fertility. Quite surprisingly, farmers of both farming systems use liquid as well as concentrated pesticides to prevent pests and diseases. But rice monoculture farmers use significantly higher amounts of liquid and concentrated pesticides compared to rice-fish farmers. Rice-fish farmers mostly use liquid pesticides. Farmers reported that there are some pesticides which do not affect fish survival rate and, consequently, they use those pesticides without knowing the residual effects. Additionally, rice-fish farmers have to expend further miscellaneous costs, which include land modification both before and after the rice-fish harvest, while some farmers produce dike vegetables which incur their own cost items, like vegetable seeds, bamboo, rope, etc.

According to our survey, the highest average annual productivity of rice per hectare is found in rice monoculture (233.19 mound equivalent to 233.19*40 = 9327.60 kg), followed by integrated farming (159.82 mound equivalent to 159.82*40 = 6392.80 kg). There is a significant difference in rice yield between the farming systems, which may be due to the differences in inputs (seed, fertilizer and pesticides) and management and technical skills. Quite a number of farmers reported that rice does not grow well in the rice-fish system and also complained that fish sometimes destroy the rice. Halwart and Gupta (2004) reported that bottom feeding carp, especially the common, herbivorous species such as the grass carp, uproot and eat whole rice plants if those species are stocked before the rice plants develop a good root system, as well as if the fingerlings of those species stocked are of the larger sizes. Thus, fingerling management is crucial for a rice-fish system, especially for rice productivity in that system. The average annual cultured and indigenous fish yield reported by rice-fish farmers is 1149.92 and 110.76 kg/ha. The rice-fish farmers who cultivate vegetables in their dike, while incurring additional costs also find additional income opportunity. Overall, the total return and gross margins differ significantly between the systems. Although the rice-fish farmer's rice yield is significantly lower than that of the rice-monoculture farmer, a rice-fish farmer's total return, as well as gross margin, is significantly higher. Thus, rice yield loss is outweighed by the higher return from fish under a rice-fish system. The resultant increase in gross margins for rice-fish technology results in a benefit cost ratio of 2.01. The result simply indicates that, holding other factors constant, for every additional Bangladeshi taka invested in rice-fish technology, the gross margin will be increased by more than two times, which is quite higher than that of investment in rice-monoculture. The results indicate that, at the farm level, rice-fish technology appears to be an economically viable alternative to rice-monoculture.

A closer look at the factor share of total revenue of these farming systems provides further insight into their economic viability. Factor shares of total revenue and total variable cost explain how the benefits shared among the production factors, as well as input intensity and input prices, influence the costs and returns of different systems. The analysis in Table 17.3 identifies labor, fingerlings, feed, and irrigation as the most costly inputs, as well most of the benefits shared among them in rice-fish systems, whereas in rice-monoculture systems, labour, fertilizer, ploughing, irrigation and seed are the most costly inputs and most of the benefits are also shared among them. Overall total return shared among the variable costa is higher in a rice-monoculture system compared to a rice-fish system. This implies that a rice-fish system and its fixed factors get a higher profit margin share than a rice-monoculture system.

Similar to Tables 17.3 and 17.4 presents the cost and return, as well as factor share, for a rice-fish system value chain with backward and forward linkage actors. As we have seen in the value chain maps, a rice-fish value chain is not very long and, consequently, not many value-added functions are carried out by the participating actors. Thus, the cost items list for backward and forward linkage actors, such as fingerling traders, fisherman, and fish traders, is not so long. Labour, transport and food are the main cost items. The quantity and value of the variable cost items are almost the same for all actors, but total return, as well as gross margin, varies among the actors. This is due to the differences in average number of days the activities run per year, the average quantity dealt with per day, and the average buying and selling price differences among the actors. For fisherman, that is largely influenced by the average catch or the average share of fish they get catching fish in a group. All the actors reported the seasonality of their activities due to irregular fish supply, water shortage (drought) and a decrease in common pull fishery sources like rivers, canals, etc. Thus, those actors cannot rely on these activities for their livelihood, which sometimes demotivates them to engage in them, or may even cause them to abandon the activities altogether. Among these, the three actors' gross margin is higher for fish traders, followed by fingerling traders and fishermen. The gross margin benefit cost ratio shows the same trend.

Analysis of factor shares of total return and total variable cost (in parenthesis) in Table 17.4 shows that labor and transport costs are the two most costly inputs in fingerling trading, fishing (fishermen) and fish trading. These two inputs also get the most share of total return from the respective business activities. These factor shares give the idea that these activities may have significant potential for credit-constrained and limited-market-access households, and even for landless marginalized households, because labour cost is the major share of total variable cost.

Partial Budgeting

The potentiality of any technological innovation can be evaluated by its private benefits and costs; a technological innovation is said to be economically feasible if the benefits from the technology outweigh the costs. Thus, to assess the relative potentiality of rice-fish technology over the performance of the rice-monoculture practices, a partial budget was constructed using the cost and benefit information derived from the interviews with farmers during field surveys. The findings of the analysis are shown in Table 17.5.

Table 17.4 Average cost/return of inputs and outputs of different backward and forward linkage actors

Items	Total value (mean in BD Taka/year)			Factor share of total revenue (total variable cost) (%)		
	Fingerling trader	Fisherman	Fish trader	Fingerling trader	Fisherman	Fish trader
Variable cost items						
Total labour[a]	36,263.84 (208.73)	34,908.27 (204.50)	35,795.68 (214.60)	25.93 (74.16)	40.50 (79.23)	25.14 (66.13)
Food cost	5186.93	2334.99	6273.66	2.94 (6.70)	2.47 (4.96)	3.46 (11.06)
Transport cost	12,127.22	9994.05	6938.33	7.95 (16.00)	7.89 (13.30)	4.46 (12.64)
Miscellaneous (e.g., ice, plastic bag, etc.)	1611.03	1152.63	4170.52	1.68 (3.14)	0.99 (2.51)	3.01 (10.18)
(A) Total variable cost (TVC)	55,189.01	48,389.94	53,178.18	38.50 (100)	51.84 (100)	36.06 (100)
(B) Total return from sales	176,806.70	94,771.93	204,471.70			
(C) Gross margin (B-A)	121,617.70	46,381.99	151,293.50			
G. Benefit cost ratio (B/A) (Undiscounted)	3.20	1.96	3.85			
Average number of days the activities run per year	87.50	82.47	126.11			
Average quantity deal with per day (Kg)	13.00	2.29	20.91			

[a]The quantity of human labour per year is indicated in parentheses

Table 17.5 Partial budgeting: net change of gross margin due to introduction of fish into the rice field instead of a rice monoculture

Costs	Tk/ha/year	Benefits	Tk/ha/year
1. Cost incurred for rice-fish	133,414.4	1. Cost saved by not doing rice mono-culture	87,477.47
2. Revenue forgone by not doing rice mono-culture	135,985	2. Revenue earned from rice-fish	258,595.9
Net change (++)	76,673.97		
Total	346,073.4	Total	346,073.4

It is evident from Table 17.5 that introduction of fish into a rice field increases cost as well as benefit, but the benefits outweigh the costs. Thus, the net change in farm income due to introduction of fish into the rice field instead of a rice monoculture is positive and can amount to 76,673.97 Bangladeshi Taka per year per hectare. This additional benefit is only at the farm level, but if we take into account the additional benefits of other rice-fish value chain actors, then the figure would be much higher. Ultimately, the net benefit of rice-fish systems, primarily, is the additional income from fish that is earned by smallholder indigenous farmers without a significant loss of income or food security from forgone rice cultivation.

SWOT of Integrated Rice-Fish Farming System Value Chain

A summary of key elements in terms of strengths, weaknesses, opportunities, and threats (SWOT) characterizing the integrated rice-fish farming system has been derived from stakeholder interactions, field observation, in-depth farm household surveys and literature review, and is presented in Table 17.6. SWOT analysis explores how the rice-fish value-chain performance could be further improved by identifying the critical factors impacting value-chain performance. Table 17.6 provides a brief summary of the key issues that impact the sector. All of the issues included in the table, and details discussed in the subsequent section, represent potential areas of action by the value-chain itself and by those relevant factors outside of it (e.g., policy-makers, research organizations and extension agents) to improve value-chain performance in this sector.

Strengths

The rice-fish farming system is feasible virtually throughout the country's irrigated and rain-fed rice areas, without the necessity for major adjustment to traditional production methods. The rice-fish farming system is not new, but there are numerous potentialities for improvement by introducing innovation to the system. Fish

Table 17.6 SWOT framework related to adoption and diffusion of rice-fish technology in Bangladesh

Strength (S)	Weakness (W)
Sustainable agricultural development is on the political agenda	No strategy is defined for the implementation of sustainable agricultural development
Ecologically and environmentally sound sustainable intensification option	Initial cost of preparation is high for poor farmers
Multifunctional agricultural system with multiple benefits	Needs continuous supervision
Can act as an important element of integrated pest management (IPM)	Needs more labour
Needs less fertilizer, pesticides and herbicides	Needs more technical knowledge
Efficient and complementary utilization of scarce land and water resources	Lack of backward and forward linkage actors and their inputs
Increases soil fertility	Lack of wider irrigation coverage
Integrated resource management options	Confusion and duplication in responsibilities of the various agencies involved in rice and fishery management at central and local levels
Use of rice field for fish seed/fingerling production	Lack of efficient and motivated expertise, resources, budget and equipment for public agencies
Rice and fish together as a sources of carbohydrate and animal protein	Lack of system thinking and coordination among the crop and fishery-related institutions like ministries, research and extension organizations
Fish for daily home consumption	
Women in the family can supervise	
Traditional importance (rice and fish) in rural Bangladeshi livelihoods	Lack of system thinking and coordination among different policies and their application
Creates employment opportunities during lean periods	Historical lack of investment in the social, economic, and policy dimensions of rice-fish system
Rice-fish technology has a pro-poor focus, and can benefit small-scale farmers, the landless, land owners, fishermen, producers, and other value chain actors	Limited or absent availability of component technologies within different rice–fish ecologies
Several direct and indirect policies support rice–fish system improvement in Bangladesh, like five-year plans, Country Investment Plan, PRSP, Protection and Conservation of Fish Act, National Fisheries Policy, National Water Policy, National Agricultural Policy, National Land Use Policy and New Agricultural Extension Policy	Less than timely availability of quality fingerlings
	Lack of post-harvest processing facilities and storage
	Weak and inadequate infrastructure
	Lack of technical knowledge needed for proper adoption and diffusion
Several agencies involved in crop and fishery management and officially concerned with developing crop and fishery technologies, specifically rice-fish technologies among farmers	Need for suitable bio-physical conditions
	Need for stronger collaboration among policymakers and development practitioners (related to rice, fish, land, water, and environment)
Availability of competent authorities BFRI (fisheries), BRRI (Rice) and WFC (Fisheries and aquaculture),	
Existence of DEA and NARS apex organization BARC to disseminate rice and fishery technologies and to provide a platform for discussion between different institutions	
Opportunities (O)	Threat (T)
Opportunity to obtain financial and technical assistance from international donors to enhance	Risk and uncertainty from climate variability, flooding, drought, poaching, poisoning, etc.

(continued)

Table 17.6 (continued)

capacities of public organizations and manpower for proper adoption and diffusion of rice-fish technology	Theft, disease and fish predators, such as snakes and *kakra*
Further increases rice and fish yield sustainably	Increasing trend towards landlessness
Uses homemade waste feed	Small farm size/land man ratio
More employment opportunities	High production cost
Increases backward and forward linkages	Increases use of fertilizer, pesticides, insecticides, herbicides and irrigation facility ownership
Increases acres to irrigated and rainfed rice field	
Food and nutrition security and self sufficiency	Significantly labor-intensive production systems
Reduces/alleviates hidden hunger problems	
Uses extensive extension system	Unfavorable property rights of land, especially for the tenant farmers
Uses scarce land and water resources optimally	Lack of supply of quality feed and high price level
Possibility of conducting successful communication campaign for public health concern about negative consequences of rice monoculture and positive benefits of rice-fish system	Low education and farmers unconcerned with the long-term environmental benefits
Sustainable development through sustainable intensification options	Increasing tendency towards tenant farmers and absentee landlords
Potential to satisfy consumption culture of the people	Poor extension service and lack of information among farmers
Potential to introduce innovation in rice-fish system, such as improvements in genetic potential and management practices, can potentially contribute to increasing agricultural productivity and food and nutrition security	Access to timely credit, high interest rate and unfavorable repayment schedules
	Higher fish mortality due to poor water quality, water pollution, turbidity, low water levels and high water temperatures
Potential to conserve nutrient rich small indigenous species (SIS)	Weak governance in extension systems
Increases dietary and crop diversity	Lack of access to land and water resources
Increases agricultural labour employment	Land fragmentation due to high population growth
Supports institutional innovation like collective management, community-based management to manage the common pool resources	Unfavorable land tenure systems and absence of successful land reforms
Develops public–private partnerships for effective implementation of rice-fish systems	Conservative societal structure due to low education, especially for women
Introduces integrated pest management (IPM) into rice-fish system	
Introduces agricultural insurance to overcome the loss associated with flood and drought	

Sources: Personal stakeholder interactions, field observation, in-depth farm household surveys and literature review of Nabi (2008), Frei and Becker (2005), Ahmed and Garnett (2011), Ahmed et al. (2011), IFPRI (2010), Dey et al. (2013), Halwart and Gupta (2004)

seed production, vegetable production, and fruit and tree production could be made easier under such a system. The main consumption item of the Bangladeshi people, especially that of the marginalized poor, is rice and fish, a fact which can drive the adoption and diffusion of rice-fish technology to keep pace with traditional food demand. The rice-fish farming system is a socio-economic and environmentally friendly sustainable intensification option for sustainable development compared to rice monoculture. It has multiple comparative benefits: improving public health by controlling rice pests, weeds, mosquitoes, and snails; reducing the use of chemical fertilizers and pesticides, insecticides and herbicides which will consequently increase biodiversity; nutrient recycling; fish as a tool of an integrated pest management (IPM) system; intensive and complementary use of land and water resources; improvement in crop diversification, which consequently improves dietary diversity; improvement in soil fertility by generating nitrogen and phosphorus, etc. In addition, rice-fish technology creates employment opportunities during the lean season in particular, and more gendered employment opportunities in general, compared to rice monoculture. As rice-fish farming land in Bangladesh tends to be situated very close to the homestead, the women in the family can supervise and attend to some of the labor, like feeding. All of these are solid strengths of rice-fish technology, which can enhance greater adoption and diffusion of the technique in Bangladesh.

In the value chain map, we have seen that the rice-fish value chain is quite short and that most of the farmers sell their fish live and fresh, typically on the same day of harvest. Thus, increasing consumer awareness about the health and quality inherent in the system, along with catering to their preference for quality food products, could provide better return to the rice-fish producers and better quality fish to the consumers. Emphasis on the paradigm of intensification of sustainability in rice-fish farming systems in several national policies such as the 5-year plans, the Country Investment Plan, the PRSP, the Protection and Conservation of Fish Act, the National Fisheries Policy, the National Water Policy, the National Agricultural Policy, the National Land Use Policy and the New Agricultural Extension Policy is a good strength for the diffusion of rice-fish technology. The traditional importance of rice and fish in the rural livelihoods and food culture (*mache vate benglai*) of the Bangladeshi people, especially poor rural households, are the major drivers for adoption and diffusion of rice-fish systems in Bangladesh. In the value chain analysis, we have seen that the rice-fish value chain creates additional backward and forward linkages compared to rice monoculture, which, as a consequence, creates additional livelihood opportunities for marginalized and extremely poor households. In Bangladesh, there are several institutions involved in crop and fishery management that are officially concerned with developing crop and fishery technologies, specifically rice-fish technologies among farmers. In addition, there are some specialized competent organizations like BFRI for fisheries, BRRI for rice and WFC for fisheries management, as well DEA and DOF, which have wide networks throughout Bangladesh to disseminate rice and fishery technologies and provide extension services to farmers. Furthermore, there is the NARS apex organization BARC, which acts as a platform for discussion between different institutions and coordinates and monitors different organization activities. Thus, Bangladesh has quite a good number of institutions related to rice-fish technology, which indicates a strong institutional framework for adoption and diffusion of rice-fish technology throughout potential areas in Bangladesh.

Weakness

There exist some weaknesses in the technology itself, and at the policy and institutional levels, that limit farmers' ability to take full advantage of the above-mentioned strengths. There are certain modifications necessary to make rice fields suitable for fish culture which involve costs that are sometimes below farmers' ability to invest, especially the marginalized, poor, indigenous, etc. Thus, initial investment cost is the major weakness for rice-fish technology to be a pro-poor innovation. The rice-fish field needs continuous supervision; otherwise, fish can easily be stolen by others. This supervision and feeding, land preparation and the catching of fish requires far more labour compared to rice monoculture. For adoption of rice-fish technology, farmers need suitable bio-physical conditions, like the water retention capacity of soil and soil quality (soil texture, topography and depth) and a guarantee that they will not be unduly hindered by neighboring farmers' behavior, such as the spilling of fertilizer and insecticide into their plots. Without water, rice-fish technology adoption will not be possible, but the water supply, especially during the irrigation season, does not cover the entire rice-producing areas of Bangladesh. Rice-fish fields require a great amount of and more continuous water than rice monoculture does. In Bangladesh, there is a well-established water market, thus, most farmers have to depend on water sellers, which sometimes works as a constraint or weakness for rice-fish technology. For successful adoption and diffusion of rice-fish technology, it is very important that there be technical knowledge of a sort which is very often lacking in the uneducated marginalized poor farmers of Bangladesh. Technical knowledge related to rice-fish technology adoption includes modification of the farm, timing of introduction of fingerling stock into the rice field, the combination of fingerling species, selection of suitable fingerling species, etc.

Depending on the rice-fish farming system characteristics, a farmer needs good quality and timely availability of fingerlings to stock the rice field, something which is very often lacking, especially in the dry season, and costly for poor smallholder farmers. There are backward and forward linkage actors involved in the rice-fish value chain, but the number of these actors is very few in Bangladesh. It has been reported that these professions are not recognized as dignified jobs in the society, which discourages people from engaging in these professions. That constrains the fingerling and fish trading businesses, as well as the overall rice-fish technology.

Policies and institutions related to rice-fish farming in Bangladesh also have certain weaknesses that inhibit rice-fish farming adoption and diffusion. A number of policies have been set forth towards the goal of sustainable agricultural development, but no definitive strategy has been established for the implementation of such development. Generally, there is a lack of system thinking and coordination among different policies and their application in Bangladesh, as in many other developing countries. Likewise, there are a number of organizations involved in rice and fishery management at central and local levels, but the duties and responsibilities of these organizations are not well defined, which leads to confusion and duplication. The organizations lack efficient and motivated expertise, resources, budget and equipment to promote rice-fish research and dissemination, and, as with the policies, there is little to no system thinking and coordination among them. Other than these policies and organizations, historically, investment in the social, economic, and policy dimensions of rice-fish systems is negligible. As a result, post-harvest processing facilities are not well developed or widespread in Bangladesh.

Opportunities

There exists good opportunity for rice-fish farming in the rural areas of the country, as most of the farmers are engaged in rice farming and there are available low-lying rice fields suitable for rice-fish farming. Growing awareness among fish consumers about quality and the huge demand for live and fresh fish, as well as increasing purchasing power, could provide the ramification for the development of rice-fish farming. Rice-fish farming requires more labour input than traditional and modern rice farming methods. Thus, Bangladesh, which has a very large amount of female unemployment and under-employment, will find rice-fish farming attractive. Moreover, the problem of unemployment during lean periods, which causes seasonal hunger like *monga*, will also be mitigated because of diversification in farming, with different stoking and harvesting schedules resulting in the requirement of a relatively high labour input. Protection of the ecosystem, flora, fauna and increased biodiversity, along with the resulting benefits to all humans and living things, are great advantages of rice-fish farming which are yet to be properly accounted for. There are numerous opportunities to obtain financial and technical assistance from international donors to enhance the capacity of public organizations and their manpower to enact proper adoption and diffusion of rice-fish technology in particular and agricultural technology in general. Introducing fish into rice fields creates opportunities for the sustainable use of scarce land water resources and produces rice (carbohydrates) and fish (animal protein) together, which can ensure food and nutrition security in Bangladesh. It also creates opportunities to increase the rice yield sustainably, which can ensure keeping pace with the soaring demand for food, chiefly rice, in Bangladesh. Bangladesh is one of the 34 countries which faces severe nutrition insecurity, especially hidden hunger problems (Ruel and Alderman 2013). The fish species grown in the rice fields, particularly SIS fish, are rich in micronutrients which can reduce or alleviate hidden hunger problems in Bangladesh and other developing countries. In the rice-fish system, and in the subsistence system most of all, farmers can use different home waste and home-made feed (e.g., waste rice, wheat and its bran, etc.) as fish feed, which can reduce feed cost. In the value chain, we have seen that rice-fish systems create additional channels, actors and networks which ultimately create considerable livelihoods and employment opportunities, especially for the marginalized poor. Although rice-fish farming is a technology with a lot of potential in terms of socio-economic profitability, gendered employment generation and food and nutrition security enhancement, its adoption and diffusion is very low. Different estimates suggest that there are huge suitable biophysical areas in Bangladesh where rice-fish technology could easily be adopted and diffused. Additionally, by introducing irrigation technologies, these areas could be further increased.

In Bangladesh, the extension systems of the DEA and DOF are quite extensive and cover almost all sub-districts, so it would be possible to use these huge extension systems for dissemination of rice-fish systems throughout potential areas in Bangladesh. By engaging these extension systems and other institutions related to rice and fish, it is possible to conduct successful communication campaigns about the public health concerns regarding the negative consequences of rice monoculture and the positive benefits of rice-fish systems, which would enhance adoption and diffusion of these systems in Bangladesh. It would also be possible to develop public–private partnerships for the effective implementation of rice-fish systems. Nutrient rich SIS fish are nearly extinct due to the introduction of green revolution technologies. Thus, rice-fish systems create the opportunity to conserve these indigenous species. In this system, it is possible to introduce integrated pest management (IPM) techniques which could reduce the cost of production, as well as being environmentally friendly and good for conserving the different fish species, SIS in particular. Rice-fish systems increase crop diversity by introducing fish into the rice fields, as well as making it possible to introduce vegetables into the dike, further increasing income and dietary diversity for poor rural farmers. Due to this diversity, a rice-fish system can be a climate resilient farming system. Although rice-fish systems are traditionally practiced in Asia, including Bangladesh, it is possible to introduce innovation into such a system, like improvements in genetic potential and management practices which can potentially contribute to increasing agricultural productivity and food and nutrition security in the country. The rice-fish system enhances institutional innovation like collective management or community-based management to manage the common pool resources, like water, especially in the low lying areas during the rainy season. As the rice-fish system is quite vulnerable to climatic shocks like drought and flood, it is possible to introduce agricultural insurance to overcome the loss associated with these shocks, which can also enhance adoption and diffusion of the systems.

Threats

Although there are tremendous strengths and opportunities associated with the adoption of rice-fish farming in Bangladesh, there are also some threatening factors that have been identified that hinder adoption and diffusion of the technology in Bangladesh. One obvious threat is the risk and uncertainty associated with climate variability, flooding, drought, poaching, poisoning, etc., all of which are very common phenomenon in Bangladesh. Other than these, theft, disease and predators such as snakes and *kakra* that eat fish from the rice field can cause huge economic losses. Poor water quality, water pollution, turbidity, low water levels and high water temperatures also cause higher fish mortality rates, which ultimately reduce motivation to continue using the system. In Bangladesh, deteriorating access to increasingly scarce natural resources (such as water and land) are also major threats for the expansion of rice-fish farming, especially among poor people. Surprisingly, in Bangladesh, landlessness, tenant farmers and absentee landlords are increasing day by day (Ahmed et al. 2013). But the land tenure system and land property rights, especially property rights for tenant farmers, are quite unfavourable towards expansion of rice-fish systems in Bangladesh. Due to mounting population growth, farm size/land-to-man ratio is declining (with land fragmentation increasing) and the increasing price of farm inputs keeps production costs in an upward trend which also threatens this type of farming, especially for the marginalized farm households in Bangladesh. To keep pace with soaring demand for food, rice in particular, Bangladeshi rice farmers have intensified their rice monoculture (up to three times in a year) by increasing use of fertilizer, pesticides, insecticides, and herbicides, which are all major threats for introducing fish into the rice. Irrigation, chiefly in the dry season, is a mandatory input for rice monoculture as well as for rice-fish systems, but in Bangladesh, not all the farmers have their own irrigation facilities. Most of them depend on the irrigation water market, which is also a threat to the adoption of rice-fish systems, because these systems need comparatively greater and more continuous supplies of water, something that can be very difficult without their own irrigation facilities.

Although labour intensive, rice-fish systems create employment opportunities; however, this too becomes a threat, because the cost of labour in Bangladesh continues to increase, and as an emerging economy's labour supply shifts from agriculture to non-agricultural sectors, labour crises in the agricultural sector have arisen in recent years. For large-scale and commercial rice-fish systems, additional feed is needed which farmers would normally have to buy from feed traders in the market. As rice-fish systems, as well as pond fish production, are not widespread, being mainly concentrated in certain regions, timely availability of quality feed also threatens greater adoption and diffusion of the technique. In addition, feed price is very high, and increasing, which increases production cost and is ultimately burdensome for poor subsistence farmers. Credit facilities, especially for agricultural purposes, are very weak, and interest rates are very high. Terms and conditions for credit are not favourable for agriculture, because farmers cannot repay within 1 week, agricultural practices like rice-fish systems needing a minimum gestation period and being seasonal. As we already mentioned, rice-fish technology is quite knowledge intensive and technical, thus, farmers need education, but the education levels in general, and those of farmers in particular, are very low. Low education is also sometimes linked with farmers' awareness of the positive and negative effects of any given technology, such as the comparative environmental benefit of rice-fish technology over rice monoculture. Good extension services could fill those gaps, but the extension services in Bangladesh are also very poor, although they have fairly wide networks. Governance in extension systems is very weak, which ultimately threatens the overall adoption and diffusion of agricultural technology and rice-fish systems in particular.

Conclusions and Policy Implications

Like many other Asian countries, Bangladesh is considered to be a "rice-fish" society, because both rice and fish are part and parcel of the Bangladeshi people's food culture, which has led to a popular Bengali saying, "Mache bhate Bangali," (in English, "Rice and fish make a Bengali.") (Dey et al. 2013). Some estimates suggest that Bangladesh has 2–3 million hectares of land with the potential for rice-fish production systems (ADB 2005; Dey et al. 2013; Ahmed and Garnett 2010; Dey and Prein 2006). A recent estimate indicates that about 0.18 million hectares of land are under use for rice-fish systems, which is much lower than the potential areas (Dey et al. 2013). Thus, one can raise the question as to its overall performance and the potential impeding factors that hinder adoption and diffusion into potential areas. This article attempts to provide a snapshot of the performance of Bangladesh's rice-fish systems by using a value chain analysis framework with detailed data from recent surveys on rice-fish value chain actors among the indigenous people of Bangladesh.

The chapter examines the financial performance of different actors in rice-fish systems and the myriad of factors determining such performance. Financial performance was measured by gross margin analysis. The article further investigated whether integration of fish into rice fields would improve profitability and justify a program for farming system improvement through the introduction of innovation. A partial budget analysis was conducted for two different rice production systems: a conventional rice production system under the green revolution regime (monoculture farm model), and an existing rice farm that diversified into aquaculture using the land and water resources of established rice farms (integrated farm model). Moreover, this chapter also explores the internal and external factors of the rice-fish system to further improve that system and encourage large-scale adoption and diffusion of rice-fish technologies. SWOT analysis was used to identify the level of rice-fish technology, as well as associated policy- and institutional-level strengths, weaknesses, opportunities and threats which can help towards future strategy building in regard to rice-fish technology adoption and diffusion.

Findings indicate that rice-fish systems offer considerable potential for increasing overall agricultural productivity and farm incomes in Bangladesh. Results also show that the rice-fish system value chain provides opportunity for landless, extremely poor households to participate in backward and forward linkage value chain activities in a profitable manner. To the best of our knowledge, a detailed cost and return survey of the rice-fish sub-sectors value chain similar to the one presented in this chapter has not previously been done in Bangladesh or elsewhere in countries with rice-fish potential. This paper has demonstrated that the rice-fish sector creates a very considerable level of profitable business opportunities at each stage of the value chain, and provides gendered employment opportunities, especially in the rice-fish production stage. Employment within the rice-fish value chain is further shaped by the social and institutional context within which it operates. Although rice-fish systems create opportunities for female labour, the affect they may have on household labour allocation decisions and womens' reproduction needs more research. Partial budgeting analysis supports the above findings that the rice-fish system is an economically sustainable competitive alternative to rice monoculture in Bangladesh. In addition to the empowerment of vale chain actors, rice-fish farming also benefits the local community and enhances rural economic growth.

However, rice-fish technology faces a number of significant challenges, and it is noteworthy that the technology exists in a rather sluggish manner within Bangladesh. Some innovative farm experimentation, private initiation by such organizations as the WFC, and motivation by NGOs like CARE, are the key impetus for bringing the rice-fish sector into mainstream agriculture in Bangladesh. There is a virtual lack of government support for the rice-fish farmers and overall value chain development. The high initial costs of rice-fish farming in terms of land, labour, fingerlings and feed, and land modification are major constraints to increasing pro-poor adoption and diffusion in the rice-fish production sector. In the short run, other rice-fish value chain actors have fewer barriers to entry and, if combined with rice-fish farming, the benefits could be significantly higher in the long run for poor farmers, despite the high initial cost outlay.

The traditional strengths of the technology are abundant water, fertile soil, strong research and extension institutions, expanded infrastructure and the encouragement of government policies to increase private-sector participation, which more than make up for its weaknesses and threats. Indeed, the enormous opportunities for further improvements in rice-fish technology and its value-chain performance provide a strong argument for action by the private sector within the value-chain, and by the government in the form of supportive policy and legislation (on issues such as property rights in land tenure, access to credit and markets, access and quality of irrigated water and feed, infrastructure, and public and private human capacity development and training). Such action would serve both to safeguard the current status of adoption and diffusion and to derive the benefits from it and further increase its adoption and diffusion, and the subsequent benefits, in the future.

Value-chain analysis has not been widely used to assess the ex-ante performance of integrated farming systems in general and integrated rice-fish technology in particular, with an aim towards further development of the sector. This paper has showed how the value chain analysis, together with SWOT analysis, helps us to understand the financial and social benefits generated by rice-fish technology, and to identify the crucial factors that hinder large scale adoption and diffusion and the overall performance of value chain development. Better understanding of these crucial factors can help to design the necessary policy and institutional actions and innovations that will increase adoption and diffusion, as well as the overall development of the sector.

Thus, this chapter emphasizes that the rice-fish-based farming system and its value chain development in potential developing countries could give momentum to the sustainable intensification paradigm, as the technology has traditional strengths and opportunities for further development, although its constraints (weaknesses and threats) should be properly acknowledged in order to make it happen. The important contribution of this study is, as this study was done among the nation's indigenous people, to help better design agricultural intervention to reduce extreme poverty and marginality in Bangladesh and possibly other countries in Asia and Africa that have similar socioeconomic, agro-ecological, and institutional settings.

Acknowledgement The study was supported by the German Academic Exchange Service (DAAD) and Dr. Hermann Eiselen's Doctoral Programm of the Foundation fiat panis research fellowship. It is a part of the first author's doctoral research at the Center for Development Research (ZEF), University of Bonn, Germany. The views and opinions expressed herein are solely those of the authors.

Open Access This chapter is distributed under the terms of the Creative Commons Attribution-Noncommercial 2.5 License (http://creativecommons.org/licenses/by-nc/2.5/) which permits any noncommercial use, distribution, and reproduction in any medium, provided the original author(s) and source are credited.
The images or other third party material in this chapter are included in the work's Creative Commons license, unless indicated otherwise in the credit line; if such material is not included in the work's Creative Commons license and the respective action is not permitted by statutory regulation, users will need to obtain permission from the license holder to duplicate, adapt or reproduce the material.

References

ADB (2005) An evaluation of small-scale freshwater rural aquaculture development for poverty reduction. Asian Development Bank, Manila

Adelman I, Taylor JE, Vogel S (1988) Life in a Mexican village: a SAM perspective. J Dev Stud 25(1):5–24

Ahmed N, Garnett ST (2010) Sustainability of freshwater prawn farming in rice fields in southwest Bangladesh. J Sustain Agric 34(6):659–679

Ahmed N, Garnett ST (2011) Integrated rice-fish farming in Bangladesh: meeting the challenges of food security. Food Secur 3(1):81–92

Ahmed N, Zander KK, Garnett ST (2011) Socioeconomic aspects of rice-fish farming in Bangladesh: opportunities, challenges and production efficiency. Aust J Agric Resour Econ 55(2):199–219

Ahmed AU, Ahmad K, Chou V, Hernandez R, Menon P, Naeem F, Naher F, Quabili W, Sraboni E, Yu B, Hassan Z (2013) The status of food security in the feed the future zone and other regions of Bangladesh: results from the 2011–2012 Bangladesh Integrated Household Survey. Project report submitted to the US Agency for International Development. International Food Policy Research Institute, Dhaka. http://www.ifpri.org/sites/default/files/publications/bihstr.pdf

Alam MF, Thomson KJ (2001) Current constraints and future possibilities for Bangladesh fisheries. Food Policy 26(3):297–313

Baecke E, Rogiers G, De Cock L, Van Huylenbroeck G (2002) The supply chain and conversion to organic farming in Belgium or the story of the egg and the chicken. Br Food J 104(3/4/5):163–174

Bahr M, Botschen M, Laberentz H, Naspetti S, Thelen E, Zanoli R (2004) The European consumer and organic food. School of Management and Business, University of Wales Aberystwyth, Aberystwyth

Barnard CS, Nix JJ (1979) Farm planning and control, 2nd edn. CUP, Cambridge, 600 pp

Barrientos S, Dolan C, Tallontire A (2003) A gendered value chain approach to codes of conduct in African horticulture. World Dev 31(9):1511–1526

Becerril J, Abdulai A (2010) The impact of improved maize varieties on poverty in Mexico: a propensity score-matching approach. World Dev 38(7):1024–1035

BER (2012) Bangladesh economic review. Economic Adviser's Wing, Finance Division, Ministry of Finance, Government of the People's Republic of Bangladesh, Dhaka

Berg H (2001) Pesticide use in rice and rice-fish farms in the Mekong Delta, Vietnam. Crop Prot 20(10):897–905

Berg H (2002) Rice monoculture and integrated rice-fish farming in the Mekong Delta, Vietnam – economic and ecological considerations. Ecol Econ 41(1):95–107

Bernard H (2006) Research methods in anthropology: qualitative and quantitative approaches. Alta Mira Press, Oxford

Bolwig S, Ponte S, Du Toit A, Riisgaard L, Halberg N (2010) Integrating poverty and environmental concerns into value-chain analysis: a conceptual framework. Dev Policy Rev 28(2):173–194

Christensen V, Steenbeek J, Failler P (2011) A combined ecosystem and value chain modeling approach for evaluating societal cost and benefit of fishing. Ecol Model 222(3):857–864

Coche AG (1967) Fish culture in rice fields a world-wide synthesis. Hydrobiologia 30(1):1–44

Coles C, Mitchell J (2011) Gender and agricultural value chains: a review of current knowledge and practice and their policy implications. ESA working paper No 11-05. Agricultural Development Economics Division, Food and Agriculture Organization of the United Nations, Rome

Courtney P, Mayfield L, Tranter R, Jones P, Errington A (2007) Small towns as "sub-poles" in english rural development: investigating rural–urban linkages using sub-regional social accounting matrices. Geoforum 38(6):1219–1232

Dey MM, Prein M (2006) Community-based fish culture in seasonal floodplains. NAGA World Fish Cent Q 29(1–2):21–27

Dey MM, Spielman DJ, Haque ABMM, Rahman MS, Valmonte-Santos R (2013) Change and diversity in smallholder rice–fish systems: recent evidence and policy lessons from Bangladesh. Food Policy 43:108–117

Dillon JL, Hardaker JB (1989) Farm management research for small farmer development, vol 41. Food and Agriculture Organization of the United Nations, Rome

DoF (2010) Fisheries statistical yearbook of Bangladesh. Department of Fisheries, Ministry of Fisheries and Livestock, Dhaka

Dolan C, Humphrey J (2004) Changing governance patterns in the trade in fresh vegetables between Africa and the United Kingdom. Environ Plan A 36(3):491–510

Doss CR (2006) Analyzing technology adoption using microstudies: limitations, challenges, and opportunities for improvement. Agric Econ 34(3):207–219

Dugan P, Dey MM, Sugunan VV (2006) Fisheries and water productivity in tropical river basins: enhancing food security and livelihoods by managing water for fish. Agric Water Manag 80(1):262–275

Fan S, Pandya-Lorch R (eds) (2012) Reshaping agriculture for nutrition and health. International Food Policy Research Institute, Washington, DC

FAO (2005a) EASYPol. On-line resource materials for policy making. Analytical tools. Module 043. Commodity chain analysis. Constructing the commodity chain, functional analysis and flow charts. Food and Agriculture Organization of the United Nations, Rome. www.fao.org/docs/up/easypol/330/cca_043EN.pdf. Accessed Nov 2009

FAO (2005b) EASYPol. On-line resource materials for policy making. Analytical tools. Module 044. Commodity chain analysis. Financial analysis. Food and Agriculture Organization of the United Nations, Rome. www.fao.org/docs/up/easypol/331/CCA_044EN.pdf. Accessed Nov 2009

FAO (2006) Strengthening national food control systems. Guidelines to assess capacity building needs. Food and Agriculture Organization of the United Nations, Rome

Fasse A, Grote U, Winter E (2009) Value chain analysis methodologies in the context of environment and trade research. Discussion paper No 429. School of Economics and Management, Leibniz University, Hannover

Feder G, Just RE, Zilberman D (1985) Adoption of agricultural innovations in developing countries: a survey. Econ Dev Cult Change 33(2):255–298

Fernando CH (1993) Rice field ecology and fish culture: an overview. Hydrobiologia 259 (2):91–113

Finnveden G, Moberg A (2005) Environmental systems analysis tools: an overview. J Cleaner Prod 13(12):1165–1173

Frei M, Becker K (2005) Integrated rice-fish culture: coupled production saves resources. Nat Res Forum 29(2):135–143

Gereffi G (1994) The organization of buyer-driven global commodity chains: how US retailers shape overseas production networks. In: Gereffi G, Korzeniewicz M (eds) Commodity chains and global capitalism. Praeger, Westport, pp 95–123

Gereffi G (1995) Global production systems and third world development. In: Stallings B (ed) Global change, regional response: the new international context of development. Cambridge University Press, Cambridge, pp 100–142

Gereffi G, Humphrey J, Kaplinsky R, Sturgeon T (2001) Globalization, value chains and development. IDS Bull 32(3):1–9

Giap DH, Yi Y, Kwei Lin C (2005) Effects of different fertilization and feeding regimes on the production of integrated farming of rice and prawn Macrobrachium rosenbergii (De Man). Aquacult Res 36(3):292–299

Gibbon P (2000) Back to the basics through delocalisation: the Mauritian garment industry at the end of the twentieth century. Working paper 00.7. Centre for Development Research, Copenhagen

Guptill A, Wilkins JR (2002) Buying into the food system: trends in food retailing in the US and implications for local foods. Agric Hum Values 19(1):39–51

Gurung TB, Wagle SK (2005) Revisiting underlying ecological principles of rice-fish integrated farming for environmental, economical and social benefits. Our Nature 3(1):1–12

Hagedorn K (2008) Particular requirements for institutional analysis in nature-related sectors. Eur Rev Agric Econ 35(3):357–384

Halwart M, Gupta MV (eds) (2004) Culture of fish in rice fields. Food and Agriculture Organization of the United Nations and The WorldFish Center, Rome/Penang

Halwart M, Borlinghaus M, Kaule G (1996) Activity pattern of fish in rice fields. Aquaculture 145 (1):159–170

Haque MM, Little DC, Barman BK, Wahab MA (2010) The adoption process of ricefield-based fish seed production in Northwest Bangladesh: an understanding through quantitative and qualitative investigation. J Agric Educ Ext 16(2):161–177

Hecht JE (2007) National environmental accounting – a practical introduction. Int Rev Environ Res Econ 1:3–66

Holland J (2007) Tools for institutional, political, and social analysis of policy reform, A source book for development practitioners. The World Bank, Washington, DC

Humphrey J, Schmitz H (2001) Governance in global value chains. IDS Bull 32(3):19–30

IFPRI (2010) Research and extension of rice-fish technology in Bangladesh: an expert opinion forum. A Cereal System Initiative for South Asia (CSISA) study on developing and eploying new technologies to smallholders in South Asia: key policies and issues. Forum report. International Food Policy Research Institute, Washington, DC Islam, A. H. M. S., Barman, B. K., & Murshed-e-Jahan, K. (2015) Adoption and impact of integrated rice–fish farming system in Bangladesh. Aquaculture 447:76-85

Islam AHMS, Barman BK, Murshed-e-Jahan K (2015) Adoption and impact of integrated rice–fish farming system in Bangladesh. Aquaculture 447:76–85

Kaplinsky R (2000) Globalisation and unequalisation: what can be learned from value chain analysis? J Dev Stud 37(2):117–146

Kaplinsky R, Morris M (2001) A handbook for value chain research, vol 113. International Development Research Centre, Ottawa

Kay RD, William ME, Patricia AD (2008) Farm management, 6th edn. McGraw-Hill Companies, New York

Kim S, Shin E (2002) A longitudinal analysis of globalization and regionalization in international trade: a social network approach. Soc Forces 81(2):445–471

Kledal PR (2006) The Danish organic vegetable chain. Report No 182. Den Kgl. Veterinarog Lantbohojskole, Copenhagen

Kohls RL, Uhl JN (2002) Marketing of agricultural products, 9th edn. Prentice-Hall, Upper Saddle River

Kurttila M, Pesonen M, Kangas J, Kajanus M (2000) Utilizing the analytic hierarchy process (AHP) in SWOT analysis – a hybrid method and its application to a forest-certification case. For Policy Econ 1(1):41–52

Lenzen M (2001) Errors in conventional and input–output–based life-cycle inventories. J Ind Ecol 4(4):127–149

Lightfoot C, van Dam A, Costa-Pierce B (1992) What's happening to rice yields in rice-fish systems? In: dela Cruz CR, Lightfoot C, Costa-Pierce BA, Carangal VR, Bimbao MP (eds) Rice-Fish research and development in Asia, ICLARM conference proceedings 24, Manila, pp 177–183

Little DC, Surintaraseree P, Innes-Taylor N (1996) Fish culture in rainfed rice fields of northeast Thailand. Aquaculture 140(4):295–321

Macfadyen G, Nasr-Alla AM, Al–Kenawy D, Fathi M, Hebicha H, Diab AM, Hussein SM, Abou-Zeid RM, El–Naggar G (2012) Value-chain analysis–an assessment methodology to estimate Egyptian aquaculture sector performance. Aquaculture 362–363:18–27

Maertens M, Swinnen JF (2010) Are African high-value horticulture supply chains bearers of gender inequality? In: Workshop on gaps, trends and current research in gender dimensions of agricultural and rural employment: differentiated pathways out of poverty, vol 31, Rome

Minten B, Murshid KAS, Reardon T (2011) The quiet revolution in agrifood value chains in Asia: the case of increasing quality in rice markets in Bangladesh, vol 1141. International Food Policy Research Institute, Washington, DC

Minten B, Murshid KAS, Reardon T (2013) Food quality changes and implications: evidence from the rice value chain of Bangladesh. World Dev 42:100–113

Nabi R (2008) Constraints to the adoption of rice-fish farming by smallholders in Bangladesh: a farming systems analysis. Aquac Econ Manag 12(2):145–153

Nhan DK, Phong LT, Verdegem MJ, Duong LT, Bosma RH, Little DC (2007) Integrated freshwater aquaculture, crop and livestock production in the Mekong delta, Vietnam: determinants and the role of the pond. Agric Syst 94(2):445–458

Nix J (2000) Farm management pocketbook, 31st edn. Wye College, London, p 244

Noltze M, Schwarze S, Qaim M (2012) Understanding the adoption of system technologies in smallholder agriculture: the system of rice intensification (SRI) in Timor Leste. Agric Syst 108:64–73

Ondersteijn CJ, Wijnands JH, Huirne RB, Kooten OV (eds) (2006) Quantifying the agri-food supply chain, vol 15. Springer, Wageningen

Pacini C, Giesen G, Wossink A, Omodei-Zorini L, Huirne R (2004) The EU's Agenda 2000 reform and the sustainability of organic farming in Tuscany: ecological-economic modelling at field and farm level. Agric Syst 80(2):171–197

Pant J, Barman BK, Murshed-E-Jahan K, Belton B, Beveridge M (2014) Can aquaculture benefit the extreme poor? A case study of landless and socially marginalized Adivasi (ethnic) communities in Bangladesh. Aquaculture 418:1–10

Rahman S (2013) Formalin–never ending woe. New Age. The Bangladesh Chronicle, 15 Jan 2013

Reardon T, Chen K, Minten B, Adriano L (2012) The quiet revolution in staple food value chains: enter the dragon, the elephant, and the tiger. Asian Development Bank International Food Policy Research Institute, Manila/Washington, DC

Rebitzer G, Ekvall T, Frischknecht R, Hunkeler D, Norris G, Rydberg T, Schmid WP, Suh S, Weidema BP, Pennington DW (2004) Life cycle assessment: part 1: framework, goal and scope definition, inventory analysis, and applications. Environ Int 30(5):701–720

Riisgaard L, Bolwig S, Ponte S, Du Toit A, Halberg N, Matose F (2010) Integrating poverty and environmental concerns into value-chain analysis: a strategic framework and practical guide. Dev Policy Rev 28(2):195–216

Roos N, Islam MM, Thilsted SH (2003) Small indigenous fish species in Bangladesh: contribution to vitamin A, calcium and iron intakes. J Nutr 133(11):4021–4026

Roos N, Wahab MA, Chamnan C, Thilsted SH (2007a) The role of fish in food-based strategies to combat vitamin A and mineral deficiencies in developing countries. J Nutr 137(4):1106–1109

Roos N, Wahab M, Hossain MAR, Thilsted SH (2007b) Linking human nutrition and fisheries: incorporating micronutrient-dense, small indigenous fish species in carp polyculture production in Bangladesh. Food Nutr Bull 28(Supplement 2):280–293

Roth S, Hyde J (2002) Partial budgeting for agricultural businesses. Pennsylvania State University, unpublished manual. http://pubs.cas.psu.edu/freepubs/pdfs/ua366.pdf. Accessed 10 Feb 2014

Rothuis AJ, Nhan DK, Richter CJJ, Ollevier F (1998a) Rice with fish culture in the semi-deep waters of the Mekong Delta, Vietnam: interaction of rice culture and fish husbandry management on fish production. Aquacult Res 29(1):59–66

Rothuis AJ, Nhan DK, Richter CJ, Ollevier F (1998b) Rice with fish culture in the semi-deep waters of the Mekong Delta, Vietnam: a socio-economical survey. Aquacult Res 29(1):47–57

Rothuis AJ, Vromant N, Xuan VT, Richter CJJ, Ollevier F (1999) The effect of rice seeding rate on rice and fish production, and weed abundance in direct-seeded rice–fish culture. Aquaculture 172(3):255–274

Ruel MT, Alderman H (2013) Nutrition-sensitive interventions and programmes: how can they help to accelerate progress in improving maternal and child nutrition? Lancet 382 (9891):536–551

Sarder R (2007) FAO fisheries technical paper, no 501. In: Bondad-Reantaso MG (ed) Freshwater fish seed resources in Bangladesh: assessment of freshwater fish seed resources for sustainable aquaculture. Food and Agriculture Organization of the United Nations, Rome, pp 105–128

Shang YC (1986) Research on aquaculture economics: a review. Aquac Eng 5(2):103–108

Sturgeon TJ (2001) How do we define value chains and production networks? IDS Bull 32(3):9–18

Thilsted SH, Roos N, Hassan N (1997) The role of small indigenous fish species in food and nutrition security in Bangladesh, Naga. ICLARM Q 20(3–4):82–84

Trifković N (2014) Governance strategies and welfare effects: vertical integration and contracts in the catfish sector in Vietnam. J Dev Stud (ahead-of-print), 50(7):1–13

Uddin R, Wahid MI, Jasmeen T, Huda NH, Sutradhar KB (2011) Detection of formalin in fish samples collected from Dhaka city, Bangladesh. Stamford J Pharm Sci 4(1):49–52

Veliu A, Gessese N, Ragasa C, Okali C (2009) Gender analysis of aquaculture value chain in Northeast Vietnam and Nigeria. World Bank agriculture and rural development discussion paper 44. The World Bank, Washington, DC

Walters D, Lancaster G (2000) Implementing value strategy through the value chain. Manag Decis 38(3):160–178

William JG, Hella JP, Mwatawala MW (2012) Ex-ante economic impact assessment of green manure technology in maize production systems in Tanzania. Res Hum Soc Sci 2(9):47–58

Williamson OE (1985) The economic institutions of capitalism: firms, markets, relational contracting. Collier Maximilian Publisher, London

Wood A (2001) Value chains: an economist's perspective. IDS Bull 32(3):41–45

World Bank (2012) Bangladesh: annual economic update. Poverty reduction and economic management, South Asia region. The World Bank, Washington, DC

Chapter 18
Technologies for Maize, Wheat, Rice and Pulses in Marginal Districts of Bihar and Odisha

P.K. Joshi, Devesh Roy, Vinay Sonkar, and Gaurav Tripathi

Abstract This chapter looks at potential technologies for marginal districts in two of the most backward states in India – Bihar and Odisha. Based on technological performance, we identified the marginal districts for four principal crops, i.e. rice, wheat, maize and pulses, and assessed the potential of the technologies in terms of their agro-ecological suitability, as well as the complementary inputs required for success. Using a primary survey, we gauge the real opportunities and constraints for technology adoption directly from the farmers, including their aspirations about crop choices and the technologies that exist to grow them. Maize and pulses turn out to be crops that farmers currently aspire to get into. Also, data distinctly reveals, in some cases, the disconnect between perceived potential of the technology among experts and the valuation of the same by likely adopters.

Keywords Marginality • Technology adoption • Aspirations • Perceptions • India

Introduction

This chapter summarizes the current state of agricultural productivity and the potential of different technologies in two of the most economically backward states in India, Bihar and Odisha, for their principal crops, rice, wheat, maize and pulses. Focusing on marginal districts in the two states, the paper assesses the suitability of different technologies to uplift the areas (districts) out of their current low level equilibrium (in terms of production performance) and thereby raise the standards of

P.K. Joshi (✉) • D. Roy
International Food Policy Research Institute (IFPRI), Washington, DC, USA
e-mail: p.joshi@cgiar.org; d.roy@cgiar.org

V. Sonkar • G. Tripathi
International Food Policy Research Institute (IFPRI), New Delhi, India
e-mail: V.K.Sonkar@cgiar.org; g.tripathi@cgiar.org

© The Author(s) 2016

F.W. Gatzweiler, J. von Braun (eds.), *Technological and Institutional Innovations for Marginalized Smallholders in Agricultural Development*,
DOI 10.1007/978-3-319-25718-1_18

living.[1] The marginal (backward) districts for these crops are identified based on current yield and its performance over time. Subsequently, the choice of technologies for marginal areas for each case is analyzed ex ante. In this approach, the potential is assessed under conditions in which a given technology might not be widely adopted currently but has a comparatively high potential to deliver upon adoption.[2]

The short listing of technologies for these crops has been done based on a clearing house approach in which, in consultation with different stakeholders, the potent technologies for districts have been chosen. Two multi-stakeholder workshops in Bihar and Odisha for identifying innovative technologies with input from farmers, the private sector and NGOs, natural resource management experts and specialists in market-linkages resulted in the short-listed technologies which have already been field tested but have yet to be adopted at all or adopted on a large scale.

Following this, through a structured survey of the households, we examine the reasons behind slow or poor adoption of available technological innovations. We look at the profile of the identified technologies in terms of their uptake over time and try to assess the role of complementary inputs that affect the feasibility for the respective areas, as well as the prospects for adopters of technology to multiply.

The paper is organized as follows. Section "Identifying Marginal Areas (Districts) in Bihar and Odisha" presents the scheme for identification of the marginal districts in the two states. The fixing of marginality is crop specific. After fixing marginality in section "Technologies for Marginal Districts in Bihar and Odisha: Findings from Multi-Stakeholder Consultations", the next section looks at the potent technologies for these areas. Subsequently, based on field surveys, we gauge the suitability of the identified technology for the marginal areas in section "Findings from the Farmer Survey". Section "Awareness of Technologies and Their Adoption in Bihar and Odisha" summarizes the findings from the survey. Section "Regression Analysis for Awareness and Adoption of Technologies in Rice, Wheat and Maize in Bihar and Odisha" presents some illustrative regression results for wheat, rice, maize and pulses. Section "Conclusions and Policy Recommendations" concludes.

---

[1] Usually, a combination of indicators is used in multi-dimensional criteria for assessing marginality. In this paper, we reduce the dimensionality problem in identification of marginal areas to just two. In particular, we look at the dynamic behavior of yields in combination with the current relative yield positions to fix marginality. Thus, the two indicators are both related to yield behavior. The first one is increasing or decreasing yields over time at different absolute levels and the second one comprises relative yield positions of preselected crops grown in the district. The filter for marginal districts is based on the intersection of these two indicators. A district with falling yields and positioned in the lowest quartile of yield distribution, for example, would clearly qualify as a marginal district in what we call Tier 1. Other levels of these two indicators determine the tier levels for marginal districts. A moderately falling yield over time with a current state of first or second quartile of yield distribution may comprise Tier 2 marginal districts and so on.

[2] Wheat is cropped in a very small area in Odisha. Hence, we do not study the case of wheat in that state.

Identifying Marginal Areas (Districts) in Bihar and Odisha

The eastern part of India where Bihar and Odisha are situated is rich in natural resources, such as water, year-round bright sunlight, fertile soil and forest, and mineral reserves. However, eastern India has not been able to capitalize upon its vast resource pool owing to different factors, such as underdeveloped basic infrastructure (like roads, power and markets), concentration of the poor population with high density in most parts, weak institutions (such as credit, insurance, education and extension) and weak governance. These bottlenecks have rendered the region unattractive for investment.

Among states, Bihar and Odisha lie at the bottom of the scale for various socio-economic indicators. The agriculture sector in these states represents their lifeline; 73.6 % and 61.8 % of the working population in Bihar and Odisha, respectively, draw their livelihood from agriculture vis-à-vis 54.6 % throughout all of India. The percentages of the working population mainly/exclusively dependent on agriculture in Bihar and Odisha are 43.1 % and 32.5 %, respectively (MHA 2011). However, the share of agriculture and the allied sector in Gross Domestic Product (GDP) of Bihar and Odisha was just 23.0 % and 16.4 %, respectively, in 2011–2012. The annual growth rates of agriculture and the allied sector in Bihar (3.9 %) and Odisha (3.0 %) were lower than the all India average (8.5 %) during the period 2004–2005 to 2011–2012. In spite of relatively better growth rates exhibited by the industry and service sectors over the last decade (albeit with a very low base), these states have not made significant headway in poverty reduction during 2011–2012, with 33.8 % and 32.6 % of the population in Bihar and Odisha respectively still below the poverty line (Planning Commission 2013).

The slow agricultural growth rate, along with its low share in state GDP, cannot raise the standard of living of a large population since inadequate infrastructure, particularly inadequacy of power, seriously limits the growth of industries. Hence, a structural shift in employment from low income farming to high value industrial and service sectors is not expected, at least in the medium term (GoI 2008). Therefore, to reduce poverty and elevate the marginal areas effectively, it is imperative to bring about sustained growth in the agriculture and allied sectors in Bihar and Odisha.

Moreover, in both states, there exists significant regional disparity. To further the discussion, we first present the criteria for the identification of marginal districts in the two states. We consider districts to be marginal based on two factors, i.e., dynamic changes in yield and the current levels of yield. Specifically, we take crop-specific marginal areas to be the ones where yields have been declining over time (or rising comparatively slowly) and are currently at levels that are subpar relative to the other districts in the state. The current yields are clubbed into four quartiles of the distribution over districts with yields in lower quartiles taken to be subpar. A district, for example, that had yield falling significantly (such as in double digit percentages, known as a double dip) for a crop and settling into the lowest quartile of the yield distribution would clearly be among the marginal districts. In more intermediate cases, it is the relative current levels of yield and relative changes in yields over time that affix the marginality of the districts.

For illustration, Fig. 18.1 presents the dynamics of rice, maize and pulse yields in Odisha. The figure plots the average yields (averaged over 6 years) of the three crops at two points in time across districts in Odisha. A few important points emerge from this figure. First, there is significant inter- district variation in the three cases. More importantly, there is distinctive variation in yields over time.[3]

Tables 18.1 (on cereals) and 18.2 (on pulses) present the status of different districts in Odisha in terms of their location in the space comprising dynamic behavior of yields of these crops and their current status in terms of relative yields. The columns present double or single dip decreases and single, double or triple digit increases in yields for the districts in the case of rice, pulses and maize, respectively.[4] The numbers next to the districts represent the quartile to which the yield of the district belongs in the yield distribution. A number 0 next to the district represents the case of lowest yield. In maize, for instance, Puri and Jagatsinghpur, with the lowest yields, have been put in the 0 category, while for rice, the lowest is Sundargarh.

Especially salient are the districts which, even after double crest increases in yields, continue to languish with their current yields in the lowest quartile. Districts such as Angul for maize and Malkangiri for rice fall into this category. Similarly, districts like Deogarh, Bolangir, Kendrapara, Angul and Keonjhar in rice, in spite of experiencing yield increases, still fall in the bottom quartile of rice yield distribution. In the case of maize, the districts that have experienced yield increases but still are in the lowest quartile of yields across districts comprise Bolangir, Boudh, Angul, Jajpur, Keonjhar, Mayurbhanj, Khurda and Balasore (Table 18.1). We select a few districts from this set of marginal areas for primary data collection to analyze the awareness about and adoption of shortlisted technologies. The districts selected in Odisha are Angul, Deogarh and Boudh, because they cover marginal areas in all of the three crops.

Analyses similar to Odisha were conducted in Bihar as well to identify marginal districts. The only difference was that, in Bihar, wheat was also included, being an important cereal crop there. Using same method as for Odisha in choosing districts for primary survey, the districts in Bihar selected for primary surveys are Araria, Muzaffarpur and East Champaran.

Though the criteria for choosing marginal districts is technology-centric, it turns out that, with principal crops, on average, the shortlisted districts are congruent with the list of poor districts in terms of expenditures based on the national sample survey in India. The list of marginal districts also overlaps significantly with the set

---

[3] Comparing rice and maize, both within as well as across, variation in Odisha is far more pronounced in the latter. This might be due to greater spatial differences in adoption of technology in maize than in rice.

[4] Single dip, double dip, single crest, double crest and triple crest imply single digit percentage drops, double digit percentage drops, single digit percentage increases, and double and triple digit percentage increases in yields over time.

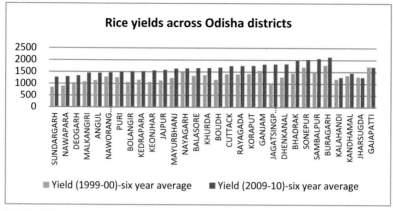

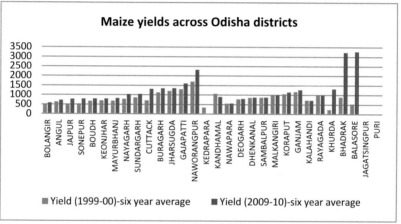

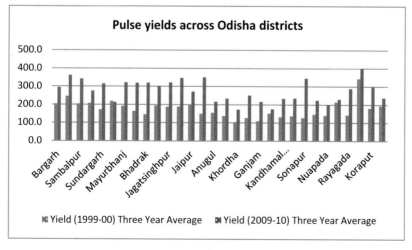

Fig. 18.1 Dynamics of yields of principal crops across districts in Odisha

328   P.K. Joshi et al.

Table 18.1 Cereals: change in productivity and current states of yields in Odisha

| Single dip | | | | | | | |
Rice	Maize	Rice	Maize	Rice	Maize	Rice	Maize
Kalahandi-2	Kedrapara-1		Naupada-1 (neutral)	Sundargarh-0	Bolangir-1	Malkangiri-1	Angul-1
Jharsugda-1	Nawapara-1			Nawapara-1	Sonepur-1	Bhadrak-3	Jajpur-1
	Kalahandi-2			Deogarh-1	Boudh-1	Kandhamal-3	Nayagarh-2
	Jagatsingpur-0			Angul-1	Keonjhar-1	Gajapatti-4	Cuttack-2
	Puri-0			Naworangpur-2	Mayurbhanj-1		Gajapatti-3
				Puri-2	Sundargarh-2		Naworangpur-4
				Bolangir-1	Buragarh-3		Kandhamal-3
				Kedrapara-1	Jharsugda-3		Deogarh-2
				Keonjhar-1	Sambalpur-2		Dhenkanal-2
				Jajpur-2	Balasore-1		Malkangiri-2
				Mayurbhanj-2			Koraput-3
				Nayagarh-3			Ganjam-3
				Balasore-3			Rayagada-3
				Khurda-3			Khurda-1
				Boudh-2			Bhadrak-2
				Cuttack-3			
				Rayagada-3			
				Koraput-3			
				Ganjam-3			
				Jagatsingpur-1			
				Dhenkanal-2			
				Sonepur-4			
				Sambalpur-3			
				Buragarh-4			

Table 18.2 Dynamics of yields and current state of pulses in Odisha

Single dip	Double crest	Double crest	Triple crest
Keonjhar-0	Gajapati-1	Mayurbhanj-3	Rayagada-3
	Nabarangapur-4	Sambalpur-4	Bhadrak-3
	Malkangiri-2	Nayagarh-2	Dhenkanal-4
	Debagarh-3	Baudh-2	Sonapur-4
	Jaipur-2	Jagatsinghpur-2	
	Anugul-1	Kandhamal Phoolbani-2	
	Nuapada-1	Sundargarh-3	
	Bargarh-3	Cuttack-4	
	Jharsuguda-4	Baleshwar-3	
	Balangir-1	Ganjam-1	
	Kendrapara-3	Puri-2	
	Koraput-3		
	Khordha-1		

of backward districts marked by the planning commission. Figures 18.2 and 18.3 present maps of districts in the two states in terms of poverty levels based on expenditure data from national sample survey data in 2006.

Technologies for Marginal Districts in Bihar and Odisha: Findings from Multi-Stakeholder Consultations

With the aim of validating identification of marginal districts, and of potent technologies for such areas, two workshops were organized in Bihar and Odisha, respectively. The workshops included experts from agricultural research institutions/universities, government officials, private sector representatives, members of non-governmental organizations, farmers and other relevant organizations, e.g., from renewable energy sources in agriculture.

The workshops followed a clearing house model in which the identification of relevant districts was put to participants and a commonly agreed-upon list of marginal areas was prepared for each crop. In addition, there were deliberations on crops and activities in the two states in terms of technologies with the potential to improve outcomes. Based on secondary data, workshop findings and interactions with scientists and technology experts, suitable technologies were identified for raising productivity in the marginal districts.

The technological solutions presented below cover a spectrum relating to improved/hybrid seed varieties, specific to particular agro-climatic ecosystems; cultivation processes; mechanization; irrigation; training and extension; and market linkages, among others. Most of these technologies have been tested in the field, both by research institutions and sometimes by innovative/progressive farmers. However, they have yet to be adopted or adopted on a large scale in the marginal

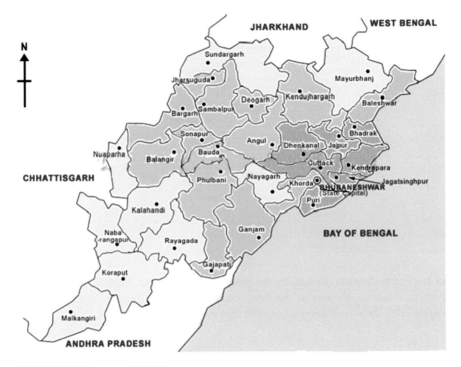

Fig. 18.3 Head count ratio Odisha

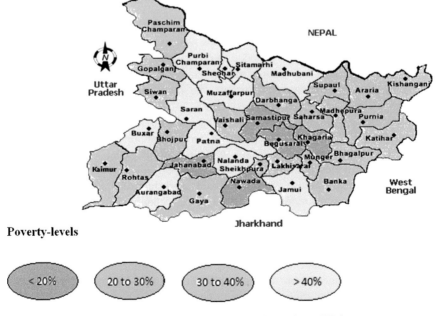

Fig. 18.2 Headcount ratios across districts in Bihar (2006) based on NSS data

districts. We thus examine the reasons behind slow or poor adoption of technological innovations through a household survey.

Technologies for Marginal Districts

Below, we present the list of technologies identified as having high potential for different crops for the marginal districts in Bihar and Odisha. Subsequently, we focus on the state of awareness of technologies and their adoption based on primary surveys in the two states.

Rice

The technologies in rice with underexploited potential in the marginal districts are as follows:

1. *Varietal substitution towards (climatic) stress-tolerant, high-yielding varieties* developed and tested for specific agro-climatic ecosystems. Some marginal districts lie in these ecosystems and adequate varieties need to be promoted for these areas. The rice cultivars that are high yielding and tolerant to climatic stress (including drought and flood) are available to try in the marginal districts of the two states. For example, the Swarna Sub 1 and Varshadhan rice varieties give high yields under flood conditions, while Sahbhagidhan cultivar (IR 74371-70-1-1) is drought-tolerant (see Yamano et al. 2013; Reddy et al. 2009). However, ready availability of the right seed cultivars is an issue.

2. *Mechanized Direct Seeded Rice (DSR) technology* for rice-cultivation among small and marginal farmers. This technology saves on cultivation costs by conserving labor and water for irrigation; it also enables timely sowing that helps achieve higher yields for rice, as well as for the succeeding winter crop (wheat) (see Pathak et al. 2011). The crop matures earlier than the traditional practice by 7–10 days. Begusarai is the most marginal rice district in Bihar (with the lowest yields currently across districts). It has started to see some success with DSR. The adoption and spread of this technology needs to be initiated and scaled up in other marginal areas as well. In Odisha, the mechanized DSR is a new phenomenon; its introduction into the state dates to 2009–2010. It has so far not been widespread. Odisha is predominantly engaged in manual DSR – 80 % of the rice cultivation area was under manual DSR in 1990.

3. *Mechanization of agriculture promoted by custom hiring centers- Specific promotion of the self-propelled paddy trans-planter machine-* This technology has been estimated to increase yields by an estimated 20 % through increased operational efficiency and facilitating uniform and optimum plant population. It is estimated to reduce seed-requirement by as much as 40 % while saving on costs of nursery preparation and transplanting. Mechanization through

establishment of custom hiring centers and machinery hubs could be tried through the system of Krishi Vigyan Kendras (Agriculture Science Centers); with the facility of training the youths, it can facilitate mechanization of agriculture for small and marginal farmers (Srinivasarao et al. 2013).

Timely transplanting of rice is facilitated through the use of the self-propelled paddy trans planter, while also reducing costs of labor, fertilizer, seed and irrigation, as well as ensuring uniform spacing and optimum plant density. Post-2004, the mechanization efforts in Bihar have met with some success; Odisha has to cover more ground in promoting mechanization. Mainstreaming custom hiring centers would be an important contributor to improving the outcomes in the marginalized districts. There are possible synergies with other forms of mechanization as well, such as the *power-tiller, the pedal thresher and the paddy reaper*. The economies of scope should be exploited among the different forms of mechanization.

4. *Use of integrated nutrient management, involving use of both organic and inorganic fertilizers* can result in superior yields and the achievement of better 'nutrient-use-efficiency', but requires many complementary inputs. The System of Rice Intensification (SRI), for example, in conjunction with organic treatment can, in principle, give significantly higher yield and superior nutrient-use efficiency to rice (see Prasad 2006).

Maize

In maize, the salient technology is the adoption of hybrid varieties, most importantly the single cross hybrids (SCH) that have become widespread in Bihar. The spread of hybrid maize is far more muted in Odisha. Within Bihar as well, hybrid varieties of maize have largely bypassed the marginal districts. Also, there has been much lower adoption of hybrid maize in the rainy season because of flooding issues. Bihar experienced severe floods in 2004–2005, 2007–2008 and 2008–2009 that were most intense in the northern and eastern districts. Waterlogging affects crop yields in an area of about 0.63 million ha (6.7 % of the total) (Chowdhary et al. 2008).

In some of the marginal districts, the moisture created under good rainfall, particularly in northern Bihar, provides for maize sown in the winter and summer seasons. Maize is less water intensive than rice and wheat. Hence, it is able to overcome the irrigation deficit during the winter and summer seasons arising from the high costs and limited availability of diesel for energizing pumps. However, there is also a greater prevalence of pests, diseases and weeds in the rainy season.

Because of these factors, the adoption pattern of hybrid maize in Bihar is mixed, with high uptake in the winter/spring seasons and relatively lower adoption in the rainy (Kharif) season. The next frontier for maize might be extending its success to those areas lagging behind and expanding the seasonal coverage of hybrid seeds by expanding hybrid maize production during the rainy season. In truth, over the last two decades, the area under maize cultivation in Bihar declined by 13.7 % during the Kharif season, but it increased by over 29 % and 58 %, respectively, in the Rabi and spring seasons.

Furthermore, in the marginal maize districts, farmers could be advised to grow maize on raised beds/ridges to reduce the risk of floods/excess water, which will in turn encourage adoption of hybrids during the rainy season as well. Some of the marginal districts for maize lie in low land areas where this technology should be promoted.

In general, there is a lack of timely availability of hybrid seeds and other agricultural inputs for farmers, along with the usual demand supply gap and lack of timely availability in the rainy season. This can be addressed through an increase in seed production and by strengthening the seed supply chains. Adequate attention is warranted from the private sector during the rainy season, similar to its role in the winter/spring seasons in heralding the hybrid maize revolution in the rainy season. Marginal districts have the potential to become the maize seed hubs of east-India owing to the favorable agro-climatic conditions prevailing mainly in the winter season, along with their fertile and plain land. Marginal districts can enjoy great commercial benefit from sales/export of seed, grain and technology for maize.

Wheat

The wheat technologies (broadly defined) identified for the marginal districts are:

1. *Surface seeding technique for rice-wheat systems*: This involves the broadcast of wheat-seeds in standing rice crops, under a condition of excess moisture (low land moist field) 15–25 days before the paddy-harvest. It helps avoid delay in wheat sowing, while also saving on tillage cost. Due to timely planting, higher yields are achieved. Further, it saves water in amounts from 35 % to 40 %.
2. *Zero tillage wheat with Resource Conserving Technologies* (RCTs): This involves different sowing practices (like equal row, paired row and control traffic). It can be done through suitable zero till (ZT) drills, double disc planters, multi-crop planters and rotary disc drills in rice residue. Immediately after the rice harvest, zero till wheat is sown through use of a ZT drill, which advances the sowing by 15–20 days, thereby helping escape the terminal heat stress prevalent in many marginal districts in Bihar and Odisha. It is estimated to save as much as Rs. 2500/ha in tillage, 20 % in seed, and 20 % in first irrigation, besides an additional yield gain of about 1.0 tons/ha (based on focus group discussions and expert elicitation).
3. *Laser land leveling (LLL)*: This saves irrigation water, increases cultivable area by 3–5 %, improves crop establishment, improves uniformity of crop maturity, increases water application efficiency up to 50 %, increases crop yield by 15 %, and improves weed control efficiency. The cost of LLL on average is Rs. 400/h and the average cost of leveling is found to be in the range of Rs. 5000–9500/ha. Like other costly machinery, this will also require a system of custom hiring to increase adoption among small and marginal farmers. LLL is thus a resource-

conserving technology. In surface-irrigated rice-wheat systems, 10–25 % of irrigation water is lost because of poor management and uneven fields. Uneven fields also lead to inefficient use of fertilizers and chemicals, increased biotic and abiotic stress, and lower yields (Lybbert et al. 2013).

Pulses

Overall, pulses have been confined to marginal environments, as comparatively resource-rich farmers have tended to prefer crops like paddy or wheat, or cash crops like cotton. Studies have attributed the low yield per hectare to various factors, mainly a lack of high-yielding and short-duration varieties and competition with other crops. Inadequate irrigation, cultivation in inferior lands, absence of fertilizer use, frequent attack by pests and diseases, dearth of extension services and poor infrastructure, and slow transfer of technology have also disadvantaged the pulse sector in states like Bihar and Odisha (Banerjee and Palke 2010). However, pulses have been coming up in Odisha since 2000, while they have continued their slide in Bihar, implying a dire need to bring in technology for pulses in that state.

In the two states, cereals continue to dominate among crops. Not much change is reflected in the case of pulses and oilseeds. Their contribution to the crop-sector fluctuated roughly 5–6 % between 2001 and 2008. In pulses, only seven districts in Bihar showed significant growth in production in the last decade. The prominent marginal districts in Bihar for pulses are Madhubani, Darbhanga, Sheohar and Vaishali. The prominent marginal districts in Odisha for pulses are Khordha, Gajapati, Nuapada, Keonjhar and Angul.

The innovative technologies identified for the marginal districts in pulses in the two states are:

1. *Stress-tolerant high-yielding varieties*- A number of stress-tolerant high-yielding varieties for all major pulses have been developed by the public sector in India in recent years. The supply of these cultivars needs to be improved in both Bihar and Odisha. Over time, pulse production has moved from eastern to western India, just as it has moved from north to south. Eastern states like Bihar and Odisha have been losing out on pulses. This interregional movement, to some extent, was driven by technology. New varieties of pulses, viz. short-run and very-short-run varieties, after they were introduced in the late 1990s, allowed for intercropping in such a way that areas less successful in pulses subsequently became quite proficient in them.

2. *Inter-cropping of pulses with other crops* can be a highly remunerative practice (Singh et al. 2009). Pulses also help in nitrogen fixation. They should thus be promoted as part of the rationalization of fertilizer usage in Bihar and Odisha.

3. *Some other technologies such as line sowing/seed drilling/zero tilling* are useful practices for improving the yield of pulses. Use of Rhizobium culture, phosphatic fertilizers and micro-nutrients like Boron, Molybdenum, Sulphur should be promoted in pulses.

Market Infrastructure

An overarching message that emerged from the elicitations in both Bihar and Odisha is that significant gains can accrue from the creation of marketing and storage infrastructure that will aid in the adoption of technology. Market development will strengthen the bargaining position of small and marginal farmers who end up selling the small surplus to middlemen or local traders at prices much lower than the government support price or the actual market price. In both states, since the central government does not procure grains, the farm gate prices are far below the minimum support price (MSP). Figure 18.4, which maps the markets across all districts in India (normalized by the number of cultivators), shows that these two states lag behind other states. With the realization of higher farm gate prices, higher incomes will strengthen the capacity of farmers to invest in new technologies.

The Factors Behind Technology Adoption

After selection of marginal districts and shortlisting of technologies, we take a snapshot of the technological options and gather details about their special features and the current state in terms of their uptake. Table 18.3 presents the details regarding technologies based on expert elicitations, focus group discussions, and a review of the literature supplemented with field visits. The main message from Table 18.3 is that technology adoption is a complex process involving several complementary inputs, and that the greater role of these complementary inputs, the less likely it is for the technology to be adopted on a larger scale.

Several technologies in Bihar and Odisha listed in Table 18.3 have not moved much beyond the introduction stage. These include technologies like laser land leveling (LLL), mechanized direct seeded rice plantation (MDSR) (also mechanized zero tillage of wheat), and furrow irrigated bed planting, all of which have moved at a sluggish pace. Even simple technologies such as mid-season drainage through wetting and drying have only expanded at a slow rate. On the other hand, technologies such as drought- and flood-resistant varieties of crops had greater uptake. Constraints in adoption of technology differ, but generally, almost all technologies have been inhibited because of lack of adequate extension services and often because of missing complementary inputs. Table 18.3 presents the basic set of existing factors. such as agro-climatic conditions, that customize technologies to specific areas.

To illustrate the role of complementary inputs, consider the case of LLL. In spite of being a promising technology, it has spread in a limited way because of its capital costs that, under liquidity constraints, screen out small farmers. Another example in which pre-conditions for the spread of technology have not been in evidence is SRI. In spite of being a promising technology, the complementary inputs and riskiness

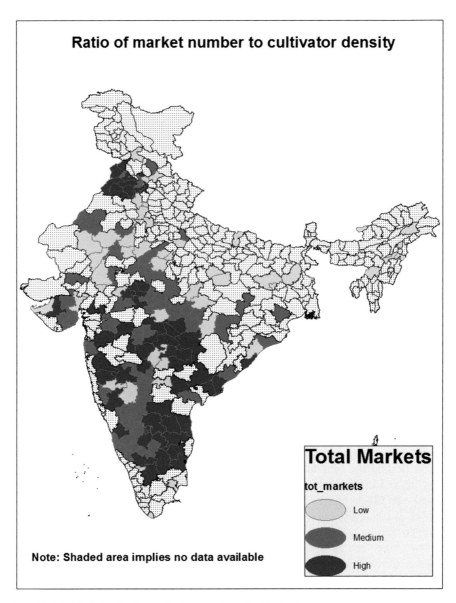

Fig. 18.4 Market density (normalized by number of cultivators)

inhibit large-scale adoption of this technology. With the basic information about technologies combined in Table 18.3, we developed the instruments for a primary survey of farmers regarding their awareness of and their adoption patterns for different technologies identified from the earlier workshops and expert elicitations.

Table 18.3 Synopsis of the technologies, including the factors for potential success

Name of the technology	Definition	Suitable for crop	Suitable area	Special features	Start/Field-experiments (year)	Current state	Comments
1. Water conservation interventions	Interventions which reduce water requirement to produce same or higher level of yield	Crops which are highly water intensive	Mostly in irrigated areas but some interventions can be applied under rain-fed situations	Reduce irrigation water requirement; Maintain water balance of the environment; Reduce GHG emissions due; Yield advantages are also observed in some cases			
1.1. Water conservation	Field water conservation through preparation of 20–25 cm dike around the field	Rice, wheat	Irrigated and rain-fed	Reduce run-off to increase effective use of water; Reduce irrigation water by 15–20 % without damaging yield	This is a traditional method which has been in use for a long time	Farmers' awareness has increased towards its adoption	
1.2. Precision water application	In this practice, farmers should determine when and how much to irrigate for optimal crop production, which can be done by irrigation scheduling	Rice, wheat, maize	Irrigated areas	Reduce excess use of water; Reduce methane emission from rice field; Around 20–52 % of water can be saved; In the case of wheat, yield can be increased up to 40 %	It has been advocated by scientists over the past decade	Awareness could spread due to poor public sector extension activity in Odisha	Many modern cultivation-methods, like direct-seeded rice and SRI, essentially involve precision-water application

(continued)

Table 18.3 (continued)

Name of the technology	Definition	Suitable for crop	Suitable area	Special features	Start/Field-experiments (year)	Current state	Comments
1.3. Mid-season drainage through AWD (alternate wetting and drying of field)	Drain out excess water to protect plants from soil and salinity effects; Effluent can be used for alternative purpose like ground water recharge	Rice, maize	Low land surface irrigated areas and rain-fed areas which are affected due to salinization or water logging	Increase crop yield; reduce salinization; Up to 42–45 % yield loss due to water logging or salinization can be protected	It has remained at the laboratory/trial level over the past decade	It has not expanded due to poor extension activity	
1.4. Laser Land-Leveling farming (LLL)	Leveling of land with laser leveler	Rice, wheat, maize	Upland areas both irrigated and rain-fed	Reduce water requirement by 20–25 % and reduce fuel use by 20 % without affecting crop yield	It is has been recently introduced	It has not spread due to lack of resources, including machinery and training facility	
1.5. Systems of Rice Intensification (SRI)	7–10-day-old seedlings are transplanted at spacing of 20 cm, with 1–2 seedlings per hill	Rice	Irrigated areas with excess labor availability for agriculture	Reduce water by 30 %; 25–30 % yield can be achieved; Poor management can reduce yield 10–20 %; Highly labor intensive	SRI was successfully field-tested in 2003 by NGO *Pradan* in remote Keonjhar and Mayurbhanj districts. Another NGO *Sambhav* has been promoting SRI since 2005; Government began focus on SRI in 2007–2008;	The government has lately begun promotion of SRI; It spent Rs. 1.6 crore in 2010–2011 and provided 6.6 crore in 2011–2012 for SRI promotion; The area under has been rising steadily in Odisha due to strong	

					Under NFSM, demos were done in the Rabi season 2007–2008 in the Kakahandi, Jajpur, Bolangir, Deogarh, Sundergarh & Dhenkanal districts; In the Kharif season 2008–2009, demos (no.) were done in Kalahandi (97), Ganjam (315), Deogarh (50), Dhenkanal (130), and Nayagarh (57)	promotional measures by government
1.6. Furrow irrigated bed planting	Growing crops on ridges or beds; Depending on each crop, height and width of the bed are maintained	Wheat, maize	Irrigated areas for wheat; Irrigated and rain-fed areas for maize	Wheat yield may increase up to 4 % and 30 % less water is required for wheat; 20 % less water is required for maize and better grain yield	Though this technology has not been focused upon, the alternate sink and bed technology was tested at the beginning of the 2000s – in this, rice is sown in a sink bed while pulses/vegetables are sown in alternate top beds	Due to lack of proper extension, the technology has not picked up

(continued)

Table 18.3 (continued)

Name of the technology	Definition	Suitable for crop	Suitable area	Special features	Start/Field-experiments (year)	Current state	Comments
2. Energy conservation	Technologies which help to reduce energy consumption during land preparation without affecting yield level; These also help to reduce water requirements for crops	Rice, wheat, maize	Irrigated and rain-fed areas	Reduce fuel and water consumption; Increase yield; Reduce GHG emission		Operational	
2.1. Direct seeded rice	Dry seed are sown either by broadcast or drilling in line	Rice	Rain-fed areas with no supplementary irrigation facility	Weeding is a critical problem which causes yield loss; Proper weeding can prevent yield loss	It has been done traditionally in Odisha – 80 % of rice-cultivated area in Odisha was under manual DSR in 1990; Mechanized DSR has been introduced in the last 2–3 years	Though manual broadcast is a popular method, particularly in the uplands, mechanized DSR has not taken root due to lack of proper planning and shortage of resources	
2.2. Zero-tillage	Crops grown in the zero till field; Happy seeder is also used for planting wheat seed; This practice also involves	Wheat, maize	Irrigated area	15–20 % water can be saved without affecting yield; Yield of wheat ranging from 7 to 12 tons/ha (or 5–7 % increase	Manual DSR is popular	Mechanized Zero-tillage has not reached the farmer level	

	incorporating residue of rice crops into rice wheat system			in yield as compared to conventional tillage); Diesel saving by 8 %; In case of Khariff, maize, 11 % yield can be increased		Operational
3. Nitrogen conservation	Technologies which supplement chemical fertilizer use for crops; Long-term benefits are greater than short-term benefits; Integrated nutrient management is key to achieving nitrogen-smart agriculture	Rice, wheat, maize	All types of area	Reduce chemical fertilizer use; Reduce on-farm and off-farm GHG emissions; Improve organic components in soil in the long run to prevent land degradation and yield loss		
3.1. Green manuring	Cultivation of legumes in a cropping system; This practice not only improves N economy but also has many other beneficial effects on soil health/quality	Rice	All types of area	Reduce fertilizer N use by 50–75 %; Integrated use of green manure increases yield 20 % in upper IGP and 8 % in lower IGP	Since start of the 2000s, the technique has been introduced along with brown manuring, i.e., cultivating Dhaincha along with rice	It is expanding but slowly

(continued)

Table 18.3 (continued)

Name of the technology	Definition	Suitable for crop	Suitable area	Special features	Start/Field-experiments (year)	Current state	Comments
3.2. Partial organic farming	Integrated use of FYM with chemical fertilizer to partially (25–50 %) compensate NPK requirements without affecting productivity	Rice, wheat, maize	All types of area	Improve soil health/quality; Reduce 25–50 % fertilizer requirement for rice and maize, 25 % for wheat	Farmers have traditionally used cow-dung as manure, but proper organic farming was introduced in the late 2000s	It has not spread in the absence of required promotion	
3.3. Complete organic farming	This technology involves use of nutrients from organic sources to 100 % compensate for chemical fertilizer use	Rice, wheat, maize	All types of area	No significant impact on yield; Improve soil health/quality	It has been demonstrated in recent years	It has not been adopted by farmers inthe absence of proper promotion and stimulus	
3.4. Leaf Colour Chart (LCC)	Techniques to estimate fertilizer requirement for crop	Rice, wheat, maize	All types of area	Reduce chemical fertilizer use by 30 %; Change in yield depends on farmers' crop cultivation practices			
4. Weather	This intervention provides services related to financial security and weather advisory for farmers.	Rice, wheat, maize	All types of area	Compensate financial loss of farmers; Help make decision about seed sowing, which has direct impact on yield			

4.1. Crop insurance	Crop specific insurance to prevent income loss of farmers	Rice, wheat, maize	All types of area	Cost of cultivation can be compensated from the losses due to climate change effects
4.2. Weather forecasting	Information and communication technology-based forecasting about the weather	Rice, wheat, maize	All types of area	Help farmers to make balance between rainfall and irrigation, which reduces irrigation water use; Proper time management gives yield advantage
5. Knowledge	This involves knowledge about intensive input use for achieving higher levels of yield and to choose seed variety to protect yield loss under difficult conditions	Rice	All types of area	Give higher level of yield; Increase input consumption, mainly fertilizer and water; Protect crop yield under sub mergence and dry situations
5.1. Fully intensive agriculture	Increase chemical fertilizer use by 100 %	Rice, wheat	Irrigated areas	Up to 20 % yield can be increased; Cost of fertilizer will increase
5.2. Partial intensive agriculture	Increase chemical fertilizer use by 50 %	Rice, wheat	Irrigated areas	Up to 10 % yield can be increased;

(continued)

Table 18.3 (continued)

Name of the technology	Definition	Suitable for crop	Suitable area	Special features	Start/Field-experiments (year)	Current state	Comments
				Cost of fertilizer will increase			
5.3. Flood-tolerant variety	Seed variety which is tolerant to flood or heavy rainfall situation	Rice	Rain-fed areas or irrigated areas when flood situation is severe	Yield loss can be prevented in a flood year	Various such varieties were developed and introduced in the 1990s and 2000s	Adoption has increased over the years; farmers have benefitted	
5.4. Drought-tolerant variety	Seed variety which is tolerant to drought or relatively dry weather situations	Rice	Drought prone areas	Yield loss can be prevented in a drought year or under dry situations	Various cultivars were introduced in last two-three decades	Adoption has increased	

Findings from the Farmer Survey

After shortlisting the technologies and summarizing the factors behind technology adoption, we conducted primary surveys in three marginal districts each in both Bihar and Odisha. The structured questionnaires were designed to assess the state of awareness about different technologies and the level of adoption directly from farmers.

Cropping Choices of Farmers

Figure 18.5 presents the distribution of farmers by crop choices. As expected, the majority of farmers are engaged in cereal production, i.e., rice and wheat in Bihar and only rice in Odisha. Importantly, more than 40 % of the farmers in Bihar are also engaged in maize production. Maize has been a revolutionary crop in Bihar in recent times and has surpassed the growth rate of production of other primary crops. In comparison, the uptake of maize in Odisha is smaller in the three districts with less than 4 % of the farmers in our survey cultivating it. Another crop that has generally been neglected in both Bihar and Odisha is pulses, but technology interventions may turn out to be quite important for this crop.

Summary Statistics from Farmer Survey

Apart from current crop choices, the survey also asked farmers which crops or activities they are not engaged in but would like to get into. This could suggest the crops/activities for which the potential of technologies should be assessed, since farmers have expressed their willingness to get into these. Maize stands out as the crop, both in Bihar and Odisha, that a large proportion of farmers want to get involved in (Table 18.4). Note that the question is directed to the farmers not currently cultivating maize. Apart from maize, the sector that farmers want to get into if not currently engaged with is pulses. For Bihar and Odisha combined, more than 18 % of farmers would like to get into pulses if they are currently not engaged with them. In looking for promising technologies for marginal districts in the two states, the ones related to maize and pulses should thus be given due attention.

Next, we analyze the status of technologies in terms of awareness and uptake from the point of the farmers from selected marginal districts in the two states.

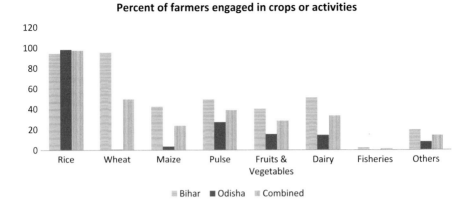

Fig. 18.5 Distribution of farmers by crop activity in Bihar and Odisha

Table 18.4 Farmer's revealed choices for activities other than the current activity in marginal districts	Bihar	Odisha	Combined
Maize	21.03	23.79	22.58
Pulses	7.68	29.59	18.50
Fruits and vegetables	13.52	42.75	27.99
Dairy	13.52	10.64	12.23
Fisheries	19.01	2.71	11.2
	3.11	4.45	3.8

Source: Field survey data

Awareness of Technologies and Their Adoption in Bihar and Odisha

Demographic, Asset and Amenities Characteristics

Characteristics such as education and experience in farming determine the likelihood of farmers being aware of certain technologies and adopting them. In addition, scale of production (reflected in size of landholdings) and social identity also seem to play a role. Access to credit, markets, information and other complimentary factors determining adoption of technology are often functions of land size and the social identity of the farmer.

About 30 % of the farmers in the Bihar sample are completely or partially illiterate and only 6 % have some college experience. Similarly, about 30 % of farmers in the Odisha sample are illiterate. These farmers, however, have significant experience in farming. There is less experience with spring and winter maize, but those cultivating them, on average, have over one and a half decade's experience with the crop (Fig. 18.6).

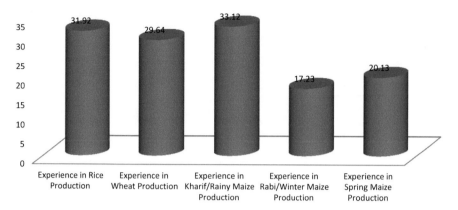

Fig. 18.6 Experience in farming in Bihar

Though a large share of farmers in the surveyed districts in Bihar own land (greater than 82 %), there is a significant share of sharecroppers in the sample (nearly 12 %). The incidence of sharecropping is lower in Odisha at less than 6 %. The rental contracts have high payment (either half or one third of the production) and risk-sharing is minimal. The ability to experiment with new technology would, thus, be quite limited for the renters. The evidence on land reforms and/or tenancy in regard to productivity has, however, been mixed (see Ghatak and Roy 2007).

In addition, the land markets are quite thin, with only 7 % of the farmers in the three districts owning purchased land. The case is starker in Odisha, with less than 2 % of the farmers having purchased or sold land. A technology that is scale dependent (for example, LLL) would face roadblocks in the current land ownership, market and tenure situation.

Another important characteristic that needs to be kept in mind about the marginal districts surveyed is that over 20 % of the land is low-lying and 50 % is medium-lying. In the case of Odisha, this figure is extremely high at nearly 40 %. Furthermore, in Odisha's surveyed districts, over 35 % of the land has sandy soil, a characteristic that has implications for choice of technology.

On the positive side, the availability of irrigation, even in these marginal districts in Bihar, is high. Nearly 95 % of the farmers surveyed have access to some form of pump irrigation, either on their own or by hiring. However, the intensity of irrigation (i.e., the number of times a plot is irrigated) is sub-optimal. In sharp contrast, more than 95 % of the land in the marginal districts surveyed in Odisha are rain-fed (Fig. 18.7). There is, however, a greater coverage of canal irrigation, at 14 %, in Odisha than in Bihar.

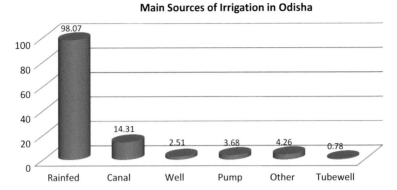

Fig. 18.7 Sources of irrigation in Odisha's surveyed marginal districts

Land Size, Social Identity, Awareness and Adoption of Technology

Based on our farmer survey, there are several salient findings regarding the association of socio-economic characteristics with technology choices, as follows.

(i) For the lowest caste group SC/ST, the proportion of farmers in both a high and a low state of landholdings (4th and 1st quartile of land distribution), there is a general lack of awareness about technology. (See Bhadauria 2013 for an account of differences in technology adoption across castes in agriculture). There are several barriers to technology adoption based on caste, some of which have been studied rigorously, for example, credit in Kumar (2013).

(ii) The level of awareness as well as implementation of technology for the other backward caste is high, at times greater than the higher caste farmers, but that could be related to the level of engagement with farming vis-à-vis non-farm activities.

(iii) Some technologies, like mechanized DSR and SRI, have low conversion from awareness to implementation. In general, translation from awareness to implementation rises with the holding size. Farmers in any social group are more likely to adopt a technology the greater their landholding is.

Fig. 18.8 presents the awareness of technologies across crops.[5] There is a definite pecking order in awareness of technology, with farmers belonging to the lowest

---

[5] The test for awareness of technology was based on the surveyed farmer being able to explain the technology in a way that revealed their knowledge of its details. For example, several farmers confused the simple traditional way of broadcasting with the mechanical direct seeding of rice. Enumerators had to ensure that the knowledge of the technology was not being confounded with similar but technically different methods.

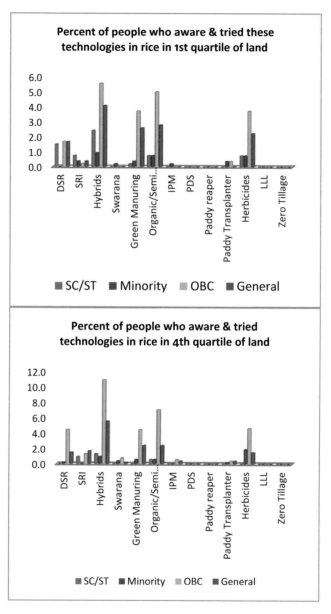

Fig. 18.8 Percentage of farmers adopting technology conditional on awareness

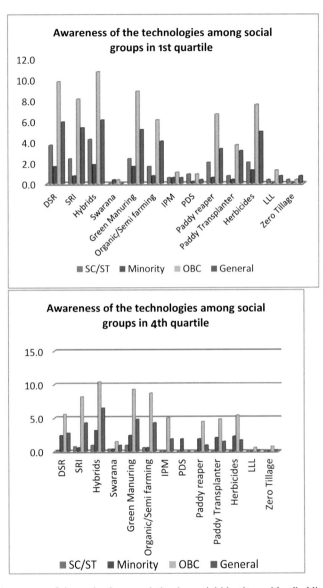

Fig. 18.9 Awareness of rice technology: variation by social identity and landholdings

caste and the minority groups being generally less aware. Moreover, some of the disadvantages based on social identity seem to be mitigated as land size increases. Larger low caste farmers are generally more aware of technologies (Fig. 18.9).

Awareness and Adoption Among Farmers in Marginal Districts: Technologies in Rice

Apart from farm and farmer characteristics, underlying the awareness and adoption of technologies are the roles of different social and institutional networks. As a source of information, expectedly, the most common networks comprise friends and neighbors. The public extension services are hardly the main sources of information, being more so in Odisha. This reinforces the possibility discussed in various circles that one of the most important factors preventing adoption of promising technologies is the lack of extension support. According to Birner and Anderson (2007), the public extension system in India has been unable to keep pace with the changes in the global technological and economic environments. With increasing demand for information, the extension system has evolved to include various information sources, including public and private, formal and informal, and traditional and modern.

In Bihar, while over 5 % of farmers got the information about their chosen technology from public sector extension services, in Odisha, it was less than 2 %. Print and electronic media account for a meager 4 % of the information farmers obtained about rice technologies (those who actually adopted them).

There are several technologies in rice for which awareness is strikingly low. Consider, for example, integrated pest management (IPM). Though shortlisted as a frontal technology, in both states, less than 7 % of farmers are aware of IPM. Also striking is the extremely low level of awareness of the Swarna submersible variety of rice. It is possible that only those farmers who have low-lying land prone to flooding would seek information on such varieties. Indeed, awareness of this variety is marginally higher (by about 2 percentage points) among farmers with low-lying land.

Also, upon awareness, adoption is not automatic. Figure 18.10 shows that, for some much publicized technologies in rice, for example, mechanized DSR and SRI, conditional on awareness, adoption is almost negligible (more so for the latter). In particular, SRI has extremely low adoption among farmers in the marginal districts, even though it has been marketed as a technology with very high potential in the scientific community. Table 18.5 above shows the pitfalls in this technology, wherein it would require several complementary inputs if it were to be adopted. Farmer surveys also show that farmers are wary of the downside risk with SRI. If, for example, there are untimely rains when young saplings are planted, the losses would be quite significant. A contrast to such situations in adoption of technology is hybrid rice in Bihar. More than 60 % of the surveyed farmers have adopted hybrid

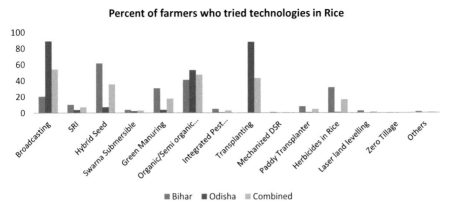

Fig. 18.10 Percentage of adopters of technologies in rice

Table 18.5 Constraints in technology adoption in rice

Constraints	Bihar	Odisha
Technology expensive and/or credit constraints	72.58	94.20
Lack of economic availability of inputs	51.37	54.55
High downside risk	43.51	46.81
Higher levels of production but markets for output not commensurate	16.27	12.57
Information constraints	35.10	33.85
Technology relatively new and I do not want to be the experimenter	10.79	40.23
Observation of several failures of technology	5.48	6.19
Technology delivers positive gains but far below the promise	1.46	4.06
Others	2.01	0.19

rice. This figure is especially striking since hybrid rice has been introduced only recently. However, the adoption of hybrid rice in Odisha is comparatively low (Fig. 18.10).

Furthermore, almost 65 % of the farmers in marginal districts who are currently not cultivating hybrid rice would like to adopt it. It is, in fact, the most aspirational technology for farmers in marginal districts of both states. Organic/semi-organic farming is another technology which the non-practitioners aspire to (30 % and 28 % farmers in Bihar and Odisha, respectively). Note that aspiration for technologies does not imply that the respective farmers have the right set of conditions to actually adopt them. Importantly, some technologies, like zero tillage and LLL, are not sought after, despite again being cases in which high potential has been accorded to them by scientists.

With awareness and aspiration in place, what are the factors that inhibit technology adoption? We compiled a list of possible inhibitors in this context. The set of constraints are given below in Table 18.5, any number of which could be working in combination. Results on the farmers' responses regarding the bind of different constraints in adoption of technologies in rice are presented below

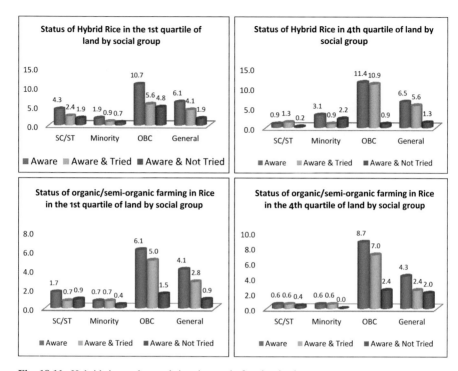

Fig. 18.11 Hybrid rice and organic/semi-organic farming in rice

(Table 18.5). In both states, the hindrances to adoption of technologies in rice are broadly similar (with only a few exceptions). Credit is a bigger constraint in Odisha than it is in Bihar. Also, the farmers in marginal districts of Odisha are comparatively averse to experimentation. Almost identical proportions of farmers in both Bihar and Odisha face information problems with regard to technologies in rice (nearly 35 %).

Similar to hybrid rice, organic and semi-organic farming is also marked by awareness commonly translating into adoption. In rice, the two technologies that could be promoted further in the marginal rice districts are hybrid varieties and organic/semi-organic farming, as there is near universal uptake across socio-economic categories (Fig. 18.11). While the awareness and uptake of organic/semi-organic rice farming is similar in Odisha, hybrid rice has extremely low uptake by farmers in the marginal districts of Odisha, at 7 % compared to over 61 % in Bihar.

Percent of farmers aware of these technologies in Wheat

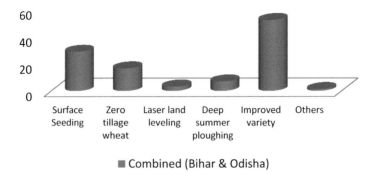

■ Combined (Bihar & Odisha)

Fig. 18.12 Percentage share of farmers aware of technologies in wheat

Percent of farmers who tried these technologies in Wheat

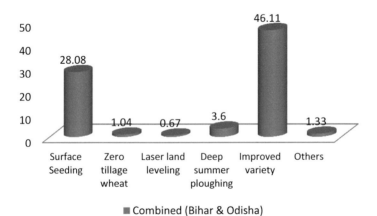

■ Combined (Bihar & Odisha)

Fig. 18.13 Share of farmers trying technologies in wheat

Awareness of Technologies in Wheat: Zero Tillage Wheat, LLL and Other Technologies

Repeating the same exercise used for rice, Figs. 18.12 and 18.13 present the awareness of wheat technologies and their adoption. Among the technologies, the highest awareness and adoption levels belong to improved varieties in wheat. Technologies like LLL have not taken off in wheat (as they have failed to do in rice) in the marginal districts of Bihar. In fact, a very small proportion of farmers are even aware of LLL. Furthermore, in contrast with the findings based on expert

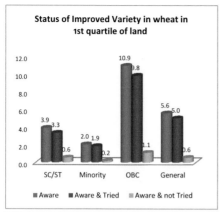
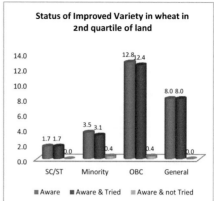
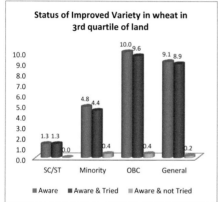
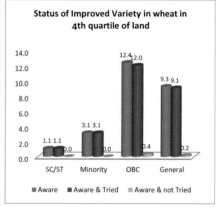

Fig. 18.14 Awareness and adoption of improved varieties in wheat

elicitation, only about 20 % of farmers are aware of mechanized zero tillage technology (MZTT). Both in the case of MZTT as well as that of LLL, data shows an extremely low share of farmers trying these methods.

In the marginal districts, what mechanized DSR and SRI are to rice, zero tillage and LLL are to wheat. Projected as highly promising, the awareness and adoption levels are extremely low for these technologies. In the lowest caste groups, there is absolutely no awareness of these technologies. Even in the higher caste categories and in the highest land size group (the 4th quartile of land distribution), the scenario is similar (data not presented here).

The most important technology for wheat that has widespread awareness and good conversion from awareness to adoption in the marginal districts is improved varieties. The farmers are more commonly updated with improved varieties and reveal a high propensity for adopting them. The awareness and subsequent adoption of improved varieties in wheat is indiscriminately high across social groups and land sizes (Fig. 18.14).

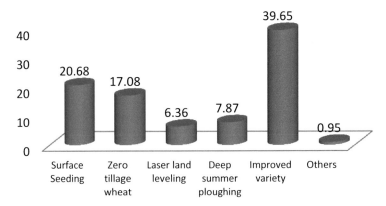

Fig. 18.15 Proportion of informed farmers aspiring for different untried technologies in wheat

Moreover, in deciding about technologies for marginal areas, one criterion can be conditional on information about a technology, namely what fraction of farmers would likely adopt it. Clearly, technologies that farmers reveal their aspirations for are more likely to be adopted. Figure 18.15 presents the percentage of farmers who find specific technologies to be most promising and would likely adopt them. The demand pull is clearly weak for LLL and deep summer ploughing.

Promoting improved varieties of wheat would likely have the greatest uptake among farmers. Other technologies such as surface seeding and ZTWT are also likely to find some acceptance. These revealed preferences are, however, based on the existing information sets of the farmers. If the information sets of the farmers themselves were to be altered, that would change the situation in regard to the valuation of the farmers.

Farmer's Choice of Technologies in Maize

An important role played by technology has been to make maize a multi-seasonal fixture, as opposed to a rainy season-specific crop. Thus, the analysis of maize has to be done by season. Across states, there are significant differences in both awareness as well as application of maize technologies between Bihar and Odisha. Based on the survey, in the marginal districts, over 42 % of farmers are engaged in maize cultivation in Bihar. The corresponding figures for Odisha are less than 3.5 %. Yet, maize remains a promising crop in Odisha, with nearly 24 % of farmers not engaged currently aspiring to get into it.

Awareness of hybrid varieties in maize is quite widespread, particularly in Bihar, where more than half of the farmers surveyed know about them. In both Bihar and Odisha, an area where a large scope remains for improving the knowledge set of farmers, their adoption is related to the high value of corn products like Quality Protein Maize for feed, in particular, baby corn and sweet corn, respectively. The percentage of farmers who think of these as profitable options is in the single digits.

Importantly, in the marginal districts of both Bihar and Odisha, we do not have a single farmer in our sample who has tried producing baby corn or sweet corn. These are high value corn items and promoting them can augment farm incomes significantly. This is true for the cases of both winter and spring maize. In terms of the constraints inhibiting adoption of hybrid maize, the survey shows the main reason is lack of availability of seeds.

Farmer's Perspective and Choice of Technologies in Pulses

More than half of the farmers in Bihar in our sample are engaged in pulses. The corresponding number for Odisha is much lower at 27 %. In both states, pulses need to fit into the cereal production cycle. Importantly, a technology that has been generally suggested to improve the outcomes in production is intercropping of pulses with different crops. In both states, the survey results show that the incidence of intercropping is limited. Less than 1 % of pulse farmers engage in intercropping, mainly with rice, maize or vegetables. This is one area in pulse production that needs to be scaled up.

Our survey did not ask specific questions about awareness regarding technologies in pulses. In terms of adoption of varieties, the dominance of local seeds is paramount, particularly in Mung, a very important pulse in Bihar and Odisha. Among pulse-growing farmers, more than 95 % of farmers use local seeds and improved varieties are rarely chosen.

Regression Analysis for Awareness and Adoption of Technologies in Rice, Wheat and Maize in Bihar and Odisha

Below, we explore the determinants of technology adoption in a more rigorous fashion through regression analysis, looking for general factors that are associated with awareness of technology, as well as its adoption. As discussed above, there are significant context specificities across technologies. Yet, there are also broad generalities in the context of technologies that can help us to understand the typology of the awareness and adoption of technologies. Regressions are aimed at identifying the target groups and characteristics that are linked with the knowledge of farmers about technologies and their propensities in adopting them.

Tables 18.6, 18.7, and 18.8 present the marginal effects of the variables in a probit regression for awareness and adoption of technologies in rice, wheat and maize in the three marginal districts of Bihar and Odisha. We pick up three frontline technologies in each crop to assess the level of association of each technology with characteristics of the farmers. Some of the associations with farmer, plot and location characteristics are crop- and technology-specific, but some are more generic. Specifically, the technologies analyzed below are, in rice, hybrid rice, SRI and DSR; in wheat, surface seeding, improved varieties and zero tillage; and in maize, hybrid maize, optimal spacing and nutrient management.

Specifically, we are interested in the characteristics of the former, such as land size and social identity, apart from demographic characteristics such as age, experience in farming and levels of education (human capital). As part of the pathway towards the uptake of technology, we also assess whether awareness, as well as adoption, is associated with different information sources, viz, public extension services, private information sources (such as input dealers) and social networks such as friends and neighbors. There is a growing body of literature that finds evidence of social learning in technology adoption (see, for example, Foster and Rosenzweig 2010; Pomp and Burger 1995).

The probit regressions in Tables 18.6, 18.7, and 18.8 control for several farmer and farm characteristics. Importantly, observed and unobserved regional characteristics are accounted for in all regressions with the inclusion of block fixed effects. Furthermore, given the possibility of correlated unobserved factors at the regional level, all standard errors are clustered at the block level. Hence, the agro climatic conditions and external factors such as distance from the markets and other time invariant location characteristics are controlled for in these regressions. Our main variables of interest are landholding characteristics, social identity variables and the main sources of information for the technology. There is very little variation in the size of landholding per plot. Yet, to take into account the differences that result from landholding, we include a number of plots owned by a farmer for which there is much greater variation in the cross-sectional data.

Several stylized facts and general features emerge from the basic regression analysis for the technologies in the three crops. These are as follows:

1. Landholding characteristic in the form of number of plots is positively and significantly associated with awareness, as well as adoption, of most technologies across the three crops. In most technologies, farmers with greater number of plots tend to be aware of technology and are also more likely to adopt. A greater number of plots likely allows for experimentation with new technology by spreading out risk. Only in the case of maize did a greater number of plots not have a distinctive effect on adoption of technologies. Land size per se does not have a significant effect on choice of technology, but that is probably because of the nature of the data, which in these marginal districts comprises very little variation in the holding sizes across households.

Table 18.6 Awareness and adoption of technologies in rice: marginal effects from probit regressions

	DSR	SRI	Hybrid	Tried DSR	Tried SRI	Tried Hybrid rice
Household size	−0.007 [0.021]	0.022 [0.018]	0.008 [0.024]	0.003 [0.02]	0.004 [0.03]	−0.02 [0.01]
Household head age	−0.011* [0.006]	−0.014*** [0.005]	−0.009 [0.008]	0.012 [0.01]	−0.01 [0.008]	0.013** [0.007]
Household head experience in farming	0.021*** [0.003]	0.010* [0.006]	0.011 [0.007]	−0.0 [0.005]	0.011 [0.01]	−0.02*** [0.007]
Number of plots	0.037*** [0.011]	0.047*** [0.008]	0.012 [0.012]	0.024** [0.012]	0.003 [0.017]	0.026*** [0.010]
Total plot area	−0.032 [0.031]	−0.032 [0.022]	0.001 [0.017]	−0.01 [0.02]	0.042 [0.029]	0.022 [0.025]
Receiving non-agricultural income	−0.44*** [0.128]	−0.118 [0.184]	−0.283** [0.115]	−0.117 [0.149]	0.553*** [0.196]	0.433*** [0.154]
Household member receiving some income from NREGA	−0.191 [0.210]	−0.213** [0.103]	0.096 [0.130]	0.214 [0.144]	0.211 [0.283]	0.117 [0.147]
Dummy for clay soil	−0.188 [0.365]	−0.095 [0.261]	−0.138 [0.273]	0.212 [0.315]	−0.328 [0.742]	−0.177 [0.248]
Dummy for loamy soil	−0.062 [0.167]	0.410* [0.243]	−0.061 [0.239]	0.259 [0.245]	0.250 [0.398]	0.029 [0.229]
Dummy for sandy soil	0.081 [0.260]	0.550** [0.234]	0.151 [0.312]	−0.185 [0.336]	0.293 [0.531]	0.349 [0.283]
Dummy scheduled caste (=1 if yes, 0 if no)	−0.009 [0.252]	−0.63*** [0.196]	−0.359* [0.207]	0.116 [0.278]	−0.243 [0.471]	−0.676** [0.288]
Dummy scheduled tribe (=1 if yes, 0 if no)	−0.056 [0.471]	−0.506* [0.275]	−0.472 [0.417]	−0.143 [0.299]	−0.241 [0.577]	−0.083 [0.287]
Dummy other backward caste (=1 if yes, 0 if no)	−0.226 [0.167]	−0.251* [0.132]	−0.095 [0.242]	−0.085 [0.169]	0.017 [0.299]	−0.081 [0.174]
Dummy minority (=1 if yes, 0 if no)	0.041 [0.359]	−0.269 [0.294]	−0.268 [0.427]	−0.234 [0.46]	−0.536 [0.543]	−0.295 [0.288]
Dummy no schooling (=1 if yes, =0 if no)	−0.02 [0.38]	0.05 [0.327]	0.28 [0.26]	−0.503 [0.359]	5.115*** [0.414]	−0.241 [0.301]
Dummy middle school (=1 if yes, =0 if no)	−0.089 [0.383]	0.329 [0.333]	0.237 [0.240]	−0.695* [0.345]	5.4*** [0.302]	0.004 [0.34]

(continued)

Table 18.6 (continued)

	DSR	SRI	Hybrid	Tried DSR	Tried SRI	Tried Hybrid rice
Dummy high school (=1 if yes, =0 if no)	−0.190 [0.423]	0.32[0.318]	0.416 [0.321]	−0.527 [0.355]	5.6*** [0.365]	0.213 [0.370]
Dummy intermediate degree (=1 if yes, =0 if no)	0.040 [0.347]	0.577* [0.343]	0.471 [0.344]	−0.73** [0.373]	5.73*** [0.347]	0.088 [0.426]
Dummy for bachelor's degree (=1 if yes, =0 if no)	0.131 [0.468]	0.701** [0.315]	1.049** [0.434]			
Dummy for information from friend/neighbor (1 = yes, =0 if no)	0.373** [0.148]	0.367** [0.149]	0.530*** [0.155]	0.254 [0.184]	−0.282 [0.248]	0.215* [0.129]
Dummy for information from public extension (=1 if yes, =0 if no)	−0.152 [0.373]	0.438 [0.388]	0.640 [0.530]	0.552 [0.482]	0.692* [0.413]	0.499 [0.483]
Dummy for information from private extension (=1 if yes, =0 if no)	−1.282* [0.757]	0.265 [0.624]	−0.489 [0.439]	−0.97*** [0.31]	0.99[0.7]	1.142*** [0.371]
Yield range in rice				−0.013 [0.071]	−0.078 [0.061]	−0.062 [0.051]
Block fixed effects	Yes	Yes	Yes	Yes	Yes	Yes
Observations	713	818	803	708	636	738

Robust standard errors in parentheses – ***$p < 0.01$, **$p < 0.05$, *$p < 0.1$. All standard errors are clustered at the district level

2. Younger farmers are more likely to be aware of the technologies. Age has no effect on knowledge about certain technologies, or, if it does affect, it does not do so negatively. In general, experience in farming also raises the probability of being aware of technologies and the likelihood of adopting them.
3. One of the broadly generalizable results is the social bias in awareness, as well as adoption, of technology. There are several technologies for which evidence shows that the lowest caste strata is disadvantaged in terms of having knowledge about technologies, as well as adopting them. For example, this is true for awareness of SRI and hybrid rice. Hybrid rice, the spread of which has been extensive in Bihar, still has a situation in which the scheduled caste and scheduled tribe households have a significantly lower probability of being

Table 18.7 Awareness and adoption of technologies in wheat: marginal effects from probit regressions

	Improved variety	Surface seeding	Zero tillage wheat	Tried improved variety	Tried surface seeding
Household size	0.071** [0.030]	−0.029 [0.023]	−0.014 [0.026]	0.008 [0.056]	−0.042 [0.026]
Household head age	−0.021** [0.010]	−0.005 [0.010]	−0.013 [0.009]	−0.007 [0.012]	−0.004 [0.007]
Household head experience in farming	0.005 [0.007]	0.013 [0.009]	0.022*** [0.008]	0.011 [0.011]	0.017** [0.008]
Number of plots	0.051*** [0.013]	0.031** [0.012]	0.043*** [0.012]	0.121*** [0.039]	0.036* [0.020]
Total plot area	−0.003 [0.031]	0.024 [0.029]	0.010 [0.032]	0.072 [0.074]	0.051 [0.036]
Receiving non- agricultural income	0.058 [0.203]	−0.116 [0.175]	−0.56*** [0.209]	0.351 [0.241]	−0.455** [0.189]
Household member receiving some income from NREGA	−0.543** [0.228]	−0.52*** [0.167]	−0.069 [0.256]	−0.092 [0.534]	−0.056 [0.736]
Dummy for clay soil	1.328*** [0.288]	0.752 [0.507]	−0.490 [0.537]	−0.092 [0.534]	−0.056 [0.736]
Dummy for loamy soil	0.530* [0.286]	0.346 [0.374]	0.213 [0.388]	0.303 [0.567]	−0.377 [0.395]
Dummy for sandy soil	−0.012 [0.344]	0.904* [0.465]	−0.011 [0.498]	−0.343 [0.579]	0.327 [0.495]
Dummy scheduled caste (=1 if yes, 0 if no)	−0.244 [0.274]	−0.285 [0.242]	−0.70*** [0.228]	0.190 [0.554]	0.075 [0.320]
Dummy scheduled tribe (=1 if yes, 0 if no)	−0.641** [0.262]	−0.532 [0.577]		−0.150 [0.700]	
Dummy other backward caste (=1 if yes, 0 if no)	0.324* [0.186]	0.192 [0.187]	−0.069 [0.192]	0.335 [0.368]	0.209 [0.250]
Dummy minority (=1 if yes, 0 if no)	0.652 [0.450]	0.270 [0.223]	−0.624 [0.420]	−0.049 [0.537]	0.020 [0.264]
Dummy no schooling (=1 if yes, =0 if no)	−1.047** [0.410]	−0.525* [0.309]	0.130 [0.528]	−6.6*** [0.64]	−1.37*** [0.424]
Dummy middle school (=1 if yes, =0 if no)	−0.856** [0.415]	−0.496* [0.276]	0.531 [0.632]	−6.67*** [0.654]	−1.24*** [0.458]
Dummy high school (=1 if yes, =0 if no)	−0.724 [0.448]	−0.500 [0.329]	0.194 [0.552]	−6.38*** [0.67]	−1.207** [0.562]
Dummy intermediate degree (=1 if yes, =0 if no)	−0.743 [0.581]	−0.294 [0.265]	0.931 [0.728]	−6.76*** [0.81]	−1.074* [0.571]
Dummy for bachelor's degree (=1 if yes, =0 if no)	−0.614 [0.512]	0.237 [0.270]	1.139** [0.569]	−6.008*** [1.055]	−0.179 [0.371]
Dummy for information from friend/neighbor (1 = yes, =0 if no)	0.352* [0.192]	0.902*** [0.163]	1.075*** [0.161]	1.03*** [0.20]	1.134*** [0.174]
Dummy for information from public extension (=1 if yes, =0 if no)	0.393 [0.657]	0.951* [0.507]	1.006*** [0.262]	1.306 [0.81]	1.100* [0.601]

(continued)

Table 18.7 (continued)

	Improved variety	Surface seeding	Zero tillage wheat	Tried improved variety	Tried surface seeding
Dummy for information from private extension (=1 if yes, =0 if no)	0.680 [0.540]	0.896 [0.604]			1.077 [0.807]
Block fixed effects	Yes	Yes	Yes	Yes	Yes
Observations	720	671	420	397	411

Robust standard errors in parentheses – ***p < 0.01, **p < 0.05, *p < 0.1. All standard errors are clustered at the district level

adopters. Recall that as much as 61 % of farmers in the marginal districts of Bihar have adopted hybrid rice, while this number is below 15 % for SC/ST farmers in Bihar. Wheat and maize are less stratified in technology adoption along caste lines. Though not presented here, the case is similar for IPM in rice.

4. Furthermore, knowledge about complex technologies is ordered not only along lines of caste but also educational qualification. Awareness of SRI, for example, is significantly associated with high school or a higher degree.

5. In terms of information sources, importantly, the private sector comprising input dealers and other non-government sources, such as the media, generally have not played a significant role. The awareness of technologies in rice has come about through strong social ties and the public sector extension services. Interestingly, in many hybrid seeds, the scaling up of adoption has happened because of an active private sector. In the adoption of hybrid rice, the private sector extension has played a significant role. This is to be expected, since in hybrid seeds, the private sector is the dominant input supplier. Surprisingly, the private sector extension does not have a significant effect on the awareness or adoption of hybrid maize whereever strong social ties with friends and neighbors alone have significant association with a farmer's knowledge about the technology.

6. Relative to the benchmark location (the omitted category in the block dummies), there are significant differences across blocks. The positive significant dummy implies that these blocks have significantly greater likelihood of being aware of/adopting the technology. With regard to the respective technologies, these blocks are in some way more marginal than the benchmark blocks. The evidence in this regard provides a basis for prioritizing such blocks which have low awareness or adoption. At the same time, to achieve maximize adoption, the blocks with greater likelihood of awareness and subsequent adoption could be targeted.

Finally, we look at the downside of respective technologies in terms of farmers' perceptions and their association with adoption (or lack of it). Here, we look merely at association rather than a causal relationship between perceptions of the flip side in a technology and its adoption. The perceived downside of a technology could be strong enough to be associated with lack of adoption. On the other side, if some

Table 18.8 Awareness and adoption of technologies in maize: marginal effects from probit regressions

	Hybrid	Spacing	Nutrient management	Tried hybrid maize	Tried spacing maize	Tried nutrient management maize
Household size	0.020 [0.014]	0.079* [0.046]	0.135* [0.075]	0.038 [0.026]	0.044 [0.064]	0.107[0.067]
Household head age	−0.006 [0.011]	−0.040** [0.017]	−0.004 [0.032]	−0.01 [0.013]	−0.038 [0.02]	−0.026 [0.028]
Household head experience in farming	0.013 [0.009]	0.022* [0.013]	−0.035 [0.038]	0.010 [0.01]	−0.022 [0.03]	−0.011 [0.03]
Number of plots	0.014 [0.012]	0.059*** [0.016]	0.032* [0.018]	0.015 [0.01]	0.058 [0.03]	0.059** [0.02]
Total plot area	0.037 [0.033]	−0.044 [0.042]	−0.120** [0.055]	0.04** [0.02]	0.12 [0.096]	0.038[0.063]
Receiving non-agricultural income	−0.221 [0.146]	0.149 [0.229]		−0.5*** [0.12]		
Household member receiving some income from NREGA	−0.297* [0.168]	0.422 [0.485]	1.452 [1.228]	−0.25 [0.378]	1.671* [0.99]	1.783[1.147]
Dummy for clay soil	−0.631 [0.395]			−0.255 [0.67]		
Dummy for loamy soil	−0.131 [0.257]	−0.498 [0.570]	5.031*** [1.272]	0.481 [0.343]	5.36*** [0.9]	5.30*** [0.97]
Dummy for sandy soil	−0.141 [0.254]	−0.195 [0.644]	4.348** [1.908]	0.676 [0.420]	5.302[.]	4.342[.]
Dummy scheduled caste (=1 if yes, 0 if no)	−0.352 [0.288]	−0.398 [0.691]		−0.34* [0.20]		
Dummy scheduled tribe (=1 if yes, 0 if no)	0.152 [0.544]			0.432 [1.206]		
Dummy other backward caste (=1 if yes, 0 if no)	0.045 [0.215]	−0.369 [0.316]	−1.084 [0.794]	0.170 [0.258]	−1.86*** [0.6]	−1.244 [0.763]
Dummy minority (=1 if yes, 0 if no)	0.700** [0.315]	−0.299 [0.885]	−0.927 [1.339]	0.71** [0.36]	−1.18 [1.43]	−0.908 [1.416]
Dummy no schooling (=1 if yes, =0 if no)	−0.179 [0.281]	−0.140 [0.415]	0.868 [0.795]	0.294 [0.420]	1.71* [0.99]	0.978[0.826]
Dummy middle school (=1 if yes, =0 if no)	0.16 [0.342]	−0.538 [0.492]	−0.379 [0.748]	0.108 [0.416]	−0.009 [0.95]	−0.963 [0.911]

(continued)

Table 18.8 (continued)

	Hybrid	Spacing	Nutrient management	Tried hybrid maize	Tried spacing maize	Tried nutrient management maize
Dummy high school (=1 if yes, =0 if no)	0.164 [0.335]	−0.657* [0.377]	−0.715 [0.885]	0.073 [0.599]	−0.017 [0.61]	−0.424 [0.587]
Dummy intermediate degree (=1 if yes, =0 if no)	0.322 [0.403]	−1.379** [0.653]		0.492 [0.460]		
Dummy for bachelor's degree (=1 if yes, =0 if no)	0.329 [0.306]	−0.810 [0.604]	0.838 [1.135]	0.250 [0.466]	2.598 [1.641]	0.782[1.321]
Dummy for information from friend/neighbor (1 = yes, =0 if no)	0.383** [0.151]	−0.306 [0.352]	0.057 [0.562]	0.119 [0.158]	−0.636 [0.65]	−0.376 [0.603]
Dummy for information from public extension (=1 if yes,0 if no)	0.880* [0.459]	0.657 [0.736]	1.062 [1.020]	−0.595 [0.42]	−0.347 [1.15]	0.371[1.012]
Dummy for information from private extension (=1 if yes, =0 if no)						
Block fixed effects	Yes	Yes	Yes	Yes	Yes	Yes
Observations	515	305	123	364	108	123

perceived constraints are also significantly associated with adoption, it would imply that the scale of adoption could be higher if those downsides of the technology were addressed.

The downside of technologies could be among the perceived constraints in adoption; in IPM, the farmers who see this as a new technology that would force them to engage in experimentation are less likely to adopt this technology. In hybrid rice, the propensity to adopt is associated with perceptions of high costs and lack of marketing opportunities for the product. Those who see downside risk in hybrid rice are significantly less likely to be adopters of hybrid rice. In mechanized DSR, lack of adoption is associated with information constraints, as well as a perception that the technology performs far below the potential.

Table 18.7 looks at the awareness and adoption of technologies in wheat. For illustration, we look at the case of two technologies, viz. surface seeding (SS) and improved varieties in wheat (IV). Again, the other backward caste turns out to be the group with greater awareness of technologies. It is also associated with greater adoption of technologies in wheat. As in the case of rice technologies, the greater number of plots the farmer has raises the likelihood of the farmer trying new technologies. Furthermore, importantly, the main providers of information regarding technologies in wheat in the marginal districts of the two states are the public extension services. Unlike other crops, such as cotton, pearl millet and maize, for which private sector seed have become widespread and sources such as input dealers have become important channels for information, the wheat case represents the case of low penetration of the private sector. Unless extension is provided as a bundled product with new technologies in wheat, systems would have to rely on public extension for introduction and spread of any new technology in wheat.

In surface seeding of rice, there are several downsides to the technology likely perceived by potential adopters. Perceptions of high input costs, high downside risk and observation of several failures of the technology are associated significantly with adoption. In wheat, both technologies are associated with information constraints, which is to be expected given the state of public extension, the primary source of information in wheat.

The analyses above point to technology-specific factors having a bearing on both awareness as well as adoption of specific technologies in different cases. In effect, each technology has different determining factors with regard to its awareness and uptake by farmers. There are three logical scenarios to consider: (i) Awareness of technology; (ii) Adoption conditional on awareness; and (iii) Non-adoption conditional on awareness. The factors explaining these three states definitely have significantly different implications. While lack of awareness mandates that information dissemination about the technology and its potential should be prioritized, lack of adoption upon awareness implies that suitability of the technology, including the need for complementary inputs, should be explored and resolved.

Conclusions and Policy Recommendations

In this paper, we conduct an ex ante assessment of technologies for crop-specific marginal districts in Bihar and Odisha. Relying on dynamic behaviour of yields and current performance for rice, wheat and maize, marginal districts in Bihar and Odisha were identified. With the yield behaviour so considered for the principal crops, the marginal districts map closely into alternative criteria that could be used to find said districts.

Based on information from several sources, the potential of different technologies was gauged for the marginal districts. The findings from stakeholder elicitation was cross-validated with a primary survey in selected marginal districts in the two states. The survey results provide important information about crops/activities that farmers aspire to. Maize and pulses are two sectors in which there could be good payoff, as a number of farmers aspire to get into producing them.

In both states, there is generally a significant lack of awareness of agricultural technology, more so in marginal districts of Odisha. Some modern technologies, like hybrid rice in Bihar, have become quite well known to the farmers, while others, like Systems of Rice Intensification, in spite of having existed for quite some time, have not yet broken the information barriers. Awareness of technologies is also stratified along socio-economic lines. Smaller farmers and farmers belonging to the lowest caste fare badly, both in awareness as well as adoption of technologies.

Translation from awareness to adoption has been quite difficult for most technologies. In general, the technologies related to varietal adoption have been comparatively successful in this regard. In many others, as they get more complex and there is a greater need for complementary inputs, adoption of certain technologies, even in the presence of awareness, has been difficult.

Policies for technology promotion in the marginal districts have to take into account the current state, as well the aspirations, of the farmers. These aspirations relate both to the crops/activities that farmers want to engage in as well as different technologies that they want to adopt but cannot because of constraints. Given the evidence of the disconnect between awareness and execution, a holistic approach taking into account the whole process of adoption from information to support in adoption will be needed. The state of the farmers dealing with illiteracy, small land sizes and social barriers mandate a tailored approach in technology choice for the lagging districts in Bihar and Odisha.

Rationalization of technologies in the context of these districts needs to be done. A demand-pull approach that takes into account a farmer's preferences and his capacity should be adopted in introducing or promoting technologies. Several of the technologies that have been much publicized, such as SRI and Laser Land Levelling, only had limited success because underlying conditions did not support the comprehensive spread of the technology.

Some technologies clearly stand out for the marginal districts and could be promoted. These constitute hybrid rice, varietal improvement in wheat, and organic/semi-organic farming, all of which exhibit high potential. On the crop front, maize and pulse technology should be made more focal in policy.

Open Access This chapter is distributed under the terms of the Creative Commons Attribution-Noncommercial 2.5 License (http://creativecommons.org/licenses/by-nc/2.5/) which permits any noncommercial use, distribution, and reproduction in any medium, provided the original author(s) and source are credited.
The images or other third party material in this chapter are included in the work's Creative Commons license, unless indicated otherwise in the credit line; if such material is not included in the work's Creative Commons license and the respective action is not permitted by statutory regulation, users will need to obtain permission from the license holder to duplicate, adapt or reproduce the material.

References

Banerjee G, Palke LM (2010) Economics of pulses production and processing in India. Occasional Paper number 51. Department of Economic Analysis and Research, National Bank for Agriculture and Rural Development (NABARD), Mumbai

Bhadauria JS (2013) Impact of caste in adoption of technology in rural India. Agric Update 8 (3):386–388

Birner R, Anderson J (2007) How to make agricultural extension demand driven? The case of India's agricultural extension policy, Discussion paper no 00729. International Food Policy Research Institute, Washington, DC

Chowdhary VM, Vinu Chandran R, Neeti N, Bothale RV, Srivastava YK, Ingle P, Ramakrishnan D, Dutta D, Jeyaram A, Sharma JR, Singh R (2008) Assessment of surface and sub-surface waterlogged areas in irrigation command areas of Bihar state using remote sensing and GIS. Agric Water Manag 95:754–766

Foster AD, Rosenzweig MR (2010) Microeconomics of technology adoption. Annu Rev Econ 2(1):395–424, 09

Ghatak M, Roy S (2007) Land reform and agricultural productivity in India. A review of evidence. Oxford Rev Econ Policy 23(2):251–269

Government of India (2008) Road map for rural industrialization in Bihar. A report of the special task force on Bihar. Government of India, New Delhi

Kumar SM (2013) Does access to formal agricultural credit depend on caste? World Dev 43:315–328

Lybbert TJ, Magnan N, Bhargava AK, Gulati K, Spielman DJ (2013) Farmers' heterogeneous valuation of laser land leveling in Eastern Uttar Pradesh: an experimental auction to inform segmentation and subsidy strategies. Am J Agric Econ 95(2):339–345

MHA (2011) Census of India. Ministry of Home Affairs, Government of India, New Delhi

Pathak H, Tewari AN, Sankhyan S, Dubey DS, Mina U, Singh VK, Jain N, Bhatia A (2011) Direct-seeded rice: potential, performance and problems – a review. Curr Adv Agric Sci 3(2):77–88

Planning Commission (2013) Press note on poverty estimates, 2011–12. Government of India Planning Commission, New Delhi

Pomp M, Burger K (1995) Innovation and imitation: adoption of cocoa by Indonesian small-holders. World Dev 23(3):423–431

Prasad CS (2006) System of rice intensification in India. Innovation history and institutional challenges. WWF Project 'Dialogue on water, Food and Environment' c/o ICRISAT, Patancheru

Reddy JN, Sarkar RK, Patnaik SSC, Singh DP, Singh US, Ismail AM, Mackill DJ (2009) Improvement of rice germplasm for rainfed lowlands of eastern India. SABRAO J Breed Genet 41, Special Supplement

Singh KK, Ali M, Venkatesh MS (2009) Pulses in cropping systems. Technical Bulletin IIPR, Kanpur

Srinivasarao C, Sreenath D, Srinivas I, Sanjeeva Reddy B, Adake RV, Shailesh B (2013) Operationalization of custom hiring centres on farm implements in hundred villages in India. Central Research Institute for Dryland Agriculture, Hyderabad, p 151

Yamano T, Rajendran S, Malabayabas M (2013) Psychological constructs toward agricultural technology adoption: evidence from Eastern India. Contributed paper at the 87th annual conference of the agricultural economics society, University of Warwick, Coventry, 8–10 Apr

Chapter 19
Technological Innovations for Smallholder Farmers in Ghana

Samuel Asuming-Brempong, Alex Barimah Owusu, Stephen Frimpong, and Irene Annor-Frempong

Abstract This chapter explores which community-based technologies have the greatest potential for reducing poverty and vulnerability among many smallholder farmers in Ghana. To this end, the stochastic dominance test was applied to rank outcomes from the different technologies used by the smallholder farmers in the study area. To show the effect of the technology on smallholder farmers' income, propensity score matching was used to test for differences in income of technology adopters and non-adopters. Based on the findings of the study, we conclude that the dominant technologies that have the potential to reduce smallholder farmers' level of poverty and marginality are: inorganic fertilizers for Afigya-Kwabre; zero tillage for Amansie-West; storage facilities for Atebubu-Amantin; marketing facilities for Kintampo South; improved varieties for Gonja East; and pesticides for the Tolon Districts.

Keywords Community-based technology • Innovations • Marginality • Smallholder farmers • Technology adoption

S. Asuming-Brempong (✉)
Department of Agricultural Economics and Agribusiness, University of Ghana, Legon, Accra, Ghana
e-mail: asumbre20@hotmail.com

A.B. Owusu
Department of Geography and Resource Development, University of Ghana, Legon, Accra, Ghana
e-mail: owusuba@yahoo.com

S. Frimpong
United Nations University-Institute for Natural Resources in Africa, Accra, Ghana
e-mail: kafrimpong@yahoo.com

I. Annor-Frempong
Forum for Agricultural Research in Africa (FARA), Accra, Ghana
e-mail: ifrempong@fara-africa.org

© The Author(s) 2016   369
F.W. Gatzweiler, J. von Braun (eds.), *Technological and Institutional Innovations for Marginalized Smallholders in Agricultural Development*,
DOI 10.1007/978-3-319-25718-1_19

Introduction

Agricultural technology and innovation are the foundations of rural economic growth and development. For this reason, many governments and aid agencies constantly introduce technological innovations to rural farmers with the view of empowering them. Farmers also innovate and develop indigenous knowledge and technologies to address their specific needs. Yet, many of the interventions introduced to farmers usually assume a top-down approach without assessing the farmers' own capabilities and skills. These wholesale technologies, which also assume that all smallholder farmers are equal in resource endowments or poverty levels, and ignore spatial variations, often lack a cutting edge approach to solving farmers' problems and may worsen farmers' plight.

This study, therefore, employed the *Technology (ex-ante) Assessment and Farm Household Segmentation for Inclusive Poverty Reduction and Sustainable Growth in Agriculture (TIGA)* approach to explore community-based technologies that have the greatest potential for reducing poverty and vulnerability among many smallholder farmers in Ghana. This will form the basis for up-scaling of community specific technologies that yield the desired results. The research also highlights important attributes or indicators which may lead to either more successful or fewer successful outcomes. Optimally, these indicators could also be used to gauge and benchmark the performance of the technologies in any implementation programme.

Poverty and marginality are common in Africa, particularly Sub-Saharan Africa, where more than 40 % of the inhabitants live on less than a dollar a day. Levels of food insecurity also remain stubbornly high, with a third of the population being undernourished (IFAD/WFP/FAO 2011). This is exacerbated by conflict, climate change, poverty-induced migration and natural disasters. Although agriculture is the mainstay of more than two-thirds of Africa's poor, and thus provides the greatest potential for pulling up the millions of people stuck in poverty, agriculture in many countries lacks the much-needed technological innovation and productivity to reduce poverty and vulnerability. While it is true that technologies may abound in many countries, smallholder farmers have not, overall, benefitted from most of these technologies. This may be due to social, cultural, political, natural and economic factors which limit their ability to successfully utilize these technologies.

In Ghana, poverty remains unacceptably high, with 19.2 % trapped in abject poverty (Ackah and Aryeetey 2012). Over 70 % of the poor engage in smallholder agriculture and cultivate less than 2 ha (MoFA 2012; WFP 2009). These farmers who reside mainly in rural areas more often use rudimentary equipment in their farming and most of the technological interventions are beyond their capacity to adopt. Agriculture is also rain-fed and most farmers lack access to financial and other productive resources. This introduces an element of risk into agriculture and exposes smallholder farmers especially to vulnerability, which in turn perpetuates poverty.

Although agricultural research in Ghana has generated a number of technologies aimed at improving the farmer's livelihood and productivity, the impact has been awful or, at best, disappointing (AdeKunle et al. 2012). This emanates from the poor involvement of farmers in research and the 'pouring on farmers' syndrome. To address this challenge of poor outcomes associated with most thwarted technological innovations, it is imperative that research is geared towards unraveling indigenous and community-specific interventions that work best for farmers. This study, therefore, explores local technologies that are best suited to farmers in the three agro-ecological zones of Ghana.

Ghana's population is around 24.6 million (24,658,823) (Ghana Statistical Service 2012a) and agriculture is the backbone of the economy. According to Al-Hassan and Diao (2007), agriculture employs more than 60 % of Ghana's labour force and contributes to about 25.6 % of its Gross Domestic Product (Ghana Statistical Service 2012b). Beyond these, agriculture is also recognized to have a greater impact on poverty reduction than any other sector in developing countries (IFPRI 2004; IFPRI 2009).

Ex-ante technology assessment refers to a forecasted estimation of the performance or outcome of an about-to-be-introduced or potential technology. Braun (1998) echoed ex-ante technology assessment as a systematic analysis aimed at foreseeing the future outcomes of a particular technology in all bases which the technology may touch. Remenyi et al. (2000) also defined ex-ante technology evaluation as predictive evaluations performed to forecast and assess the impact of future technology. Technology assessment should not be muddled with technology evaluation. Yet, technology assessments and discussions at the national and international levels have often been infused with ideological, theoretical and value-based beliefs by people of different technological blocs, techno-optimist, techno-skeptic or non-allied groups. Ruben et al. (1998) add that research on agricultural production technologies takes place from different viewpoints. This partitioning of people into different technological factions eventually leads to social debate and political conflicts between opposing teams (Jamison and Baark 1990).

The first section of this chapter, therefore, provides an explanation of ex-ante methodologies for assessing technology with the overall aim of achieving a unifying front from all fractions of the technological divide. The rest of the paper has been arranged to facilitate readership by technocrats or experts, as well as bureaucrats. Section two deals with methodological issues, while section three deals with outcomes of the research. The last section is dedicated to recommendations from the research.

Overview of Technological Development in Ghana

Efforts to modernize and improve agriculture in Ghana date back to the pre-colonial era. According to Rodney (1984), these emanated from the exploitative colonial system, which was mainly a conveyance system for carrying minerals and other goods from the hinterland to urban areas for onward transport to the west to feed their industries. Long stretches of link roads were constructed which brought many rural farming households into the national focus. This helped in the distribution of farm inputs, as well as the transportation of rural commodities, and thus helped to reduce post-harvest losses. Rural electrification was also pursued to some extent.

The periods after independence also saw governments initiating programmes with the aim of improving agricultural growth and productivity. For instance, the Convention Peoples Party, the then-ruling government, introduced the State Farms System to serve as modules for farmers in Ghana. The Block Farming Systems were another innovative means of enhancing farmers' productivity in such similar modular programmes, particularly for farmers in the transition and savannah zones. Also, fertilizers were introduced into Ghana and were subsidized. Several agricultural colleges were established in Ghana during this period to train extension agents. In addition, there was the establishment of many food crop and agro-processing firms. However, the major agricultural technological breakthrough in post-independent Ghana was the introduction of innovations in the cocoa sector in the 1970s which affected the livelihoods of millions of Ghanaians.

Periods of intermittent military rule wiped out some of these programmes, but nonetheless, some agricultural growth and technological interventions were observed. Particularly, during the Acheampong regime, rural electrification, as well as rural communication facilities, mainly post offices, were vigorously pursued which had indirect linkage to agricultural growth and development. Tractors were also introduced in Ghana in large numbers during those military periods. But the dominant programme in the military days that led to food sufficiency in Ghana was the Operation Feed Yourself (OFY) Programme. As the name suggests, this programme encouraged workers in the public sector, particularly in the cities, to cultivate farms on patches of lands around their houses. These periods are also closely linked with the structural adjustment programme (SAP) of 1985.

Periods after the post-structural adjustment have also seen the introduction of many interventions, either by the state, aid agencies or even individuals. For instance, the establishment of the Ministry of Food and Agriculture and other allied agricultural institutions has helped to co-ordinate activities in the agricultural sector. Ghana has also developed an agricultural sector policy document, FASDEP, and an implementation Plan, METASIP. Other programmes, such as the Root and Tuber Improvement Programmes (RTIMP), Millennium Development Authority (MiDA) Programmes, Youth in Agriculture Programmes, Savannah Accelerated Improvement Programmes and District Tractor Services Programmes, have all helped to improve farmers' capacity and enhanced growth in the agricultural sector. But two programmes appear to have had the greatest impact. These were the re-introduction of fertilizer subsidies and the National Cocoa Diseases and Pest Control (CODAPEC) programme, popularly known as "Mass Spraying" to assist all cocoa farmers

From the foregoing, it can be deduced that several technological innovations and agricultural productivity growth programmes have been introduced in Ghana since independence. Whereas some had limited impact, others were of great success. But these successess unequivocally did not affect many rural poor farming households, who lack many resources and the capacity to adopt these technologies. Ex-ante assessment of technology and innovation for poverty and marginality reduction, therefore, provides a more effective tool for exploring interventions that have the potential to reduce farmers' poverty and marginality but also within the farmers' limit of possible adoption.

Assessment of Technological Innovations in Agriculture

Several scholars and researchers have attempted to holistically undertake an assessment (including ex-ante assessment) of technologies introduced into many poor rural farming communities using a variety of different approaches and models (Ruben et al. 1998; Ruben and van Ruijven 2001; van Keulen et al. 1998; Berkhouta et al. 2010; Ruben et al. 2006). These models generally span from normative decision-making and accounting techniques, such as benefit-cost ratios (BCR), internal rate of returns (IRR) and the net present value (NPV), to econometric models, such as multi-market models and supply response models, continuous production functions and efficiency measures, farm household models (FHMs), economic surplus models, general equilibrium models (GCE), policy analysis matrix (PAM) procedures, farming system research (FSR) procedures, and statistical simulation models such as mathematical programming, linear programming or measures of welfare dominance (Veeneklaas et al. 1994; Ruben et al. 1998; Ruben and van Ruijven 2001).

Multi-market models often deal with various agricultural sub-sectors and market distortions, considering interactions in both the product and the factor market, and the impact of price changes on incomes, expenditures and production. These models require a detailed specification of supply and demand elasticities. Farming systems research (FSR) provides a framework for classification of farm households into marginality groups or spots, and a detailed analysis of farm household resource-use decisions. FSR helps to explain the basis of technology choice, and the identification of resource constraints at the farm household level (Steenhuijsen Piters 1995). Mathematical programming procedures are usually applied to analyse optimum allocative choice (Ruben and van Ruijven 2001). They provide insights into the optimal agro-ecological production possibilities for a farm or region and are useful for indicating physical trade-offs between different (long-term) objectives (Ruben et al. 1998). These are based on utility maximization principles and usually use optimization approaches, such as profit maximization or risk minimization, as the objective function.

Multimarket, farming system research and mathematical programming models simulate either a farm household's behavior, such as technological (non) adoption and input choices, or agro-ecological processes separately, and cannot be used directly for ex-ante analysis, because the relationship between technological options and behavioural driving forces is not adequately specified. To address this problem, new research programmes focusing on the integration of economic simulations within biophysical simulation models offer important opportunities for the appraisal of the attractiveness of technological options from the farmers' viewpoint, and the identification of incentives to make their adoption feasible.

The NPV has also been traditionally used to assess the economic benefit of technology implementation in various sectors of the economy. For instance, in assessing vehicle safety technologies in European Union countries in 2006, the European Commission estimated the NPV as the difference between discounted stream of benefits and the required cost:

$$NPV = \sum_{t=0}^{T} NB_t * \frac{1}{(1+r)^t}, \tag{19.1}$$

where NPV is the net present value of the stream of net benefits from year t to T; T is the time horizon of the evaluation; NBt is the net benefits (benefits minus costs) incurred in year t; and r is the rate of discount. Benefit-cost analysis (BCR), through valuation of physical inputs and outputs, can be applied to assess the minimum conditions for technology change or profit. It is expressed as the present value of benefits divided by the present value of costs. BCR has been used to assess technologies in the public sector of many countries. Wulsin and Dougherty (2008), and Garrido et al. (2008) used BCR in assessing health technology in the United States of America and Europe, respectively.

Supply response models (SRM) use (expected) prices as a major explanatory variable for adjustment of agricultural production (Askari and Cummings 1976). Supply response models only consider the production side of the farm household, and linkages between production and consumption decisions, ignoring the characteristics for farm households operating under imperfect markets. But to effectively assess potential impact of technology, economic models (which identify the behavioural reasons for crop or livestock and technology choice) and agro-ecological models (used to select feasible technologies and cropping options for specific agro-ecological conditions and to assess their consequences in terms of sustainability of the resource base) should be combined (Ruben et al. 1998). Combining both approaches into a single analytical framework greatly assists policy-making, enabling the identification of possible trade-offs between economic and environmental objectives, as well as assessment of the impact of government interventions in markets for land, inputs, products, technology and infrastructure on farmers' decisions and the consequences for farm household welfare and sustainability of the resource base. The integration of agro-ecological and socio-economic information, therefore, takes place at the farm household level. Farm household models (FHMs) offer another perspective for the analysis of production and consumption decisions at the farm household level (Singh et al. 1986). Differences in risk behaviour (Roe and Graham-Tomasi 1986), market failures or missing markets (de Janvry et al. 1991), and inter-temporal choice (Deaton 1990; Fafchamps 1993) can also be taken into account. Due to the possibility of analysing both production and consumption decisions, the FHM approach represents a useful starting point for analysis of the effectiveness of potential technologies.

Another new or emerging technology assessment tool of great importance in recent times is that which combines both econometric and biophysical models, generally referred to as bio-economic models. These models incorporate technical input–output coefficients derived from agro-ecological simulation models into econometrically-specified farm household models (FHM) (Ruben and van Ruijvenvan 2001). These models usually involve functional integration of four models, namely biophysical crop growth simulation models, mathematical programming models that reveal the resource allocation implications of alternative crop and technology choices, FHMs that capture farmers' behavioural priorities, and aggregation procedures to address the effectiveness of policy instruments. A common feature of bio-economic models is that they usually originate from two sources – production and consumption models – and can generally be put into three categories. These are: biological process models with an economic analysis component, integrated or meta-bio-economic models (commonly referred to as meta-modelling), and economic optimization models, often used when new and potential technologies have to be included; the process involves the use of mathematical programming approaches with integrated or biological process models. Bio-economic models are used to analyse the impact of different types of economic incentives on farmers' resource allocation decisions, as well as their implications for the natural resource base (Ruben and van Ruijvenvan 2001; Sullivan 2002).

According to Ruben et al. (1998), for an ex-ante assessment of potential or new technologies, modelling and simulation approaches are required. Van Keulen et al. (1998) used a bio-economic model comprising linear programming with constraint optimization and farm household models in their study of sustainable land use and food security in developing countries: DLV's approach to policy support. The possibilities of introduction of more sustainable land use systems and their consequences for socio-economic indicators were analyzed. Indicators included in the sustainability model were use of biocide and soil nutrient loss, with farm income being the optimization constraint. The results showed that more sustainable land use systems can be introduced. Farm households' responses to specific policy instruments were also analyzed with the farm household model (FHM). The model was used to identify those price instruments that affect improvement of the competitiveness of agricultural production in the Atlantic Zone and improved natural resources management, which are two regional development objectives. These objectives were transformed into four clear goal indicators at the farm household level, namely income (utility) and plantain and cassava production were used as indicators for improved competitiveness, while biocide and fertilizer use served as indicators for natural resource management. The model results showed that higher product prices, lower fertilizer prices, and reduced transaction cost favour substitution of actual production activities with alternatives, leading to more sustainable land use. Increased biocide prices, on the contrary, resulted in a decrease in biocide use, mainly as a result of a reduction in cultivated land area.

The authors undertook similar studies in Mali and Costa Rica. An interactive multiple goal linear programming technique was applied to analyze options for rural development. Technical innovations used included more effective integration of arable farming and animal husbandry, based on the use of crop residues and fodder crops to provide high quality forage, the use of animal manure for nutrient cycling in cropping systems, and improved access to animal traction. The authors introduced various constraints reflecting different kinds of market imperfections, such as the possibility of hiring outside labour, availability of chemical fertilizers and price-setting for inputs and outputs. The results indicated that, with full knowledge of alternative (agro-ecologically sustainable) production techniques, the values of sustainability indicators such as soil nutrient, organic matter (O.M) depletion or soil mining can be improved up to 55−80 % by introducing these production techniques without sacrificing required incomes.

The FHM identified microeconomic supply reactions to various policy measures. Production and consumption decisions were jointly analyzed. Four household types were distinguished according to resource endowment and their objective functions to account for straight directions of supply response (SR). Savings and investment were included through the savings and investment model, while different time discount rates accounting for subjective time preferences by type of household food and labour balances were identified for the appraisal of market interactions and exchange among farm types. The agro-ecological sustainability indicators used were the balances of the macro plant nutrients (N, P, K) and soil O. M. content. The results showed that, given a farm household's resources, their goals and aspirations, and their subjective time discount rate, non-sustainable technologies resulting in soil nutrient depletion remain to be practiced. They added that low supply response causing price policies to be largely ineffective is a major constraint for stimulating agricultural intensification. The authors concluded that structural policies such as improving rural infrastructure, credit systems and land policies are required to promote adoption of technological innovation. They stressed that the impact of policy instruments depends on the market, the institutional environment and overall resource availability. They added that, in low income countries like southern Mali, where factor markets for land and capital are not very well developed, instruments of price policy appear to have limited influence on resource allocation, and market and institutional development are the required instruments.

By contrast, in highly commercialized regions, like the Atlantic Zone of Costa Rica, modification of input prices and lower transaction costs appear to be suitable instruments for promoting sustainable land use while maintaining household consumption prospects. The authors recommended a further refinement of the methodology to cope with the absence of the increasingly recognized role of non-agriculture income in farm household decision- making, and the incomplete aggregation of procedures between the farm and regional levels. The authors further recommended a multi-market model which includes migration and agricultural factor use. Roetter et al. (2000), in their synthesis of methodology and case studies, employed simulation models, geographic information systems (GIS) and optimization techniques. Characterization of resources of farmers was done using GIS and a spatial database.

In a case study, a trade-off between cereal production and environmental impact in Haryana, India, primary productivity and milk was estimated using crops and livestock modeling. Multiple goal linear programming was used for optimizing land, water, capital and labour, with objective variables such as food, milk, income, land use, irrigation, N-fertilizer, employment, capital, N loss and biocide index. The preliminary results indicated current availability of water is a major constraint to increasing food production in Haryana. They also developed a technical coefficient generator (TCG) for describing the input–output relations of the various production activities and technologies based on the concept of production ecology. Berkhout et al. (2010), in their study, asked: Does heterogeneity in farmer goals and preferences affect allocative and technical efficiency? A case study in Northern Nigeria fitted a Tobit regression model of the form $E_s = \beta_0 + \sum_{i=1}^{N} \beta_i K_i + \sum_{j=1}^{M} \gamma_j Z_j + \varepsilon_{is}$ first to quantify heterogeneity in farm production attributes among smallholder farmers in a rural African setting; and secondly, to investigate whether heterogeneity in these attitudes and goals indeed results in different production strategies. In the notation, E_s is the score of three efficiency measures – technical efficiency, profit and food allocative efficiency – obtained through data envelopment analysis (DEA); K_i is a vector of household characteristics such as age, level of education and distance to markets; and Z_j is behavioral variables. To arrive at a measure of profit allocative efficiency (E_3), the authors used the linear decomposition proposed by Ray (2004), illustrated as:

$$E_3 = \frac{\left(\Pi^* - \Pi^A\right)}{C_A} = \frac{\left(\Pi^* - \Pi^{A''}\right)}{C_A} + \frac{\left(\Pi^{A''} - \Pi^A\right)}{C_A}, \qquad (19.2)$$

where Π^* is the profit-efficient production point; C_A is the actual cost level ($C_A = wL^A$); Π^A is the actual level of profit based on the observed level of output and observed use of labor $\left(\Pi^A = pQ^A - wL^A\right)$; and is the level of profit when input-oriented technical inefficiency is eliminated $\left(\Pi^{A''} = pQ^A - wL^{A''}\right)$. The first part of the expression is the allocative efficiency. The latter part of the expression, which is the technical efficiency, equals: $\Pi^{A''} = pQ^A - E^I wL^A$, with E^I being a measure of input-oriented technical efficiency. Then, the last term in the equation reduces to $(1 - E^I)$. Food efficiency was estimated similar to the more widely used concept of revenue efficiency, albeit using nutritional content of crops instead of output prices.

Three surveys consisting of a survey on general household characteristics, production, farmer goals and preferences were undertaken. The third survey consisted of two parts: a fuzzy pair-wise ranking and a set of Likert scale questions. Data was collected from 155 farmers in seven villages based on differences in market access, population pressure and differences in soils and climate. Farmers indicated their preference for five different goals presented in the fuzzy pair-wise goal ranking, such as getting the highest net benefits from farming; getting the highest subsistence food production; minimizing the risks of farming; safeguarding the soil for future generations; and minimizing labor use in agriculture. Principal component analysis was used to reduce the data from the rankings and the additional questions separately and jointly. Factor analysis was used to reduce the dimensionality of the data, such that z is the minimum set of variables describing most of the variance observed.

In order to increase the efficiency of the DEA approach, the authors aggregated outputs of the 22 crops into three main groups: cereals, legumes and high-value crops (roots, tubers and vegetables), adding rice and sugarcane as separate crops, and using eight different kinds of inputs. The results of the fuzzy ranking suggested that staple food production and sustainability are the most important attributes to farmers in the area of study, followed by risk aversion, while gross margins and labor use minimization are relatively unimportant. The researchers also found that, on average, farmers are relatively food efficient but far from profit efficient. They further added that this not only results from household characteristics directly, but also from personal goals and preferences. They concluded that both socio-economic characteristics and goals and preferences have direct effects on efficiency levels, in addition to some indirect effects of household characteristics through changes in goals and preferences. They stressed that, since village dummies qualify as potential instruments for behavioral factors, it suggests local conditions are strongly related to expressed attitudes and preferences. Hence, they recommended that further studies should be undertaken to identify the causal relationships between the different behavioral factors and socio-economic characteristics and focus on how rural agricultural policies should account for this effectively.

Ruben and van Ruijven (2001) analyzed technical coefficients for bio-economic farm household models: a meta-modelling approach with applications for Southern Mali fitted with a bio-economic model of two components – production and consumption. The authors applied a meta-modelling approach for the production side of a bio-economic farm household simulation model in order to generate continuous production functions on the basis of discrete production data that can be derived from agro-ecological simulation results. In the study, a typical farm household in the 'Koutiala' region of Southern Mali was used, composed of 25 people, with 12 active people that supply 1800 labour days, and have at their disposal 18 ha of land with defined soil quality characteristics, three pairs of oxen and four ploughs. The production side of the model included a set of 1443 technical coefficients for cropping activities (maize, cotton, millet, sorghum, cowpea and groundnuts) and 96 technical coefficients for livestock activities (milk and meat production).

A range of input–output coefficients for potential production (technological) activities that guarantee higher levels of input efficiency, such as control of crop losses, making use of improved input applications, crop residue management strategies, better timing of operations (soil preparation, weeding, grazing) and the implementation of soil erosion control measures, and lower levels of soil nutrient depletion were estimated from agro-ecological simulation models. The consumption side of the model was based on a cross-section budget survey regarding expenditures for cereals, meat, milk, vegetables and non-agricultural commodities.

This survey data was used to estimate marginal utility of consumption for different expenditure categories, making use of a continuous farm household utility function. Expected prices for produced commodities and inputs (labour, traction, implements, fertilisers and manure) were derived from local surveys. Expected utility of consumption (corrected for nutrient losses) under given market conditions and defined resource constraints was optimized.

The meta-modelling approach was applied to the series of several hundreds of data points for all crop and livestock activities to derive continuous production functions for each activity, making use of the Battese (1996) procedure to account for zero input use. For arable cropping, the authors estimated the following Cobb–Douglas production function:

$$lnY = \beta_0 + \beta_1 \ln(L) + \beta_2 \ln(T) + \beta_3 \ln(N) + \beta_4 \ln(P) + \beta_5 \ln(M), \qquad (19.3)$$

where Y represents the quantity of the different harvested crops (in monetary units); L and T are the total amounts of labour and traction (in working days); N and P are the amounts of active ingredients of nitrogen and phosphorous fertilisers applied to the crop (in kg/ha); and M is the amount of manure applied (in kg/ha). Livestock activities were defined for meat and milk production under different regimes of animal feeding. For livestock, a linear specification of the production function was estimated as:

$$lnY = \beta_0 + \beta_1(q1) + \beta_2(q2) + \beta_3(q3) + \beta_4(q4) \ldots + \beta_{10}(q10), \qquad (19.4)$$

where $q(1) \ldots q(10)$ represent feed sources available during the wet and dry seasons that correspond to different levels of energy intake and digestible organic matter. The estimated functions for crop and animal production were incorporated into a non-linear bio-economic farm household model, which was optimized for the objective of expected utility of consumption, given the availability of resources (land, labour, traction):

$$\begin{aligned} Max \; EU &= \sum \left(u \cdot C \middle| \left(Y^* - p_e \cdot E \right) \right), \\ \text{s.t } Y^* &= p_i \cdot I + p_c \cdot C + p_l \cdot L, \end{aligned} \qquad (19.5)$$

where C represents a vector of consumption goods; Y^* represents income derived from production; I represents the different inputs; L is labour force; and p are their respective prices. The vector E includes environmental externalities (e.g., nutrient losses) valued against their replacement costs.

The household model was first optimised under the assumption of perfect markets, allowing for separability, and thus, sequential optimization. This base run of the model was used as a reference point. Subsequently, the authors imposed constraints on the labour, capital and animal traction market by limiting the use of these inputs to the quantities owned by households. The model specifications with different market imperfections were optimised in a non-separable way, which meant that the production and consumption parts were estimated simultaneously.

The standard Gams software was used for the optimisation. The results showed that the coefficients for labour were positive and significant, and especially in cotton and cowpea production, the elasticity of labour was high. The traction elasticities for cereals and cotton were estimated between 0.06 and 0.20, while for cowpea and groundnut, these were estimated to be about 0.7. The (valid) coefficients for different types of fertiliser were lower than 0.3; for sorghum, cowpea and groundnut, the fertiliser coefficients were not significant. The authors, however, were unable to explain the negative coefficients for manure in millet and cowpea production. The study also found that all functions for crop production have increasing returns to scale.

The livestock results showed a negative constant, implying that cattle needs feed for maintenance, which does not contribute directly to production, and only above a certain level of food intake do cattle start producing milk and meat. The results of the optimization also indicated that, with market imperfections, utility decreases compared to the situation with perfect markets. The results for the consumption side of the model suggested consumption of all categories of goods is lower when market constraints are taken into consideration, while also indicating a shift from meat consumption towards cereals if per capita income falls, which is consistent with consumer demand theory where meat is normally considered to be a luxury good. Consequently, a decrease in income will cause a more than proportional fall in meat consumption. Cereals are considered to be basic requirements for food security, and therefore, cereal consumption does not decrease as much as meat consumption. The researchers concluded that decision-support systems for policymakers should be able to address issues related to the implications of technological change for farmers' welfare and sustainable resource management, and could be helpful in identifying feasible policy instruments to induce farmers towards the adoption of these technologies. They added that behavioural aspects of farmers' choice and available options for technological change must be combined within a single and consistent modeling framework. They further stressed that the meta-modelling approach provides a useful tool for exploring the characteristics of the discrete technical input–output coefficients that are subsequently incorporated into the framework of a dynamic and continuous bio-economic farm household model. They stressed that these procedures enable improvement of the specification and the robustness of meta-models based on data sets derived from different disciplines, and, therefore, policy simulations based on such integrated models could provide consistent estimates of response elasticities based on income, substitution and scale effects (Foster et al (1984); Frimpong and Asuming-Brempong (2013).

Considering the agricultural systems and agro-ecological conditions in Ghana, and based on the foregoing discussions, the meta-modelling approach involving econometric analysis of ex-ante technologies and FHM for estimating potential adoption rate of the various technologies were used. The FHM provided causality analysis in order to assess socio-economic factors influencing technology use. The aggregation procedures, the fourth component of the bio-economic approach, with which the effectiveness of policy intervention at the regional level has to be addressed, was taken into account within the common technology assessment framework design (TIGA).

Methodology

Ex-ante Assessment of Suitable Technological Innovation

Following Scaillet and Topaloglou (2010) and Davidson and Duclos (2000), in identifying which commonly practiced technology gives superior welfare outcome, the stochastic dominance test was applied to rank outcomes from the different technologies used by the smallholder farmers in the study area. Consider the distribution of independent samples of welfare measures such as per capita expenditure, Y of smallholder farmers using any two technologies, A and B with cumulative density functions F_A and F_B with the lower bound of the common support fixed at zero and the upper bound fixed to any acceptable poverty line, that is, $(0, Z)$. Then,

$$F_A(x) = \frac{1}{N} \sum_{i=1}^{N} 1\left(X_i^A \leq Z\right), \tag{19.6}$$

$$F_B(x) = \frac{1}{M} \sum_{i=1}^{M} 1\left(X_i^B \leq Z\right), \tag{19.7}$$

where X_i^A and X_i^B represent distribution of per capita expenditures of smallholder farmers who use technology A and B respectively, N and M represent sample sizes of the two technologies, and $1(\cdot)$ takes a value of 1 when the farmer is equal to or above the poverty line and zero otherwise. The stochastic dominance for A can be expressed as:

$$D_A^1(x) = F_A(x) = \int_0^x dF_A(Y). \tag{19.8}$$

For any integer, $S \geq 2$, here S is the order of the stochastic dominance, $D_A^s(x)$ takes the form:

$$D_A^s(x) = \int_0^x D_A^{s-1}(Y) dy. \tag{19.9}$$

Technology A is said to be dominant over B at order S if $D_A^s(x) \leq D_B^s(x)$, for all $X \in [0, Z_{\max}]$. Following Davidson and Duclos (2000), the general form is specified as:

$$D_{(x)}^s = \frac{1}{(S-1)!} \int_0^x (Z - Y_i)^{S-1} dF(Y). \qquad (19.10)$$

For a random sample of N independent observations, the natural estimator of $D_{(x)}^s$ is specified as:

$$\hat{D}_{(x)}^S = \frac{1}{(S-1)!} \int_0^x (Z - Y_i)^{S-1} dF(Y) = \frac{1}{N(S-1)!} \sum_{i=1}^N (Z - Y_i)^{S-1} (X_i^A \leq Z). \qquad (19.11)$$

The main hypothesis for testing dominance at order $S = 1, 2, 3$ (first, second and third order dominance) is stated as:

$$H_0 : D_A^s(x) - D_B^s(x) = 0 \; for \; all \; Z \in [0, Z], \qquad (19.12)$$
$$H_1 : D_A^s(x) < D_B^s(x) = 0 \; for \; all \; Z \in [0, Z]. \qquad (19.13)$$

This was done using a t-test. The variance for the test was specified as:

$$\left[\hat{D}_A^S(x) - \hat{D}_A^S(x) \right] = Var\left(\hat{D}_A^S(x) \right) + Var\left(\hat{D}_A^S(x) \right). \qquad (19.14)$$

The t-statistic on the basis for which H_0 is tested is stated as:

$$\frac{D_A^s(x) - D_B^s(x)}{\sqrt{Var\left(\hat{D}_A^S(x) \right) + Var\left(\hat{D}_A^S(x) \right)}}. \qquad (19.15)$$

Four outcomes are possible: A dominates B; B dominates A; no dominance because $A = B$; or no dominance because A crosses B. When A crosses B, the second and third order dominance are used to test for differences. For A to be said to be dominant over B, the null hypothesis must be rejected.

Potential Adoption Rate

The Farming Systems Research approaches were used to predict the maximum adoption rate of the recommended technologies in the zones. According to Hildebrand and Russell (1996), the likelihood that a farmer will adopt a technology depends on farmer categories, production goals and the environment. This is mathematically expressed as:

$$PAR = \left(\frac{F*G*E}{10000}\right). \tag{19.16}$$

In the notation, F = Frequency of farmer categories (%), G = Frequency of production goals (%) and E = frequency of production environments (%).

From the study, four farmer categories, very poor, poor, rich and very rich, were identified. Factors that influence these farmers' production goals are access to production resources and institutions (credit, FBOs, extensions), age, and gender. An indicator used for the production environment is whether households are settlers or natives. A composite indicator for farmers' goals (G) and environment (E) was estimated using weighting values. Males were given a weight of 1 and females 0.5, while respondents 50 years or younger were given a weight of 1 and those older than 50 were given a weight of 0.5. The rest of the indicators were given a weight of 1. The composite indicator was estimated as:

$$C_i = \frac{\sum_{i=1}^{n} w_i X_i}{\sum_{i}^{n} w_i}. \tag{19.17}$$

In the notation, C_i is the composite indicator value for G or E and w_i is the weight of indicator X_i.

Income Effect of Technological Innovations

To show the effect of the technology on smallholder farmers' incomes, propensity score matching was used to test for differences in income of the treated (adopters of the technology) and the untreated (non-adopters of the technology). The income differential was then expressed as a percentage change of the income of adopters had they not adopted the technology (counterfactual).

Data Collection

Characteristics of the Study Area

Ghana is located between latitudes 4.5°N and 11.5°N and longitudes 3.5°W and 1.3°E (Antwi-Agyei et al. 2012), and can be distinguished into five main agro-ecological zones of fairly homogeneous climate, landform, soil, vegetation and land use systems (MoFA 2012) typical of West Africa. Ghana's population is around 24.6 million (24,658,823) (Ghana Statistical Service 2012a) and agriculture is the backbone of the economy. Generally, annual average temperatures range from 26.1 °C in places near the coast to 28.9 °C in the extreme north, with the highest

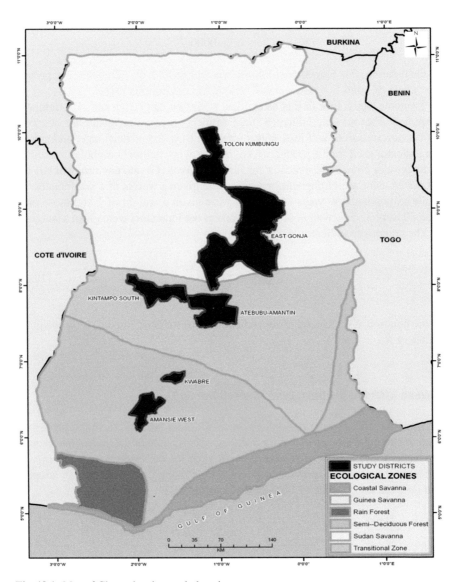

Fig. 19.1 Map of Ghana showing study locations

temperatures recorded in the Upper East Region (MoFA 2012). The topography of Ghana is predominantly undulating, with slopes less than 1 %. Average annual rainfall is about 11,796 mm, according to the Ghana Meteorological Service (2010), as cited in MoFA (2012), and less than 0.5 % of agriculture is under irrigation (World Bank 2010). Figure 19.1 shows a Map of Ghana with the study locations.

Selection of Study Area

An ex-ante assessment of the potential benefits of new technologies in communities and households provides the most reliable means of gauging a household's acceptance of the new technology (Wood 2003). It is, therefore, appropriate to select communities with populations for whom the technology is intended. Even then, it is unlikely that such studies can be carried out in all the potential communities where the new technology might provide significant impact. Therefore, the overall focus of sampling is to reach all strata of the people living in potential communities, particularly poor and small-scale farmers living in marginal or less-favoured areas (LFA's). Following Wood (2003), Stoorvogel et al. (2004), and Smale et al. (2003), the study adopted a multistage sampling procedure in selecting respondents. This involved zoning or stratification of Ghana into three parts, namely savanna, transition and forest zones, on the basis of differences in vegetation, income and livelihood activities. Two districts were selected purposively from each strata or zone using the crop type produced. Three communities within each district were randomly selected using the lottery approach. A simple random sampling technique was employed to select farmers within the communities. Following Yamane (1967), the sample size for each community was estimated as:

$$n = \frac{N}{1 + N(e^2)},$$ (19.18)

where n = the sample size, N = population, and e = significance level.

The sampling frame for each community was established with village elders, District Assemblies and the Ministry of Food and Agriculture District directorates. The survey included poverty or marginality hotspot mapping using a Global Positioning System (GPS), a collection of primary data on household and farm level factors, and agro-ecological variables using structured questionnaires (Simelton 2012). Key informant and expert interviews were also conducted. In all, 402 smallholder farmers were interviewed for the study. This comprised 139 respondents from the forest zone, 156 from the transition zone, and 107 from the savannah zone. The proportion of the respondents from the various zones in the total sample is shown in Fig. 19.2.

Results and Discussions

Trend Analysis of Technological Interventions

The trend analysis of technological innovations provides project implementers with the historical overview of major technological interventions in the intended project sites and the purpose or reason for introducing the interventions. This section,

Fig. 19.2 Distribution of smallholder farmers in the study area

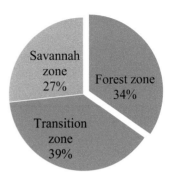

therefore, presents a brief temporal overview of interventions in the study zones, based on expert opinions, key informants and focus group discussions.

Since Ghana's independence in 1957, the then-ruling government, the Convention Peoples Party (CPP), made a steady effort to modernize agriculture in Ghana. According to the key informants, the CPP government introduced an early maturing rice variety, the 'Red rice', into the savannah zone. Also, during the same period, cowpeas, improved maize varieties and special yam seeds were introduced into the zone. The period under CPP also saw the establishment of state farms and block farms in the forest and transition zones, respectively. These farms served as models to train farmers.

The CPP government was closely followed by periods of rule by military regimes, which adversely affected progress. However, the Acheampong regime stands out in terms of agricultural development and technological advancement. For the first time, under the Acheampong-led National Redemption Council (NRC) government, Massey-Ferguson tractors were introduced into Ghana in 1974. Marketing Standards Boards were also established to ensure and maintain standardization in selected crops. The government also introduced local breeds of banana into the savannah zone in 1973. To improve food security, the Acheampong government introduced cassava into the savannah zone for the first time, cassava having hitherto been grown mainly in the forest zone. In the south, particularly among public workers, the Operation Feed Yourself (OFYS) Programme was also introduced. This programme made Ghana self-sufficient in the production of several crops. Silos were constructed across the country to reduce post-harvest losses. However, this period was also followed by other military regimes which virtually wiped out the successes achieved during the period.

After Ghana returned to civilian rule in 1979, under the leadership of Dr. Hilla Liman, attempts were made once again to boost agricultural production. The government introduced fertilizers into the country and also subsidized them as a means of improving crop yields. These fertilizer subsidies and other agro-input subsidies were wiped out during the Structural Adjustment Programmes (SAP). Periods after the Peoples National Party (PNP) and the Liman-led government saw more military rule. In 1992, when Ghana turned into a democratic state, the

National Democratic Congress (NDC) government introduced an improved maize variety, Dobidi, through MoFA, in conjunction with IFAD.

Under the New Patriotic Party (NPP), which won power from the NDC government in 2000, fertilizer subsidies were re-introduced. The Cocoa Mass Spraying Programme was also introduced, which boosted yields of cocoa farmers, along with the cocoa certification programme. New cassava varieties with high starch content, improved oil palm seedlings, and improved soybean varieties were introduced under the NPP-Kuffuor government. Through the Millennium Development Authority, yam minisetts technology, as well as improved maize seeds, were also introduced into selected districts across the nation.

The NDC government, after their re-installation in 2008, re-introduced the Block Farm Programme and the District Tractor Services Concept. In 2009, the Village Mango Project was introduced into the transition zone, offering farmers, on the average, five improved mango seedlings to maintain and nurture. This was followed by the Root and Tuber Improvement and Marketing Programme (RTIMP), introduced across the country with the overall aim of providing improved planting materials such as yam minisetts and cassava cuttings.

The foregoing discussions show that there have been a number of intermediations in technological programmes introduced into Ghana by various governments. However, most of these programmes failed to identify the specific local challenges of the farmers. Monitoring and supervision were poor and the required resources were either not available or were not provided at the time that they were required. Corruption and bureaucracy also crippled most of these interventions and prevented the targeted farmers from benefitting from the available resources. For instance, some politicians and bureaucrats hauled fertilizer or bought the subsidized fertilizer and re-sold to farmers at higher prices. Also, in some cases, only farmers known to be allied with the government in power benefitted from the interventions.

Technologies That Will Work for Poor Smallholder Farmers

Agricultural technologies are often locational, since they are affected by environmental changes. Technologies work best when they are adapted to the specific conditions of the intended beneficiaries and have optimum adoption rates. Therefore, technologies that give the greatest potential welfare benefit to the intended user group and increase the beneficiary's utility are to be chosen IFPRI (2009). Also, cultural/economic support systems and political or administrative conditions surrounding the target area may influence the scaling up of a technology. The stochastic dominance test provides a measure of the welfare benefits of a technology for smallholder farmers. From the results of the test, six main technologies are suggested for up-scaling in each district of the study zones. This is of particular importance, as farmers in the study zones and districts were involved in production of different crops and, hence, faced different technological challenges.

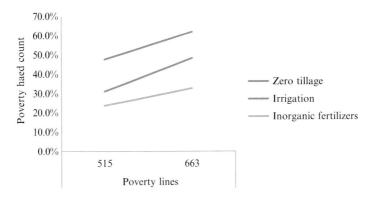

Fig. 19.3 Stochastic dominance test results for Afigya-Kwabre District

Smallholder farmers in the Afigya-Kwabre District of the forest zone were mainly involved in vegetable production, such as okro, tomatoes and garden eggs. Farmers in this district were mainly constrained by fertilizer and irrigation facilities for dry season farming and erratic rainfall conditions which affected their production. From the first order stochastic dominance test (see Fig. 19.3), the inorganic fertilizers provided the greatest welfare benefits to households. Fertilizer technology has the greatest dominance, as fewer of the smallholders who applied the technology have their income below the poverty line. However, application of fertilizer without adequate soil moisture will not lead to the intended benefit. Therefore, the technology best fitted for such communities will be one which combines irrigation facilities and fertilizer application.

Smallholder farmers in the Amansie West District have cocoa as their main crop, but also engage in food crops such as maize and cassava on a subsistence basis. As such, during the cocoa off-seasons, households are faced with food insecurity, which pushes many of them into illegal mining and other coping mechanisms. Since most of the farmers grow cocoa, they do not use weedicides to control weeds, as they claim the practice has implications for some useful flora and fauna. However, farmers who engaged in off-season farming using zero tillage were better off, as shown in the results in Fig. 19.4. Therefore, it would be useful to provide farmers in this district with resources such as weedicides and fertilizers to enter into off-season farming.

In the transition zone, although the technologies intersect, the best results after testing for first, second and third order dominance are marketing and irrigation in the Kintampo South District (see Fig. 19.5), and storage facilities and the use of inorganic fertilizers for smallholder farmers in the Atebubu-Amantin District (Fig. 19.6).

This observed difference in the zones emanates from differences in crop types produced by smallholder farmers, even within the same zone. Whiles ginger is the major crop produced by farmers in the Kintampo South District, maize is the major crop produced by smallholder farmers in the Atebubu-Amantin District. Also, in the

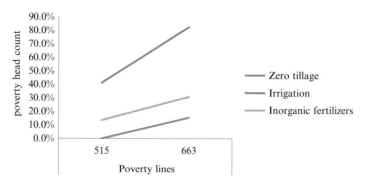

Fig. 19.4 Stochastic dominance test results for Amansie-West District

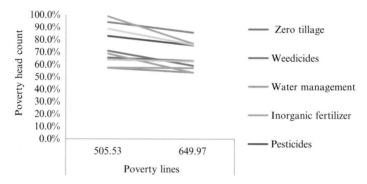

Fig. 19.5 Stochastic dominance test results for Kintampo South District

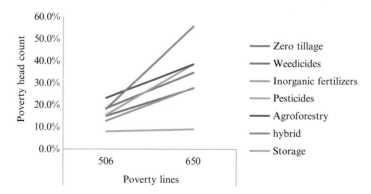

Fig. 19.6 Stochastic dominance test results for Atebubu-Amantin District

Tolon District of the savanna zone, pesticides and improved seeds were identified as the dominant technologies for reducing poverty among smallholder farmers in the district (Fig. 19.7). Similarly, in the Gonja-East District, improved seeds and agro-forestry were identified as the two dominant technologies (Fig. 19.8).

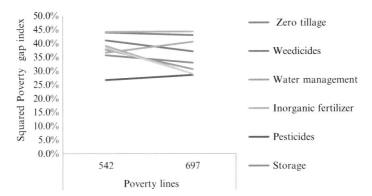

Fig. 19.7 Stochastic dominance test results for Tolon District

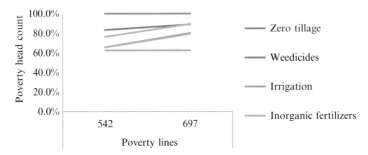

Fig. 19.8 Stochastic dominance test results for East-Gonja District

Potential Adoption Rate

The results of the potential adoption rate (PAR) of the technologies are shown in Table 19.1. The table presents the four farmer categories, very poor, poor, rich and very rich, aggregate value of the production goals index (G), and the production environment (E) for each of the farmer categories. From the results, the PAR for the forest zone is 23.3 %. This means that, other things remaining constant, the rate at which the technologies in the forest zone will be adopted or diffused is 23.3 %. This is, however, different from maximum adoption, which is the percentage of farmers who will adopt the technology. Predicted adoption rate for the transition and savannah zones are 22.5 % and 18.0 %, respectively.

Income Effect of Technologies

Table 19.2 shows the income effect of the recommended crop technologies for the study locations. The results of the *t*-test showed significant differences in income

Table 19.1 Potential adoption rates in the zones

Farmer categories	Forest Zone				Transition zone				Savannah zone		
	Very poor	Poor	Rich	Very rich	Very poor	Poor	Rich	Very rich	Very poor	Poor	Rich
Farmer categories (%)	14.4	46	38.1	1.4	9.6	65.4	22.4	2.6	31.8	42.1	26.2
Production goals (%)	50.6	55.1	53.9	68.8	48.3	53.4	49.5	51.6	43.9	48.3	44.2
Production environment (%)	35.0	45.8	42.1	50.0	48.9	43.5	41.9	33.3	39.2	39.3	39.3
PAR (Zone)	23.3				22.5				18.0		

Table 19.2 Income effect of crop technologies in study locations

District	Technology		N	Per capita expenditure	Mean difference	df	t	Sig. (2-tailed)	Income effect (%)
Afigya Kwabre	Inorganic fertilizers	Adopters	55	1171.12	914.18	66	3.420	.001	355.8
		Non-adopters	13	256.95					
Amansie-West	Zero tillage	Adopters	13	2313.32	1046.13	69	2.169	.034	82.6
		Non-adopters	58	1267.19					
Atebubu-Amantin	Storage	Adopters	47	7390.29	6258.03	65	1.917	.060	552.7
		Non-adopters	20	1132.26					
Kintampo South	Marketing	Adopters	2	891.65	70.85	87	.154	.878	8.6
		Non-adopters	87	820.80					
Gonja East	Improved seeds	Adopters	8	1018.67	660.85	55	3.388	.001	184.7
		Non-adopters	49	357.82					
Tolon	Pesticides	Adopters	17	513.42	173.64	48	1.701	.095	51.1
		Non-adopters	33	339.78					

between adopters and non-adopters of the technologies, except for the Kintampo South District. This insignificant difference may be due to the small sample size of adopters. The results further indicate income change from 8.6 % to 552.7 % for adopters of the various technologies.

Conclusions

Based on the findings of the research, the following conclusions are drawn:

- Dominant technologies that have the potential to reduce smallholder farmers' level of poverty and marginality are: inorganic fertilizers for Afigya-Kwabre; zero tillage for Amansie-West; storage facilities for Atebubu-Amantin; marketing facilities for Kintampo South; improved varieties for Gonja East; and pesticides for the Tolon Districts.
- Potential adoption rate varied among the various poverty segments.
- The technologies have significant effects on the incomes of adopters.

Policy Recommendations

The following recommendations are made from the study:

- Government should strengthen, resource, and build the capacity of institutions to train and offer support to smallholder farmers. These institutions should have a separate wing to see to the needs of smallholder farmers in helping them adopt innovations.
- The Ministry of Food and Agriculture (MoFA) and other partner organisations and ministries should also provide routine workshops and training for smallholder farmers.
- Government, through its Extension Services Directorate, should help disseminate, diffuse or up-scale technologies that have greater potential of reducing poverty and marginality.
  - For the Afigya-Kwabre District, technologies that enhance soil fertility, such as use of inorganic fertilizers, has a greater likelihood of reducing poverty in the district.
  - In the Amansie-West District, activities that will provide income to smallholder farmers during the off-cocoa season, such as use of zero tillage for short duration cropping, specifically vegetables, will help.
  - In the Atebubu-Amantin District, storage facilities for maize and a warehouse credit facility that will provide some income for farmers as they look for a good market price is the dominant strategy.

- In the Kintampo South District, marketing farmer-based organisations gives the farmers higher income.
- Improved yam minisetts is recommended for the Gonja East District
- Pesticides for controlling yam borers is the best technology for smallholder farmers in the Tolon District.

Open Access This chapter is distributed under the terms of the Creative Commons Attribution-Noncommercial 2.5 License (http://creativecommons.org/licenses/by-nc/2.5/) which permits any noncommercial use, distribution, and reproduction in any medium, provided the original author(s) and source are credited.

The images or other third party material in this chapter are included in the work's Creative Commons license, unless indicated otherwise in the credit line; if such material is not included in the work's Creative Commons license and the respective action is not permitted by statutory regulation, users will need to obtain permission from the license holder to duplicate, adapt or reproduce the material.

References

Ackah C, Aryeetey E (2012) Introduction and overview. In: Ackah C, Aryeetey E (eds) Globalization, trade and poverty in Ghana. Sub-Saharan Publishers and International Development Research Centre, Accra

Adekunle AA, Ellis-Jones J, Ajibefun I, Nyikal RA, Bangali S, Fatunbi O, Ange A (2012) Agricultural innovation in sub-Saharan Africa: experiences from multiple-stakeholder approaches. Forum for Agricultural Research in Africa (FARA), Accra

Al-Hassan R, Diao X (2007) Regional disparities in Ghana: policy options and public investment implications. International Food Policy Research Institute, Washington, DC

Antwi-Agyei P, Fraser EDG, Dougill AJ, Stringer LC, Simelton E (2012) Mapping the vulnerability of crop production to drought in Ghana using rainfall, yield and socioeconomic data. Appl Geogr 32:324–334

Askari H, Cummings S (1976) Estimating agricultural supply response in the Nerlove model: a survey. Int Econ Rev 18:257–292

Battese G (1996) A note on the estimation of Cobb-Douglas production functions when some explanatory variables have zero values. J Agri Econ 48:250–252

Berkhouta ED, Schippera RA, Kuyvenhovena A, Coulibalyb O (2010) Does heterogeneity in goals and preferences affect efficiency? A case study of farm households in northern Nigeria. Paper presented at the international association of agricultural economists 2009 conference, Beijing, 16–22 Aug 2009

Braun E (1998) Technology in context: technology assessment for managers. Routledge, London

Davidson R, Duclos JY (2000) Statistical inference for stochastic dominance and measurement for the poverty and inequality. Econometrica 68:1435–1464

de Janvry A, Fafchamps M, Sadoulet E (1991) Peasant household behaviour with missing markets: some paradoxes explained. Econ J 101(409):1400–1417

Deaton AS (1990) Savings in developing countries: theory and review. World Bank Econ Rev 3:183–210

European Commission (2006) Cost-benefit assessment and prioritisation of vehicle safety technologies: final report. European Commission Directorate General Energy and Transport, Brussels

Fafchamps M (1993) Sequential labor decisions under uncertainty: an estimable household model of West-African farmers. Econometrica 61(5):1173. doi:10.2307/2951497

Foster J, Greer J, Thorbecke E (1984) A class of decomposable poverty measures. Econometrica 52(3):761. doi:10.2307/1913475

Frimpong S, Asuming-Brempong S (2013) Comparative study of determinants of food security in rural and urban households of Ashanti region, Ghana. Int J Econ Manag Sci 2(10):29–42

Garrido MV, Kristensen FB, Nielsen CP, Busse R (2008) Health technology assessment and health policy-making in Europe: current status, challenges and potential. Observatory studies series No 14., European Observatory on Health Systems and Policies, Brussels

Ghana Statistical Service (2012a) Population and housing census (2010). Ghana Statistical Service, Government of Ghana, Accra

Ghana Statistical Service (2012b) Revised gross domestic product 2011. Statistical Service, Government of Ghana, Accra

Hildebrand PE, Russell JT (1996) Adoptability analysis. A method for the design, analysis and interpretation of on-farm research-extension. Iowa State University Press/Ames, Iowa City

IFAD/WFP/FAO (2011) The state of food insecurity in the world 2010. How does international price volatility affect domestic economies and food security? International Fund for Agricultural Development, World Food Programme and Food and Agriculture Organization of the United Nations, Rome

IFPRI (2004) Ending hunger in Africa: prospects for the small farmer. Policy brief from smallholder African agriculture: progress and problems in confronting hunger and poverty. IFPRI development strategy and governance division discussion paper 9. International Food Policy Research Institute, Washington, DC

IFPRI (2009) Estimating the impact of agricultural technology on poverty reduction in rural Nigeria, IFPRI discussion paper 00901. International Food Policy Research Institute, Washington, DC

Jamison A, Baark E (1990) Modes of biotechnology assessment in the USA, Japan and Denmark. Technol Anal Strat Manag 2(2):111–128

MoFA (2012) Agriculture in Ghana: fact and figures 2010. Statistics, Research and Information Directorate, Ministry of Food and Agriculture, Accra

Ray SC (2004) Data envelopment analysis: theory and techniques for economics and operations research. Cambridge University Press, Cambridge

Remenyi D, Money A, Sherwood-Smith M, Irani Z (2000) The effective measurement and management of IT costs and benefits, 2nd edn. Elsevier/Butterworth/Heinemann, Oxford

Rodney W (1984) How Europe underdeveloped Africa. Howard University Press, Washington, DC

Roe T, Graham-Tomasi T (1986) Yield risk in a dynamic model. In: Singh I, Squire L, Strauss J (eds) Agricultural household models: extension, application and policy. Johns Hopkins University Press, Baltimore, pp 255–276

Roetter RP, Van Keulen H, Van Laar HH (eds) (2000) Synthesis of methodology development and case studies. Makati City (Philippines), SysNet research paper series No 3. International Rice Research Institute, Los Baños

Ruben R, Menk H, Kuyvenhoven A (1998) Integrating agricultural research and policy analysis: analytical framework and policy applications for bio-economic modeling. Agr Syst 58 (3):331–349

Ruben R, Van Ruijven A (2001) Technical coefficients for bio-economic farm household models: a meta-modelling approach with applications for Southern Mali. Ecol Econ 36:427–441

Ruben R, Kruseman G, Kuyvenhoven A (2006) Strategies for sustainable intensification in East African highlands: labour use and input efficiency. Agr Econ 34:167–181

Scaillet O, Topaloglou N (2010) Testing for stochastic dominance efficiency. J Bus Econ Stat 28:169–180

Singh I, Squire L, Strauss J (1986) Agricultural household models: extensions, applications and policy. John Hopkins University Press, Baltimore

Simelton E (2012) Mapping the vulnerability of crop production to drought in Ghana using rainfall, yield and socioeconomic data. Appl Geogr 32(2012):324–334

Smale M, Nkonya M, Nkonya E (2003) 'Information about farmers use of banana varieties in IFPR's uganda policy database' in assessing the social and economic impact of improved banana varieties in East Africa. EPTD workshop summary paper No 15. Uganda Strategy Support Program/IFPRI, Kampala

Steenhuijsen Piters B (1995) Diversity in fields and farmers: explaining yield variations in Northern Cameroon. PhD thesis, Wageningen Agricultural University, Wageningen

Stoorvogel JJ, Antle JM, Crissman CC, Bowen W (2004) The tradeoff analysis model: integrated bio-physical and economic modeling of agricultural production systems. Agr Syst 80(1):43–66

Sullivan C (2002) Calculating a water poverty index. World Dev 30:1195–1210

van Keulen H, Kuyvenhoven A, Ruben R (1998) Sustainable land use and food security in developing countries: DLV's approach to policy support. Agr Syst 58(3):285–307

Veenenklaas FR, van Keulen H, Cisse S, Gosseye P, van Duivenboden N (1994) Competing for limited resources: options for land use in the fifth region of Mali. In: Fresco LO, Stroosnijder J, Bouma J, van Keulen H (eds) The future of the land: mobilising and integrating knowledge for land use option. Wiley, Chichester, pp 227–247

WFP (2009) Comprehensive food security and vulnerability analysis, Executive brief. World Food Programme, Rome

Wood (2003) Strategic ex ante evaluation of the potential economic benefits of improved banana productivity' in assessing the social and economic impact of improved banana varieties in East Africa. EPTD workshop summary paper No 15. Uganda Strategy Support Program/IFPRI, Kampala

World Bank (2010) The World Bank annual bank conference on land policy and administration Washington, DC April 26 and 27, 2010; Government's role in attracting viable agricultural investment: experiences from Ghana, Accra. Number 133. The World Bank, Washington, DC

Wulsin L Jr, Dougherty A (2008) A briefing on health technology assessment. California Research Bureau, California State Library, Sacramento

Yamane T (ed) (1967) Statistics: an introductory analysis. Harper and Row, New York

Chapter 20
Potential Impacts of Yield-Increasing Crop Technologies on Productivity and Poverty in Two Districts of Ethiopia

Bekele Hundie Kotu and Assefa Admassie

Abstract Ethiopian agriculture is characterized by low productivity which has contributed to the persistence of food insecurity and poverty in the country. Reports indicate that several yield-increasing technologies are available but have not yet been adequately utilized. This chapter assesses the potential impact of yield-increasing crop technologies on productivity and poverty based on the data collected from two districts in Ethiopia. We focus on the use of improved seeds, together with appropriate agronomic packages such as chemical fertilizers and row planting technique. Results suggest that the resulting monetary gains would be enough to lift the "better-off" poor households out of poverty, but they would not be enough to lift up the ultra-poor out of poverty, implying that other livelihood strategies are desirable for improving the well-being of the latter.

Keywords Agricultural productivity • Food insecurity • Yield-increasing technologies • Ex-ante assessment • Poverty

Introduction

In his lecture when accepting the 1979 Nobel Prize in Economics, T.E. Shultz posited "...[m]ost of the world's poor people earn their living from agriculture, so if we knew the economics of agriculture, we would know much of the economics of being poor."[1] This is a worthy statement for scholars, practitioners, and

---

[1] see http://www.nobelprize.org/nobel_prizes/economic-sciences/laureates/1979/schultz-lecture.html.

B.H. Kotu (✉)
International Institute of Tropical Agriculture, Tamale, Ghana
e-mail: bekelehu@yahoo.com

A. Admassie
Ethiopian Economic Association Yeka, Addis Ababa, Ethiopia
e-mail: aadmassie@yahoo.com

© The Author(s) 2016
F.W. Gatzweiler, J. von Braun (eds.), *Technological and Institutional Innovations for Marginalized Smallholders in Agricultural Development*,
DOI 10.1007/978-3-319-25718-1_20

policymakers focusing on poverty reduction, and is quite relevant to be considered in regard to Ethiopia because of the fact that agriculture is the dominant sector in the Ethiopian economy and because poverty is mainly a rural phenomena in Ethiopia.[2] In fact, agriculture has a strong multiplier effect on the current economy and, if nurtured to grow fast, will have a better impact on poverty than other sectors (Diao et al. 2010). Such a comparative advantage of agriculture has been recognized by the governments of Ethiopia at different times since the 1980s (and even before), although policy interventions couldn't always produce positive outcomes as desired. In fact, policy interventions prior to 2001 couldn't bring substantial changes in agriculture and, hence, in poverty levels in rural areas. The total production of cereal crops was stagnant during the 1980s due to stagnant yield levels (Hundie 2012). While the growth in production was substantial (i.e., 6.6 %) in the 1990s, the source of growth was expansion of agricultural lands to marginal areas. However, average yield levels have been rising since 2003, resulting in a rapid growth of output (i.e., 7.5 % of average annual growth in cereal production between 2003 and 2012) (ibid). This is a desirable result which implies the success of the national policies during this period.[3]

Nevertheless, the average yield of staple food crops in Ethiopia is still low, which has contributed to the persistence of poverty in the country's rural areas. For instance, wheat yield at 2 Mt/ha is 65 % below the average of the best African region (i.e., Southern Africa) and 260 % below the average of the best world region (i.e., Western Europe) (FAOSTAT 2013). The low yield observed among the staple crops is mainly attributed to low use of improved technologies. Evidence shows that only 7.3 % of the area under cereals was planted with improved varieties in 2010/2011 (CSA 2011a, b). While maize is far better than the other crops in terms of percentage of area under improved varieties (28 %), the adoption rate is still low as compared to several countries in eastern and southern Africa, such as Zimbabwe (80 %), Zambia (75 %), Kenya (72 %), and Uganda (35 %) (ATA 2012).

On the other hand, a large number of improved varieties are available which can be used to increase productivity. More than 700 improved crop varieties are ready for use together with their agronomic packages (MoA 2011). About 70 were released recently (i.e., in 2011). Grain crops are dominant in terms of the total number of improved seeds, constituting about 68 % of the technologies. Among the grain crops, cereals account for about 56 % of the total varieties corresponding to that category, followed by pulses (30.6 %) and oil crops (13.8 %). Cereals are dominated by wheat,[4] constituting about 31 % of the total varieties in that category, whereas

---

[2] For instance, considering that poverty rates in rural and urban areas are 30.4 % and 25.7 %, respectively (MoFED 2012), and given that more than 80 % of the Ethiopian population live in rural areas, it is possible to conclude that rural areas account for the majority of poor people in Ethiopia.

[3] The two national development strategies associated with this period are the Plan for Accelerated and Sustainable Development of End Poverty (PASDEP) (2004/2005–2009/2010) and the Growth and Transformation Plan (GTP) (2010/2011–2014/2015).

[4] Wheat constitutes bread wheat, durum wheat, emmer wheat, and buck wheat.

maize (16.2 %), barley[5] (14.7 %), sorghum (12.9 %), tef (11.5 %), rice[6] (7.2 %), millet[7] (4 %), and others (1.8 %) take consecutive ranks in that order (MoA 2011).

While agricultural growth is an important means for reducing poverty in rural areas of developing countries, marginality is cited as a root cause of extreme poverty (Gatzweiler et al. 2011; von Braun et al. 2009; Ahmed et al. 2007). This is because of the fact that marginality implies the presence of a set of constraints which need to be lifted in order to recognize the capabilities of people and transform them into functioning actors (Gatzweiler et al. 2011). Thus, focusing on marginal areas would be a good strategy to be followed in order to bring a substantial reduction of extreme poverty in rural areas of developing countries, which might also apply to the Ethiopian case. However, the level of contributions of agriculture to poverty reduction may not be evenly distributed across all marginalized areas where agriculture is practiced as a means of livelihood. Intuitively speaking, its contribution would be high where the potential is high and could decline as potential declines due to relatively low/high costs of production in high/low potential areas.

Therefore, this research was initiated with the purpose of assessing the potential contribution of the adoption of yield-increasing crop technologies (YICT) to household poverty in marginalized areas characterized by high potential of agriculture. While several technologies could increase crop yield, we consider here the use of improved seeds together with appropriate agronomic packages, such as chemical fertilizers and row planting technique. The remainder of the paper is divided into five sections. Sections "Impacts of Agricultural Technologies" and "Time and Adoption" review literature on the impacts of YICT and the time dimension of adoption. Section "Methods of the Study" describes the methods used for the study, including selection of the study areas and households, sources of data, and methods of data analysis. Section "Household Strata" describes the household strata with regards to income poverty. Section "Potential Impacts of the Introduction of the Technologies" presents the potential impacts of existing YICT on poverty among different strata of households. The last section concludes the paper.

Impacts of Agricultural Technologies

Technologies are important sources of productivity growth in agriculture, thereby leading to better income and lower poverty.[8] This was observed practically in Asia and parts of South and Central America during the so-called "Green Revolution" era in the

---

[5] Barley constitutes food barley and malt barley.

[6] Rice constitutes irrigated type and upland type.

[7] Millet constitutes finger millet and pear millet.

[8] See, for instance, Nomaan Majid, Reaching Millennium Goals: How well does agricultural productivity growth reduce poverty? ILO Employment Strategy Department, 2004/2012.

1960s and 1970s. An important manifestation of the green revolution was the adoption of YICT such as improved seeds, chemical fertilizers and pesticides, along with expansion of infrastructure such as irrigation, roads and electricity. The widespread adoption of scientific agricultural techniques during this period resulted in a rise in labor productivity, thereby increasing income and reducing poverty (Hazell 2009).

Studies on post GR impacts in different countries also show that improved agricultural technologies have positive impacts on yield level and variability, income, and food security, as well as poverty (Macharia et al. 2012; Krishna and Qaim 2008; Napasintuwong and Traxler 2009; Hareau et al. 2006, Qaim 2003). Macharia et al. (2012) assessed the potential economic and poverty impacts of 11 improved chickpea varieties in Ethiopia using the economic surplus approach. They found that the new technologies can generate a total of $111 million (US) for 30 years, which could lift more than 0.7 million people (both producers and consumers) out of poverty. Krishna and Qaim (2008) studied the potential impacts of Bt eggplant on economic surplus and farmers' health in India. The results show that the technology can significantly reduce insecticide applications and increase effective yields while generating an economic surplus of about $108 million (US) per year, which could be harnessed by diverse economic groups, including resource-poor farmers. The ex-ante analysis on the benefits of herbicide-resistant transgenic rice in Uruguay using a stochastic simulation technique show that the technologies would generate a benefit of $1.82 million (in terms of mean net present value) for producers and $0.55 million for the multinational corporations who develop the technologies (Hareau et al. 2006). Napasintuwong and Traxler (2009) estimated that total economic surplus of the adoption of GM papaya in Thailand is in the range of $650 million to $1.5 billion, which would be generated within the first 10 years of adoption. The primary beneficiaries would be small-scale papaya farmers, who would benefit even with the loss of export markets.

Time and Adoption

Adoption of technologies may not take place overnight. Rather, it is a process that occurs over time. Davies (1979) posits that adoption starts with innovators and expands via early adopters, early majority, and late majority, finally ending with laggards. Studies by Mosher (1979), Rogers (1983), Mahajan and Peterson (1978), and Bera and Kelley (1990) associate variations observed among households in adoption decisions to variations in the capacity of households to acquire and process information, as well as differences in resource constraints. The distribution of total net benefit from adoption of the new technology depends on the adoption path. However, adoption rates are highly uncertain in ex-ante analysis and empirical results are rarely available for consumption. The difficulty of fixing ex-ante values arises from the fact that many factors affect both the adoption path and the maximum rate of adoption. Despite variable adoption patterns that might exist, Alston et al. (1995) suggest sigmoid curves for adoption paths in ex-ante studies, which is what we have adopted in our analysis.

The aggregate net benefit generated from adoption of improved technologies depends on the maximum achievable rate of adoption and the speed of adoption. Indeed, these are important for assessing the impacts of the technologies, taking into account the time dimension of adoption. The first challenge to this exercise is fixing the maximum achievable adoption rate. While results are variable depending on the nature of technologies and other contexts, empirical studies indicate that adoption rate of improved technologies of cereals can be 90 % or more (Ephraim and Featherstone 2001; Motuma et al. 2010; Hailu 2008). For instance, Ephraim and Featherstone (2001) reported a 90 % adoption rate for improved maize varieties in Tanzania, while, more recently, Motuma et al. (2010) reported an adoption rate of 92 % in one district of central Ethiopia. Similarly, Hailu (2008) reported that nearly 90 % of the farmers in central Ethiopia adopted improved wheat varieties and associated technologies.

A few studies are available to fix ideas regarding the speed and pattern of adoption of crop technologies. Tesfaye et al. (2001) showed that it took about 30 years to reach an adoption rate of about 90 % in northwestern Ethiopia. That study revealed a sigmoid pattern of adoption whereby about 80 % of the total change was attributed to the last 10 years. However, the speed of adoption can be influenced by external intervention such as good extension services and, hence, the maximum length of time to reach the maximum rate of adoption can be shortened (Tesfaye et al. 2001; Hundie et al. 2000; Motuma et al. 2010). For example, Tesfaye et al. (2001) noted that, while the rate of adoption of improved wheat varieties has increased from about nil to 72 % within 20 years in Ethiopia, most of the changes occurred within 6 years after the extension system had been strengthened. Similarly, other studies (e.g., Hundie et al. 2000; Motuma et al. 2010; Hailu 2008) show that extension services have significant effect on adoption.

Methods of the Study

Selection of Marginality Hotspot Districts (Woredas)

Marginality hotspots are rural areas in which high prevalence of poverty and high agricultural potential overlap (Graw and Ladenburger 2012). Based on Gatzweiler et al. (2011), marginality is defined as "an involuntary position and condition of an individual or group at the margins of social, political, economic, ecological and biophysical systems, preventing them from access to resources, assets, services, restraining freedom of choice, preventing the development of capabilities, and eventually causing extreme poverty" (2011, p. 3). We used a two-step procedure to identify the woredas for study. First, marginality hotspots were identified based on the work of Graw and Ladenburger (2012), which classifies areas in Ethiopia with respect to marginality levels. Out of seven levels of marginality identified in that paper,[9] we selected areas marginalized in terms of four dimensions or more based on a visual assessment of the marginality hotspot map.[10] These include 37 woredas (15 in the Amhara region and 22 in SNNPR). These woredas represent a number of the country's different agro-ecologies, and, hence, they are diverse in terms of farming systems, i.e., cereal-based, perennial-crop-based, and livestock-based pastoral areas. We focused on cereal-based farming systems, specifically on those areas producing the four major cereals in Ethiopia, teff, maize, sorghum, and wheat. This was done for the sake of maximizing the benefit of a focused analysis.[11] Moreover, cereals were focused on due to the fact that they are dominant sources of calories and income among Ethiopian smallholders and, hence, their contribution to food security is quite substantial. A total of 17 cereal-based woredas were identified, out of which 15 are from the Amhara region and 2 are from SNNPR.

The second parameter considered to identify the study areas was agricultural potential. Precipitation was used as a proxy variable to measure agricultural potential. Thus, all drought-prone woredas were excluded, which resulted in 11 woredas.[12] The woredas lacking information on precipitation level were also excluded. The remaining six woredas (4 from the Amhara region and 2 from SNNPR) were prioritized based on their market access.[13] Two woredas from the Amhara region and 1 woreda from SNNPR were dropped since they have better market access, which resulted in 2 woredas from the Amhara region (Debre Sina/ Borena in the South Wollo zone and Baso Liben in the East Gojam zone) and one woreda from SNNPR (Halaba special woreda). These woredas are characterized by high prevalence of poverty and marginality, as characterized by Graw and Ladenburger (2012) and by high agricultural potential. Finally, Baso Liben and Halaba were selected for our study while Debre Sina was dropped because of its lower potential in cereal production than Baso Liben. Figure 20.1 displays the location of the study areas.

Selection of Subdistricts (Kebeles)

Both marginality and agricultural potential were considered in selecting kebeles. However, our sampling at this stage was dependent on local knowledge and less

---

[9] The dimensions considered in Graw and Ladenburger (2012) are: (1) economy, (2) demography, (3) landscape design, land use and location (spatial variables), (4) behavior and quality of life, (5) ecosystem, natural resources and climate, (6) infrastructure, (7) public domain and institutions.

[10] The detailed maps used to identify marginality hotspots were obtained from ZEF (courtesy of Christine Hausmann).

[11] However, cereal-based systems are not devoid of livestock and perennial crops and, hence, the possibility that farm households can be directed toward the latter options can be assessed.

[12] The data on moisture status at the woreda level was obtained from IFPRI-ESSP II (courtesy of Dr. Alemayehu Seyoum Tafesse).

[13] The data on market access at the woreda level was obtained from IFPRI-ESSP II.

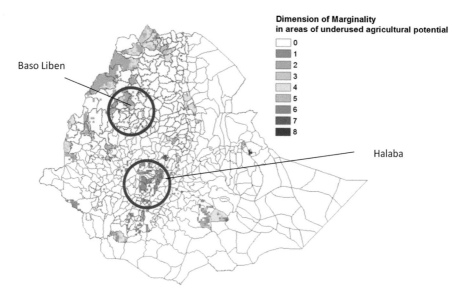

Fig. 20.1 Location of the study woredas in Ethiopia

sophisticated variables, since we couldn't find a readymade dataset to locate marginality hotspots at the kebele level. Marginality was proxied by the distances of kebeles from woreda towns, whereas agricultural potential was proxied by a composite of parameters such as amount and reliability of rainfall, irrigation potential, soil fertility, and topographic characteristics. Assessments of the kebeles against these parameters were made by agricultural experts from the selected woredas. The sampling involved several steps. First, all cereal-based kebeles were identified and put under three categories (nearby, medium, and far) based on their distance from woreda towns. Kebeles more than 10 km away from woreda administrative centers were put under the "far" category, and, hence, were considered to be marginalized. Second, kebeles were put under three groups based on agricultural potential (i.e., high potential, medium potential, and low potential); the categorization was done based on assessments made by experts at woreda offices of agriculture, taking into account the factors listed earlier. Third, a new list of kebeles was developed constituting marginalized kebeles and those having high agricultural potential. Last, three cereal-based kebeles (mainly growing wheat, teff, and maize) were purposively selected based on their accessibility for conducting the household survey.

Selection of Households

Households were stratified based on two parameters, namely landholding and gender of household head. Land is traditionally used as a stratifying variable due to the fact that it is a crucial asset among Ethiopian smallholder farmers.[14] Gender was used because one can easily obtain gender disaggregated lists from secondary sources, and gender is one of the key factors related to poverty. The selection process involved three steps. First, households were grouped into three categories based on landholding: bottom category (<34 %), medium category (34–66 %), and upper category (>66 %). Second, each stratum was stratified again into two sub-strata based on the gender of household heads. Third, sample households were drawn from each sub-stratum using a proportionate-to-size sampling technique such that the total number of sample households in each stratum would be 20. The entire process of sampling resulted in a total of 360 sample households.

Post-survey Re-stratification of Households

The sample households were re-stratified after the survey based on income level. Income level is used as a measure of living standard. Basically, the households earn income from three different sources, namely: crop production, livestock production, and off-farm activities. Expenses for commercial inputs such as fertilizer, improved seeds, chemicals, machinery rents, and hired labor were deducted from income from crop production. The purchasing power parity (PPP) exchange was used to convert the income into USD.[15] The total income of each household was converted into per capita income (PCI) per day, which was used to stratify households into four strata, namely: (1) better-off households (those who earn a PCI of at least $1.92/day); (2) subjacent poor (those with income between $1.54/day and $1.92/day); (3) medial poor (those with income between $1.15/day and $1.54/day); and (4) ultra poor (those who earn less than $1.15/day).[16]

In addition to the above method, stratification was done by using self-reported data gathered from the sample respondents on their households' wealth status. In this regard, respondents were asked to position their households within the above four strata, taking

---

[14] The other important asset is livestock, but this doesn't lend itself for use in the sampling process of rural household surveys due to the absence of secondary data on livestock ownership at the household level. Thus, we did not consider it in our sampling process.

[15] According to the IMF World Economic Outlook, the Implied PPP conversion rate for Ethiopia at the end of 2012 was 7.04 (http://www.quandl.com/IMF-International-Monetary-Fund/MAP_WEO_IMPCR_ETH-World-Economic-Outlook-Implied-PPP-Conversion-Rate-Ethiopia) (accessed on 28/03/2013). A slightly different rate (i.e., 7.2) is implied in http://www.indexmundi.com/ethiopia/economy_profile. We adopted the rate implied in the World Economic Outlook.

[16] According to the project manual, the stratification goes like this: subjacent poor are those with incomes between $1 and $1.25/day, medial poor: between 75¢ and $1/day, and ultra-poor: below 75¢/day. We adjusted these cut-off values by considering the national poverty line, which was set to be 3,781 birr (or $702) per annum in 2011 (PPP rate for 2011 = 5.389). We also added the better-off category to include households with daily PCI greater than the poverty line.

into account other households in their kebeles.[17] Such an introspective approach may be useful for capturing the objective reality of the households' conditions and their capabilities to mitigate or cope with various kinds of risks, since it gives room for the households to reflect on their household situation. Thus, results based on this stratification are supposed to supplement the findings regarding the income-based stratification.

Methods of Data Analysis

Adoption of the recommended technologies can be considered as a partial adjustment to households' crop enterprises. Thus, a partial budget approach was used to elicit the potential income impacts of adopting the technologies. Four important adjustments are expected to occur due to the introduction of new technologies. These are: (1) increased benefits, (2) reduced costs, (3) increased costs, and (4) forgone/reduced income. The first two constitute changes in revenue as a result of adopting the technologies, whereas the last two constitute changes in costs associated with the technologies.

The average net benefit per hectare from adopting a new YICT is given by

$$NB_{jt} = P_{jt}\Delta Y_j - \Delta VC_{jt}, \tag{20.1}$$

where NB_{jt} is the net benefit per hectare from crop j in year t, P_{jt} is the price of crop j in year t, ΔY_j is the yield gap of crop j to be filled by applying the technology, and ΔVC_{jt} is the change in variable cost of production of crop j per hectare due to the introduction of the technology.

$$\Delta VC_{jt} = VC_{jtN} - VC_{jtT}, \tag{20.2}$$

where VC_{jN} is the variable cost of production of crop j per hectare under a new technology and VC_{jT} is the variable cost of production of crop j per hectare under the existing technology in year t.

Computing net benefits per hectare for each year requires forecasted output and input prices for each year, which may be difficult to acquire. In that case, a simpler approach may be adopted, i.e., computing the net benefit per hectare for the base year based on the actual price information and adjusting it over time based on a forecasted average inflation rate. This may help to adjust for possible changes in costs of living over time. This can be done as follows:

$$NB_{jt} = NB_{j0}(1+r)^t, \tag{20.3}$$

where r is an average annual inflation rate.

---

[17] For the sake of convenience during interview, the above four strata were defined as rich, intermediate, poor, and very poor.

The total net benefit in year t can be computed as:

$$NBT_{jt} = A_{jt}NB_{jt},\qquad(20.4)$$

where NBT_{jt} is the total net benefit of all famers applying the new technology to produce crop j in year t, and A_{jt} is the area under crop j planted with the new technology.

The distribution of total net benefit over time from adoption of the new technology depends on the adoption path. That is, we require the change of area under crop j planted with the new technology and the rate of adoption over time. However, adoption rates are highly uncertain in ex-ante analysis and empirical results are rarely available for consumption. The difficulty of fixing ex-ante values arises from the fact that many factors affect both the adoption path and the maximum rate of adoption. Despite variable adoption patterns that might exist, Alston et al. (1995) suggest logistic curves for adoption paths in ex-ante studies. Following their suggestion,

$$A_{jt} = f(t) = \frac{L}{1 + be^{-kt}},\qquad(20.5)$$

where L is the expected maximum land to be allocated to improved technologies, t is time, and b and k are constants.

Therefore, the total net benefit of all farmers adopting the technologies in a given year can be given by:

$$NBT_{jt} = NB_{j0}\frac{L}{1 + be^{-kt}}(1 + r)^t.\qquad(20.6)$$

The total net benefit over the entirety of adoption years (NBG_j) is given by the cumulative of the above distribution:

$$NBG_j = \int_0^T NB_{j0}\frac{L}{1 + be^{-kt}}(1 + r)^t dt,\qquad(20.7)$$

where T is the maximum number of years it takes to reach the maximum adoption level. The net present value of the aggregate net benefit over the years (NBG_{pj}) can be computed as:

$$NBG_{pj} = \int_0^T NB_{j0}\frac{L}{1 + be^{-kt}}(1 + r)^t(1 + \rho)^{-t} dt,\qquad(20.8)$$

where ρ is the nominal discount rate.

The total net present value of net benefits from adoption of YICT for the production of the three crops is, therefore,

$$NBG_{pT} = \sum_{j=1}^{3} NBG_{pj}. \qquad (20.9)$$

Measuring Components of Net Benefit

Crop Price

The average prices for each of the study districts were extracted from the household survey data of this study, which was conducted in February 2013. These prices were used as a base for the three crops. The average prices of maize, tef, and wheat are, respectively, 5.2, 10.5, and 6 birr/kg in Baso Liben, whereas the average prices of maize and tef are, respectively, 4.5 and 11.35 birr/kg in Halaba.

Measuring Yield Gap

We defined two levels of potential yields which led to two levels of yield gaps.[18] The first one is based on the average grain yields of selected improved varieties, as reported in the national crop variety register, which is published by the Ministry of Agriculture every year (MoA 2011). In this regard, a variety known as BH 540 was considered to compute yield gaps of maize in Halaba, while BH 660 was used for Baso Liben. These varieties are under distribution in these areas. With regards to tef, two varieties, namely Quncho /Dz-Cr-387 (RIL-355)/ and Tseday (Dz-Cr-37), were considered in both areas. Since both varieties have not been distributed yet in the study areas, but are suitable varieties for distribution, we defined the potential yield as the average yield of these two varieties in both areas. A variety known as Kakaba (Picaflor) was considered for potential yield of wheat. The second type of yield gap was computed taking into account the average yields of the top 10 % performing households. In this case, the average of top 10 % of yield was computed for each of the selected crops and districts and used to assess the yield gaps.

Yields based on on-farm variety trials are superior to average yields attained by the top 10 % performing farmers. On-farm variety trials are conducted under the

---

[18] Yield gaps can be defined as the difference between what is attainable and what is actually attained by the farmers. However, what is attainable (i.e., yield potential) can vary depending upon the level of definition yielding different types of yield gap. The first type of yield gap is the difference between what is theoretically conceived by scientists and what is attained at experimental stations. The second type of yield gap is the difference between yield at the experimental station and potential yield at farmers' yield, perhaps due to environmental conditions and technological differences between experimental stations and farms. The third (last) type is the difference between potential on farm yield and actual farmers' yields. We considered the third type of yield gap in our analysis.

highest care of researchers and, hence, may be difficult to achieve in the short run by many poor farmers who lack capacities.[19] Thus, such high yield levels can be taken as the upper boundary for comparing local achievements. On the other hand, the "top 10 %" average yield has already been achieved by some farmers. However, some of these farmers might not have used recommended practices. Thus, the "top 10 %" yield can be taken as the lower boundary for comparing achievements. Therefore, the "top 10 %" average yield can be taken as the second best, while the "on-farm trial" yields can be taken as the first best targets for policymakers and practitioners at different levels.

Computing Variable Costs

Several variable costs are incurred in the process of crop production; costs incurred from labor, chemical fertilizers, oxen power, pesticides, seeds/planting materials, and rents of farm machineries are the major ones. However, not all of these are important in our partial budget analysis. Actually, we were interested in those costs which are affected due to the introduction of the new technologies. In this regard, we assumed that only seed and fertilizer costs would be affected due to the introduction of the new technologies under consideration.[20] Prices of fertilizer and seeds prevailing in 2013 were used as a base.

Household Strata

Farming households were put under four strata based on the criteria discussed in section "Time and Adoption". Considering the income-based approach, about 10 % of the households were categorized as better-off while the rest were put under the three "poor" categories, i.e., subjacent poor, medial poor and ultra-poor. Ultra-poor households constitute 71.4 % of the total households, whereas the subjacent poor and the medial poor account for about 5.3 % and 13.1 %, respectively. The mean per capita income for all households is $1.26/day, which is below the national poverty line.[21] The mean values for better-off, subjacent poor, medial poor, and ultra-poor households are $3.57, $1.71, $1.30, and 66¢, respectively. The F-test indicates that there is a significant difference among the four strata of households with regards to per capita income per day (p = 0.000). Post hoc multiple

---

[19] Most of them are equivalent to average yields of developed countries.

[20] Row planting is expected to increase labor input during planting, but experts comment that it reduces labor input during weeding. Since its net effect has not yet been studied, we assumed that row planting wouldn't affect the aggregate labor input.

[21] The national poverty line is estimated at 3781 birr per person per annum (equivalent to $1.92 per person per day using the PPP rate of 2012).

comparisons show that better-off households are significantly better off than all other strata of households in terms of per capita income per day; ultra-poor households are also significantly different from all others from the lower side. However, the two strata in the middle of the poverty spectrum (i.e., the subjacent poor and the medial poor) do not differ significantly from each other.

Results based on the self-reported stratification show that 5.8 % of households are within the category of better-off, while the rest of the households fall in the remaining three strata. The subjacent stratum takes the largest share, constituting 50.6 % of the total households, which is followed by the medial poor (36.9 %). The ultra-poor constitute only 6.7 % of the total number of households. The mean daily per capita income significantly varies among households in different strata. Better-off households could earn about $2.05 per day on a per capita basis. This is above the national poverty line set by the Ministry of Finance and Economic Development in 2010/2011[22] (MoFED 2012). The mean daily per capita income for the ultra-poor is about 53¢, which is about one fourth of the income of the better-off. The figures corresponding to the subjacent poor and the medial poor are $1.42 and $1.05. A post hoc multiple comparison test (using Tamhane's test) shows that better-off households are significantly different from the medial poor and the ultra-poor, but not different from the subjacent poor. On the other hand, the ultra-poor households are significantly lower than households in all other strata in terms of income. The two strata of poor households in the middle are not different from each other.

Potential Impacts of the Introduction of the Technologies

The direct potential effect of the intervention is that the yields of the target crops will grow substantially, resulting in a rise in total production that can be consumed and/or sold by the households. Here, we present the potential benefits disaggregated by the three target crops and by the household strata we have defined so far. In both cases, the analyses for the two locations were done separately.

Potential Net Benefits by Crop Type

The total net benefits for farmers from adopting the YICT by crop type and study district are presented in Table 20.1. The figures reported under the higher target case (HTC) are based on the assumption that smallholders would attain the average yields of the target crops equivalent to the average yields computed from on-farm

---

[22] The poverty line was set based on the data from Household Income and Consumption Expenditure Survey of 2010/2011 conducted by Central Statistical Agency of Ethiopia. The HICE survey covered 27,830 rural and urban households in the country.

Table 20.1 Net benefit per hectare of land from YICT

	Halaba				Baso Liben			
	NB, HTC (Birr/ha)	NB, LTC (Birr/ha)	IRR, HTC (%)	IRR, LTC (%)	NB, HTC (Birr/ha)	NB, LTC (Birr/ha)	IRR, HTC (%)	IRR, LTC (%)
Maize	18,024	3831	947	280	28,283	13,484	1844	931
Tef	13,880	6820	827	457	6,513	8,319	434	527
Wheat	–	–	–	–	22,601	12,335	3648	2036

Note: *NB* net benefit, *IRR* internal rate of return. *HTC* higher target case, *LTC* lower target case

variety trials. Those figures, which are reported under the lower target case (LTC), are based on the assumption that the current yield gaps between typical farmers and the most productive farmers (or the top 10 % of productive farmers) would be eliminated. This latter target is in line with the current target of the Ethiopian government to scale up best practices (MoFED 2010).

The additional net benefit is highest for maize in both districts under the HTC. Under the HTC, smallholders could get as high as 18,000 birr per hectare of additional net benefit in Halaba; the figure in Baso Liben is higher by nearly 60 %. While wheat is not dominantly produced in Halaba, it ranks second in Baso Liben in terms of additional net benefit per hectare. The additional net benefit in the case of tef is lower than that of maize, as well as wheat. It is lower by about 23 % in Halaba and by about 77 % in Baso Liben as compared to maize. However, the benefit from adopting YICT is quite high even for tef, as indicated by the Internal Rate of Return (IRR).[23]

Benefits corresponding to the LTC are substantially lower than that of the HTC. The exception is the case of tef in Baso Liben.[24] In Halaba, the net benefit under the LTC is about one-fifth of the figures corresponding to the HTC for maize, while it is about one half for tef. The net benefits corresponding to the LTC are about one half of the net benefit under the HTC for wheat and maize in Baso Liben. Nevertheless, the potential net benefits from YICT of the target crops are high in both districts, even under the LTC. The values of IRR reported in the table may confirm this assessment. Maize is the most rewarding crop in Halaba if technologies are adopted, followed by tef. Wheat takes the first rank in Baso Liben, while tef takes the second.

Potential Net Benefits by Household Strata

The distribution of the potential net benefit among households depends on the total size of land allocated for the target crops. Results are displayed in Table 20.2. The

---

[23] An IRR between 50 % and 100 % is supposed to be enough to adopt improved varieties and associated packages (CIMMYT 1988).

[24] The average of top 10 % yield is greater than the average yield of on-farm variety trials.

Table 20.2 Net benefit per household from YICT (Birr), by household strata

	Halaba		Baso Liben	
	Higher target case	Lower target case	Higher target case	Lower target case
Better-off	24,273	4497	30,236	11,171
Subjacent poor	18,131	3363	22,564	7976
Medial poor	24,022	4457	18,982	6653
Ultra-poor	14,262	2656	13,221	4618
Total	17,216	3198	17,962	6386

average net benefit per household under the HTC is about 17,000 birr in Halaba and nearly 18,000 birr in Baso Liben. The HTC yields more than five-fold higher benefits than that of the LTC in Halaba, while it yields about three-fold higher benefit than the LTC in Baso Liben. Under both targets, the benefits are not uniformly distributed among the four strata of households, i.e., better-off households would receive the highest benefit, while ultra-poor households would receive the lowest. The daily net per capita benefit from the adoption of YICT follows the same pattern as that of the net benefit per household, i.e., the highest benefit was computed for better-off households and the lowest for ultra-poor households in both districts and with respect to both targets.

Potential Impacts on Poverty

The average additional net benefit per household for each stratum was used to compute the potential impacts of the technologies on poverty after changing it to its dollar equivalent and computing per capita figures. Figures 20.2 and 20.3 display the potential impacts of the adoption of YICT on poverty reduction. If higher targets are achieved, households under the subjacent poor and the medial poor strata would be lifted up to the non-poor category in both districts. However, ultra-poor households remain under the poverty line, though the poverty gap substantially declines. Achievement of the lower targets has different effects in the two districts with respect to the medial poor households, i.e., it would enable them to be above the poverty line in Baso Liben, but it wouldn't do the same in Halaba. Again, households in the ultra-poor stratum would remain under the poverty line.

Time Considerations

Based on the pieces of information discussed in section "Time and Adoption", we made three important assumptions in conducting our analysis: (1) the maximum

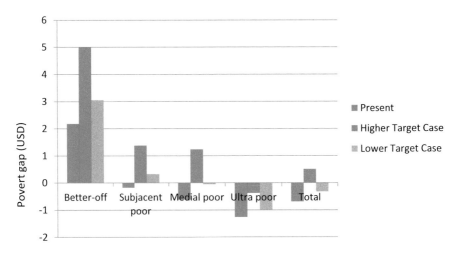

Fig. 20.2 Potential impacts of adoption of YICT on poverty reduction in Halaba

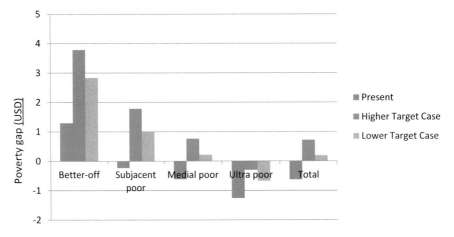

Fig. 20.3 Potential impacts of adoption of YICT on poverty reduction in Baso Liben

adoption rate is 90 %, (2) the maximum rate of adoption would be reached within 20 years after first introduction, and (3) about 80 % of the change would occur during the last 10 years. In regards to the speed of adoption, the above assumption can be considered as a "typical case scenario", as we dub it. We also considered another scenario in our computation, which could be called an "accelerated case scenario", whereby the maximum adoption would be reached within 10 years. The accelerated case scenario assumes that an efficient extension system would exist, while it requires a higher commitment of the government and non-state actors to realize the targets. Given the current ambition of the Ethiopian government to increase crop productivity by about 30 % in 2014/2015 (MoFED 2010) and the

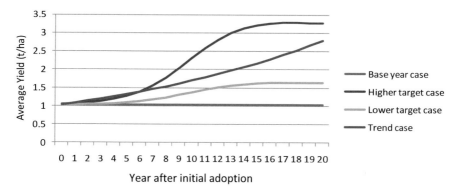

Fig. 20.4 Average grain yield of targeted crops in Halaba, typical case scenario

massive movements going on at grassroot levels to enhance adoption of improved agricultural technologies, it seems that the accelerated case scenario would be a plausible option for analysis.

The simulation results are displayed in Figs. 20.4, 20.5, 20.6, and 20.7. The results further assume that (1) all farmers will have access to extension services, (2) cropping patterns, and thus, land allocation among smallholders for the target crops, remain unchanged over time, (3) the growth rate of area under cereal production will be similar to that of the past 10 years (i.e., 1.26 %); and (4) the growth rate of the rural population will be similar to that of the past 10 years (i.e., a declining trend starting from 1.7 % in 2014).

The average yield of the target crops would potentially increase from about 1 t/ha at present to about 3.3 t/ha in Halaba and from about 1.6 t/ha to about 3.8 t/ha in Baso Liben under the typical scenario. This would happen when the HTC is realized, which will also lead to a rise in per capita grain production from 0.3 to 0.7 t in Baso Liben and from 0.2 to 0.6 t in Halaba. If the LTC is to be achieved, the average yield would rise from about 1 to 1.6 t/ha in Halaba and from about 1.6 t/ha to about 2.4 t/ha in Baso Liben. In this case, the per capita grain production would increase from 0.2 to 0.3 t in Halaba and from 0.3 to 0.5 t in Baso Liben. This would happen within two decades. The growth in grain productivity is expected to be faster under the accelerated scenario. In this case, the average yield would increase from about 1.2 to 3.9 t/ha in Halaba and from 1.7 to 4.1 t/ha in Baso Liben within a decade, provided that the higher target is achieved. This would also result in a rise of per capita grain production from 0.2 to 0.6 t in Halaba and from 0.3 to 0.8 t in Baso Liben. If the lower target is to be realized, the average yield would increase from 1 to 1.8 t/ha in Halaba and from 1.7 to 2.5 t/ha in Baso Liben. This would increase per capita production of the targeted crops from 0.2 to 0.3 t in Halaba and from 0.3 to 0.5 t in Baso Liben.

The figures also display projected actual yields based on the trend of average yield of the target crops in the past 10 years. The latter case shows what will happen to the average yield of the target crops if the current trend continues until the end of the projection period. The LTC can be reached within 10 years in Halaba and within 9 years in Baso Liben if average yield grows at the pace of the past 10 years. This

B.H. Kotu and A. Admassie

414

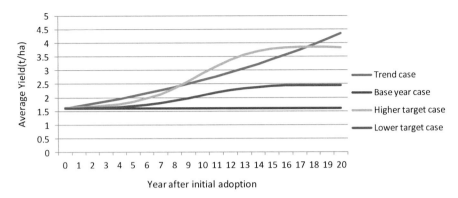

Fig. 20.5 Average grain yield of targeted crops in Baso Liben, typical case scenario

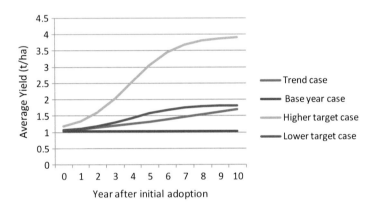

Fig. 20.6 Average grain yield of targeted crops in Halaba, accelerated case scenario

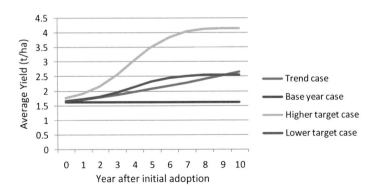

Fig. 20.7 Average grain yield of targeted crops in Baso Liben, accelerated case scenario

implies that no additional effort is needed to achieve the LTC under the typical case scenario, further implying that the extension system is doing well. However, sustaining the current productivity growth for the coming decade requires additional efforts by itself, which may include promoting the YICT proposed in this paper. The growth paths produced by our model to achieve the higher yield targets deviate considerably from the forecasted trend line, implying that more efforts should be made (than are being made at present) to realize these targets, even within two decades. There would be a better possibility of reaching the targets within the stated time in Baso Liben than in Halaba. Under the accelerated case scenario, it is assumed that the targets would be reached within a decade and the impacts of the YICT on the livelihoods of smallholders would be realized faster. If the growth rate of the past decade is sustained in the coming 10 years, only 47 % of the target average yield in Halaba and about 64 % of the target average yield in Baso Liben will be achieved, which implies that the higher target cannot be reached unless the pace of the past decade is substantially improved. Thus, innovative strategies need to be implemented to achieve the higher targets. However, the lower targets can be reached given the existing pace of growth.

The mean discounted net benefits were analyzed taking into account a time horizon of 20 years regarding the use of target YICT.[25] The main results are displayed in Figs. 20.8, 20.9, 20.10, and 20.11, whereas the patterns of change over the period considered are displayed in Figs. A1 to A2 in the annex. The results further assume that (1) the costs of dissemination of the technologies are sunk costs which will not change because of the introduction of these technologies, and (2) there will be adequate demand for outputs, and increasing production will not have negative effects on output prices.

If we consider the "typical" scenario of adoption, the mean net benefit per year for all farmers in both study areas over the stated period is 135.2 million birr, provided that the higher target is achieved; the net benefit would be about 71 million birr if the lower target is to be achieved (Fig. 20.8). Under the "accelerated" scenario, the mean figure would be 181.7 million birr for the HTC and 94.9 million birr for the LTC (Fig. 20.9). There is a visible difference between the two districts, Baso Liben taking a better position. The mean per capita net benefit corresponding to the HTC is 9.8 birr/day under the typical scenario, while it is 13.4 birr/day under the accelerated scenario (Fig. 20.10). The figures corresponding to the LTC are 5.1 and 7 birr/day under the typical and accelerated scenarios, respectively (Fig. 20.11). Overall, the adoption of the YICT will have a total net benefit of about 2.8 and 3.8 billion birr under the typical and accelerated scenarios, respectively, over the entire planning horizon, provided that the higher target case is realized. The figures corresponding to the two scenarios would be 1.5 and 2 billion birr if the lower target is to be achieved.[26]

---

[25] An annual inflation rate of 12.3 % was considered to simulate future prices. This figure is an average figure for the last 16 years.

[26] The net benefits from the YICT under consideration may decline because some varieties may not cope well with new pest out-breaks or because of other reasons. However, it is expected that farmers would keep the momentum of existing high yield or increase it by adopting better varieties and, thus, the trend will at least level off after the maximum point.

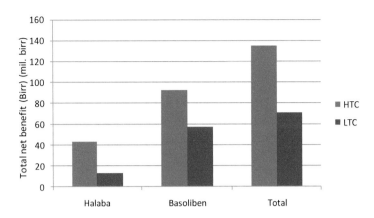

Fig. 20.8 Average discounted total net benefit (for all farmers), typical scenario (in million birr)

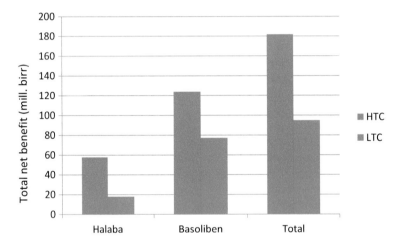

Fig. 20.9 Average discounted total net benefit (for all farmers), accelerated scenario (in million birr)

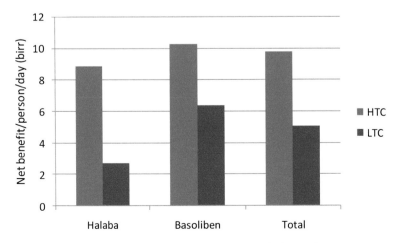

Fig. 20.10 Discounted net per capita benefit per day (birr), typical scenario

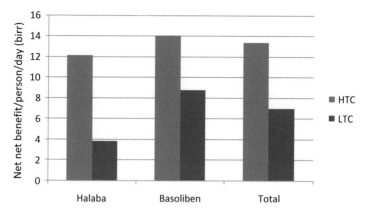

Fig. 20.11 Discounted net per capita benefit per day (birr), accelerated scenario

Conclusions

This study assessed the potential impact of existing yield-increasing crop technologies on productivity and poverty. Results are based on household survey data collected from two districts of Ethiopia, namely: the Halaba special district in the Southern Nations, Nationalities, and Peoples Region (SNNPR) and the Baso Liben district in the Amhara Region. The two districts were selected for showing high agricultural potential while being marginalized.

Results indicate that about 90 % of farm households are poor and food insecure in these areas. Adoption of yield increasing crop technologies is low, resulting in low productivity and income. Our results show that adoption of agricultural technologies (namely improved seeds with appropriate agronomic packages) would increase yields and incomes substantially, thereby reducing poverty. The average net benefit per household is about 17,000 birr in Halaba and nearly 18,000 birr in Baso Liben, assuming that technologies produce yields as indicated by reports of on-farm yield trials. The total net benefit per district ranges from 71 to 182 million birr per year, while the net additional benefit per person per day ranges from 5 to 13 birr. These amounts of additional benefit would be enough to lift up the subjacent poor and the medial poor out of poverty in both districts. However, the amounts would not be sufficient for the ultra-poor to shift above the poverty line considered in this analysis, which implies that other options (such as promoting non-farm rural businesses) would be required to lift these households out of poverty. These are rough results to the extent that the future is uncertain in terms of prices, technologies, climate. Nevertheless, the overall implication of the study is that the benefits from proposed crop technologies would be high, with the potential of reducing poverty significantly.

Open Access This chapter is distributed under the terms of the Creative Commons Attribution-Noncommercial 2.5 License (http://creativecommons.org/licenses/by-nc/2.5/) which permits any noncommercial use, distribution, and reproduction in any medium, provided the original author(s) and source are credited.

The images or other third party material in this chapter are included in the work's Creative Commons license, unless indicated otherwise in the credit line; if such material is not included in the work's Creative Commons license and the respective action is not permitted by statutory regulation, users will need to obtain permission from the license holder to duplicate, adapt or reproduce the material.

Annex

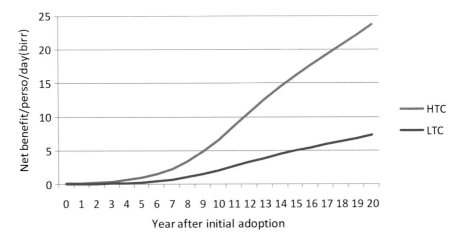

Fig. A.1 Per capita additional net benefit per day from adopting YICT in Halaba, typical case scenario

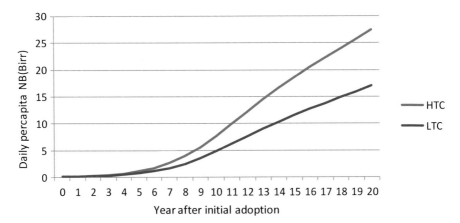

Fig. A.2 Per capita additional net benefit per day from adopting YICT in Baso Liben, typical case scenario

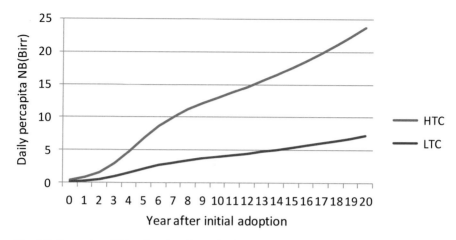

Fig. A.3   Per capita additional net benefit per day from adopting YICT in Halaba, accelerated case scenario

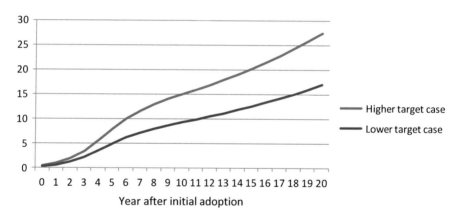

Fig. A.4   Per capita additional net benefit per day from adopting YICT in Baso Liben, accelerated case scenario

References

Ahmed AU, Vargas Hill R, Smith LC, Wiesmann DM, Frankenberger F (2007) The world's most deprived: characteristics and causes of extreme poverty and hunger. International Food Policy Research Institute, Washington, DC

Alston JM, Norton GW, Pardey PG (1995) Science under scarcity. Cornell University Press, Ithaca

ATA (2012) Five year seed sector strategy document, draft. Ethiopian Agricultural Transformation Agency, Addis Abeba

Bera AK, Kelley TG (1990) Adoption of high yielding rice varieties in Bangladesh: an econometric analysis. J Dev Econ 33:263–285

CIMMYT (1988) From agronomic data to farmers recommendation: an economics training manual, completely revised. International Maize and Wheat Improvement Center, Mexico

CSA (2011a) Report on farm management practices (private peasant holdings, meher season), statistical bulletin 505. Central Statistical Agency, Addis Abeba

CSA (2011b) Report on area and production of major crops (private peasant holdings, meher season). Central Statistical Agency, Addis Abeba

CSA (2012) Agricultural sample survey 2011/12, Report on land utilization, private peasant holdings, meher season, Bulletin No 532. Central Statistical Agency, Addis Abeba

Davies S (1979) The diffusion of process innovations. Cambridge University Press, Cambridge

Diao X, Seyoum Taffesse A, Yu B, Nin Pratt A (2010) Economic importance of agriculture for sustainable development and poverty reduction: the case study of Ethiopia, Global forum on agriculture. Organisation for Economic Co-operation and Development, Paris

Ephraim M, Featherstone AM (2001) Cross pollinated crop variety adoption studies and seed recycling: the case of maize in Tanzania. East Afr J Rural Dev 17(1):25–34

FAOSTAT (2013) http://faostat.fao.org, accessed Jul 22, 2014

Gatzweiler F, Baumüller H, Ladenburger C, von Braun J (2011) Marginality: addressing the root causes of extreme poverty, ZEF Working paper series No. 77. Center for Development Research, Bonn

Graw V, Ladenburger C (2012) Mapping marginality hotspots: geographical targeting for poverty reduction, ZEF Working paper series No. 88. Center for Development Research, Bonn

Hailu BA (2008) Adoption of improved tef and wheat production technologies in crop-livestock mixed systems in northern and western Shewa zones of Ethiopia thesis. University of Pretoria, Pretoria

Hareau GG, Mills BF, Norton GW (2006) The potential benefits of herbicide-resistant transgenic rice in Uruguay: lessons for small developing countries. Food Policy 31:162–179

Hazell PBR (2009) The Asian green revolution. IFPRI Discussion Paper 00911

Hundie B (2012) The current state of agriculture in Ethiopia: productivity, technology innovations, technology reach and food security. A background paper prepared for TIGA project kick-off meeting. Center of Development Research, Bonn

Hundie Kotu B, Verkuijl H, Mwangi W, Tanner D (2000) Adoption of improved wheat technologies in Adaba and Dodola Woredas of the Bale Highlands, Ethiopia. International Maize and Wheat Improvement Center and Ethiopian Agricultural Research Organization, Mexico/Addis Abeba

Krishna VV, Qaim M (2008) Potential impacts of Bt eggplant on economic surplus and farmers' health in India. Agri Econ 38:167–180

Macharia I, Orr A, Simtowe F, Asfaw S (2012) Potential economic and poverty impacts of improved chickpea technologies in Ethiopia. A selected paper prepared for presentation at the International Association of Agricultural Economists (IAAE) Triennial Conference, Foz do Iguaçu, 18–24 Aug 2012

Mahajan V, Peterson RA (1978) Innovation diffusion in a dynamic potential adopter population. Manag Sci 24:1589–1597

MOA (2011) Crop variety register, vol 14. Ministry of Agriculture, Addis Ababa

MoFED (2010) Growth and transformation plan (2010/11–2014/15), vol I. Ministry of Finance and Economic Development, Addis Ababa

MoFED (2012) Ethiopia's progress towards eradicating poverty: an interim report on poverty analysis study (2010/11). Ministry of Finance and Economic Development, Addis Ababa

Mosher TA (1979) An introduction to agricultural extension. Singapore University Press for the Agricultural Development Council, Singapore

Motuma T, Dejene A, Wondwossen T, La Rovere R, Girma T, Mwangi W, Mwabu W (2010) Adoption and continued use of improved maize seeds: case study of Central Ethiopia. Afr J Agric Res 5(17):2350–2358

Napasintuwong O, Traxler G (2009) Ex-ante impact assessment of GM papaya adoption in Thailand. AgBioForum 12(2):209–217

Qaim M (2003) Bt cotton in India: field trial results and economic projections. World Dev 31 (12):2115–2127

Rogers E (1983) Diffusion of innovations. Free Press, New York

Tesfaye Z, Girma T, Tanner D, Vekuijl H, Aklilu E, Mwangi W (2001) Adoption of improved bread wheat varieties and inorganic fertilizers by small-scale farmers in Yilmana Densa and Farta Districts of Northwestern Ethiopia. Ethiopian Agricultural Research Organization and International Maize and Wheat Improvement Center, Addis Abeba/Mexico

von Braun J, Hill RV, Pandya Lorch R (2009) The poorest and hungry: a synthesis of analyses and actions. In: The poorest and hungry. Assessment, analyses, and actions. IFPRI 2020. International Food Policy Research Institute, Washington, DC, pp 1–61

Index

© The Author(s) 2016
F.W. Gatzweiler, J. von Braun (eds.), *Technological and Institutional Innovations
for Marginalized Smallholders in Agricultural Development*,
DOI 10.1007/978-3-319-25718-1